CLIMATE CHANGE AND AGROFORESTRY SYSTEMS

Adaptation and Mitigation Strategies

CLIMATE CHANGE AND AGROFORESTRY SYSTEMS

Adaptation and Mitigation Strategies

Edited by

Abhishek Raj

Manoj Kumar Jhariya

Dhiraj Kumar Yadav

Arnab Banerjee

AAP | APPLE ACADEMIC PRESS

Apple Academic Press Inc.
4164 Lakeshore Road
Burlington ON L7L 1A4
Canada

Apple Academic Press Inc.
1265 Goldenrod Circle NE
Palm Bay, Florida 32905
USA

International Standard Book Number-13: 978-1-77188-822-6 (Hardcover)
International Standard Book Number-13: 978-0-42928-675-9 (eBook)

Library and Archives Canada Cataloguing in Publication

Title: Climate change and agroforestry systems : adaptation and mitigation strategies / edited by Abhishek Raj, Manoj Kumar Jhariya, Dhiraj Kumar Yadav, Arnab Banerjee

Names: Raj, Abhishek, editor. | Jhariya, Manoj Kumar, editor. | Yadav, Dhiraj Kumar, editor. | Banerjee, Arnab (Professor of environmental science), editor

Description: Includes bibliographical references and index.

Identifiers: Canadiana (print) 20190240237 | Canadiana (ebook) 20190240253 | ISBN 9781771888226 (hardcover) | ISBN 9780429286759 (ebook)

Subjects: LCSH: Agroforestry. | LCSH: Agroforestry—Environmental aspects. | LCSH: Climate change mitigation.

Classification: LCC S494.5.A45 C63 2020 | DDC 634.9/9—dc23

CIP data on file with US Library of Congress

Apple Academic Press also publishes its books in a variety of electronic formats. Some content that appears in print may not be available in electronic format. For information about Apple Academic Press products, visit our website at **www.appleacademicpress.com** and the CRC Press website at **www.crcpress.com**

About the Editors

Abhishek Raj, PhD
Research Scholar, Department of Forestry, Indira Gandhi Krishi Vishwavidyalaya, Raipur, Chhattisgarh, India

Abhishek Raj, PhD, is a Research Scholar in the Department of Forestry at Indira Gandhi Krishi Vishwavidyalaya, Raipur, Chhattisgarh, India, and is the author or co-author of more than 20 research and review papers in a peer-reviewed journals, 10 book chapters, and several extension articles. Dr. Raj acquired his BSc (Forestry) from SHIATS Allahabad (U.P.), MSc (Forestry), and PhD (Forestry) degrees from Indira Gandhi Krishi Vishwavidyalaya, Raipur, Chhattisgarh, India. He was awarded a Young Scientist Award for outstanding contribution in the field of forestry in the year 2017 from the Science & Tech Society for Integrated Rural Improvement (S&T SIRI)-INDIA. He has qualified ICAR/ASRB NET (National Eligibility Test) in Agroforestry for lectureship. He is an editorial board member of many reputed journals. He has proved himself as an active scientist in the area of forestry.

Manoj Kumar Jhariya, PhD
Assistant Professor, Department of Farm Forestry, Sarguja University Ambikapur, Chhattisgarh, India

Manoj Kumar Jhariya, PhD, is an Assistant Professor in the Department of Farm Forestry at Sarguja University Ambikapur, Chhattisgarh, India, and is the author or co-author of more than 60 research papers in peer-reviewed journal, four books, 20 book chapters, and several extension articles. Dr. Jhariya acquired his BSc (Agriculture), MSc (Forestry), and PhD (Forestry) degrees from Indira Gandhi Krishi Vishwavidyalaya, Raipur, Chhattisgarh, India. He won the University Gold Medal for securing first class first position in his PhD examination. He was awarded the Chhattisgarh Young Scientist Award in the year 2013 from the Chhattisgarh Council of Science and Technology. He was awarded an UGC-RGNF Fellowship, New Delhi, India. He is an editorial board member of several journals. He is a life member of The Indian Science Congress Association, Applied and Natural Science Foundation, Society for Advancement of Human and Nature, Medicinal and Aromatic Plants Association of India, and ISDS Society & International Journal of Development and Sustainability. He has supervised 41 MSc (Farm Forestry) students to date. He is dynamically involved in teaching (post-graduate) and research. He has proved himself as an active scientist in the area of forestry.

Dhiraj Kumar Yadav, PhD
Assistant Professor and Head, Department of Farm Forestry, Sarguja Vishwavidyalaya, Ambikapur, Chhattisgarh, India

Dhiraj Kumar Yadav, PhD, is an Assistant Professor and HOD (Incharge), Department of Farm Forestry, Sarguja Vishwavidyalaya, Ambikapur, Chhattisgarh, India. He completed graduation, postgraduation, and PhD in Forestry from JNKVV, IGKVV, and Kumaun University Nainital, respectively. He has pursued an MBA in Human Resource Management from Sikkim Manipal University, and a diploma in environmental management. He has been awarded a Chhattisgarh's Young Scientist Award in the year 2008 from the Chhattisgarh Council of Science and Technology. He won a merit scholarship during his BSc curriculum. He also worked as JRF, SRF, RA, and Project Coordinator in various forestry institutes. He has published 30 research papers articles, two books, and 10 book chapters with various national and internationally reputed publishers. He is also a member of several academic societies in India. He has supervised 41 MSc (Farm Forestry) students to date. He is dynamically involved in teaching (post-graduate) and research. He has proved himself as an active researcher in the area of forestry.

Arnab Banerjee, PhD
Assistant Professor, Department of Environmental Science, Sarguja Vishwavidyalaya, Ambikapur, Chhattisgarh, India

Arnab Banerjee, PhD, is an Assistant Professor in the Department of Environmental Science, Sarguja Vishwavidyalaya, Ambikapur, Chhattisgarh, India. He has completed MSc and PhD (Environmental Science) from Burdwan University and MPhil in Environmental Science from Kalyani University West Bengal. He won the University Gold Medal for securing first class first position in the MSc examination. He has been awarded the Young

Scientist Award for the best oral presentation at the International Conference held at the University of Burdwan, West Bengal, India. He was a project fellow under UGC sponsored major research project. He has published 64 research papers in reputed national and international journals. To his credit, he has published five books and 10 book chapters. He is a life member of the academy of environmental biology. He has supervised 16 PG students. He is dynamically involved in teaching (post-graduate) and research. He has proved himself as an active scientist in the area of environmental science.

Contents

Contributors

A. O. Akanwa
Chukwuemeka Odumegwu Ojukwu University (COOU), Uli Campus, Anambra State, Nigeria, Mobile: +234-8065813596, E-mail: angela.akanwa1@gmail.com

C. C. Anukwonke
Chukwuemeka Odumegwu Ojukwu University (COOU), Uli Campus, Anambra State, Nigeria

Arnab Banerjee
Assistant Professor, University Teaching Department, Department of Environmental Science, Sarguja Vishwavidyalaya, Ambikapur – 497001, Chhattisgarh, India, Mobile: +00-91-9926470656, E-mail: arnabenvsc@yahoo.co.in

B. P. Bhatt
ICAR-Research Complex for Eastern Region, Patna, Bihar–800014, India

Vinita Bisht
College of Forestry, BUAT, Banda (UP), India

Sumit Chakravarty
Professor, Uttar Banga Krishi Viswavidyalaya, Pundibari–736165, West Bengal, India, Mobile: 9434082687, E-mail: c_drsumit@yahoo.com

Jayanta Kumar Datta
Retired Professor, Department of Environmental Science, The University Burdwan, West Bengal–713104, India

Zia Ur Rahman Farooqi
Doctoral Student, Institute of Soil and Environmental Science, University of Agriculture, Faisalabad–38040, Pakistan, Mobile: +923156040622, E-mail: ziaa2600@gmail.com

Muhammad Mahroz Hussain
Doctoral Student, Institute of Soil and Environmental Science, University of Agriculture, Faisalabad–38040, Pakistan, Mobile: +923217251329, E-mail: hmahroz@gmail.com

Manoj Kumar Jhariya
Assistant Professor, University Teaching Department, Department of Farm Forestry, Sarguja Vishwavidyalaya, Ambikapur–497001, Chhattisgarh, India, Mobile: +00-91-9407004814, E-mail: manu9589@gmail.com

S. M. Paul Khurana
Amity Institute of Biotechnology, Amity University Haryana, Manesar, Gurgaon–122413, India, E-mail: smpaulkhurana@gmail.com

Dhiraj Kumar
ICAR-Central Agroforestry Research Institute, Jhansi (U.P)–284003, India

Narendra Kumar
Amity Institute of Biotechnology, Amity University Haryana, Manesar, Gurgaon–122413, India, E-mail: narendra.microbiology@rediffmail.com

Ubalt Lepcha
PhD Scholar, Uttar Banga Krishi Viswavidyalaya, Pundibari–736165, West Bengal, India

K. Abha Manohar
PhD Scholar, Uttar Banga Krishi Viswavidyalaya, Pundibari–736165, West Bengal, India

H. C. Mba
University of Nigeria Nsukka (UNN), Department of Urban and Regional Planning, Faculty of Environmental Sciences, Enugu Campus, Enugu State, Nigeria

Keerti Mishra
Post Graduate Student, University Teaching Department, Department of Environmental Science, Sarguja Vishwavidyalaya, Ambikapur, Chhattisgarh–497001, India

Buddhadev Mukhopadhyay
Research Scholar, Department of Environmental Science, The University Burdwan, West Bengal–713104, India, Mobile: +919635878593,
E-mail: buddhadevmukhopadhyay@gmail.com

M. U. Nwachukwu
University of Nigeria Nsukka (UNN), Department of Urban and Regional Planning, Faculty of Environmental Sciences, Enugu Campus, Enugu State, Nigeria

E. B. Ogbuene
Centre for Environmental Management and Control (CEMAC), University of Nigeria Nsukka, Enugu State, Nigeria

Prabhat Ranjan Oraon
Junior Scientist-cum-Assistant, Professor, Department of Silviculture and Agroforestry, Faculty of Forestry, Birsa Agricultural University, Ranchi–834006, Jharkhand, India,
Mobile: +00-91-9431326222, E-mail: prabhat.ranjan.oraon@gmail.com

Nazir A. Pala
Assistant Professor, Uttar Banga Krishi Viswavidyalaya, Pundibari–736165, West Bengal, India

Vinita Parte
College of Agriculture, Ganjbasoda, Vidisha, Jawaharlal Nehru Krishi Vishwavidyalaya, Jabalpur, Madhya Pradesh, India

Anju Puri
Assistant Professor, Baring Union Christian College, Batala–143505, Punjab, India

Prakash Rai
Uttar Banga Krishi Viswavidyalaya, Pundibari–736165, West Bengal, India

Abhishek Raj
PhD Scholar, Department of Forestry, College of Agriculture, Indira Gandhi Krishi Vishwavidyalaya, Raipur–492012, Chhattisgarh, India,
Mobile: +00-91-8269718066, E-mail: ranger0392@gmail.com

Vijay Rajwade
Post Graduate Student, University Teaching Department, Department of Environmental Science, Sarguja Vishwavidyalaya, Ambikapur, Chhattisgarh–497001, India

Madhur Mohan Ranga
Professor, University Teaching Department, Department of Environmental Science, Sarguja Vishwavidyalaya, Ambikapur, Chhattisgarh–497001, India

K. K. Rao
ICAR-Research Complex for Eastern Region, Patna, Bihar–800014, India

Muhammad Sabir
Assistant Professor, Institute of Soil and Environmental Science, University of Agriculture, Faisalabad–38040, Pakistan, Mobile: +923336545518, E-mail: cmsuaf@gmail.com

Gopal Shukla
Assistant Professor, Uttar Banga Krishi Viswavidyalaya, Pundibari–736 165, West Bengal, India

Nongmaithem Raju Singh
ICAR-Research Complex for Eastern Region, Patna, Bihar–800014, India,
E-mail: rajuforestry@gmail.com

Vishnu K. Solanki
Department of Agroforestry, College of Agriculture, Ganjbasoda, Vidisha,
Jawaharlal Nehru Krishi Vishwavidyalaya, Jabalpur, Madhya Pradesh, India,
E-mail: rvishnu@hotmail.com

Pratap Toppo
Department of Forestry, College of Agriculture, Indira Gandhi Krishi Vishwavidyalaya,
Raipur–492012, Chhattisgarh, India

Vineeta
Assistant Professor, Uttar Banga Krishi Viswavidyalaya, Pundibari–736165, West Bengal, India

Dhiraj Kumar Yadav
Assistant Professor, University Teaching Department, Department of Farm Forestry,
Sarguja Vishwavidyalaya, Ambikapur–497001, Chhattisgarh, India,
Mobile: +00-91-9926615061, E-mail: dheeraj_forestry@yahoo.com

Muhammad Zia-ur-Rehman
Assistant Professor, Institute of Soil and Environmental Science,
University of Agriculture, Faisalabad–38040, Pakistan,
Mobile: +923216637127, E-mail: ziasindhu1399@gmail.com

Abbreviations

ADB & IFPR	Asian Development Bank and International Food Policy Research
ADB	Asian Development Bank
ADD	attention deficit disorder
ADHD	attention-deficit/hyperactivity disorder
AFs	agroforestry systems
AICRAF	All India Coordinated Research Project on Agroforestry
C	carbon
CAFRI	Central Agroforestry Research Institute
CAST	China Association for Science and Technology
CBNRM	community based-natural resource management
CC	climate change
CDM	clean development mechanisms
CEE	Center for Environmental Education
CESR	Center for Economic and Social Regeneration
CFCs	chlorofluorocarbons
CH_4	methane
CIFOR	Center for International Forestry Research
CLEAN	Community-Led Environmental Action Network
CNG	compressed natural gas
CO_2	carbon dioxide
CPRE	Campaign to Protect Rural England
CS	carbon sequestration
CTLA	Council of Tree and Landscape Appraisers
Cv	cultivar
DA	development alternatives
DID	Department of International Development
ECCA	Environmental Camps for Conservation Awareness
EE	Environmental Education
EESD	environmental education and sustainable development
EOSE	Environmental Orientation to School Education
ESD	education for sustainable development
EU	European Union

EU-ETS	European Union Emissions Trading System
FAO	Food and Agriculture Organization
FNS	Food and Nutritional Security
GCS	geological carbon sequestration
GHGs	greenhouse gases
GHI	global harvest initiative
GLOBE	global learning and observations to benefit the environment
GoI	Government of India
Hg	mercury
IASC	Inter-Agency Standing Committee (UN)
ICE	internal combustion engines
ICRAF	International Center for Research in Agroforestry
IDMC	Internal Displacement Monitoring Center
IDRC	International Development Research Center
IFAD	International Fund for Agricultural Development
IFS	integrated farming systems
IITA	International Institute of Tropical Agriculture
IPCC	Intergovernmental Panel on Climate Change
ISWM	integrated solid waste management
IUCN	International Union for Conservation of Nature
JUNP	Joint United Nations Programme
LPG	liquefied petroleum gas
LULUCF	land use, land-use change, and forest
MHRD	Ministry of Human Resource Development
MLD	millions of liters per day
MOEF	Ministry of Environment and Forest
MOEFCC	Ministry of Environment, Forests, and Climate change
MPTs	multipurpose tree species
N	neutral
N_2O	nitrous oxide
NAAEE	North American Association of Environmental Educators
NCERT	National council for educational research and training
NGO	non-governmental organization
NH_4	ammonium
N_2O	nitrous oxide
NO_3	nitrates
NO_x	nitrogen oxides

NRC & IDMC	Norwegian Refugee Council and Internal Displacement Monitoring Center
NRM	natural resource management
NTFPs	non-timber forest products
NWFPs	non-wood forest products
OECD	Organization for Economic Cooperation and Development
PAR	photo-synthetically active radiation
PDL	previously developed land
Pg	petagram
PG	post-graduate
PHRAME	plant health risk and monitoring evaluation
REDD	reduction of emissions from deforestation and degradation
RGGI	regional GHGs initiative
RGI	rate of germination index
RUBISCO	ribulose 1,5-bisphosphate carboxylase-oxygenase enzyme
SDF	seminar talk along with documentary film
SEPA	State Environmental Protection Administration
SHGs	self-help groups
SIDA	Swedish International Development Authority
SOC	soil organic carbon
SOCS	soil organic carbon stocks
SOM	soil organic matter
SO_x	sulfur oxides
SUD	Sustainable Urban Drainage
SUDS	sustainable urban drainage system
UG	undergraduate
UHI	Urban Heat Island
UK	United Kingdom
UN-DESA	United Nations, Department of Economic and Social Affairs
UNDP	United Nations Developmental Program
UNEP	United Nations Environmental Program
UNEP-JUNP	United Nations Environment Programme and Joint United Nations Programme
UNESCO	United Nations Educational, Scientific, and cultural Organization
UNFCCC	United Nations Framework Convention on Climate Change

UNICEF	United Nations International Children's Emergency Fund
UNRISD	United Nations Research Institute for Social Development
USA	United States of America
VCD	video compact disc
VHS	video home system
WB	World Bank
WCFSD	World Conservation on Forests and Sustainable Development
WFP	World Food Program
WHO	World Health Organization
WRI	World Resource Institute

Preface

Undoubtedly, climate change is a major curse and a burning issue around today's globe. Their effects are not only confined to human health but also affects morphology, phenology, reproduction, health, and overall productivity of the vegetation and farming system. It destroys our environment and ecosystem structure and functions. It has serious recuperations for a number of sectors and resources, including agriculture, forest, environment, etc. Emission of carbon dioxide, methane, chlorofluorocarbons (CFC), and nitrous oxide are identified as potent greenhouse gases (GHGs) causing global warming and affecting agro-ecosystem production and economy. Events such as outbreaks of plant pathogens, change in land-use, and alteration of structure and composition of vegetation is a very problematic issue for entire ecosystem functioning, and associated biodiversity loss are the negative outcome of the evil. Very minor changes in temperature can have a major impact on systems on which human livelihoods depend, including water availability and crop productivity. The lives and livelihood of many communities will be at risk. Adaptation and mitigation are complementary to each other. The integration of better crops with other tree components and management practices could efficiently reduce and mitigate climate change worldwide. Moreover, transforming agricultural sectors, including crop and livestock production, forestry, fishery, etc., in a climate-smart way, can address the challenges of global warming and current food security. In this context, adoption of different practices of climate-smart agriculture, viz., agroforestry, conservation agriculture, organic farming, improved crop management, livestock management and soil and water conservation, environmental awareness, waste disposal, etc., have the capacity to build up carbon accumulation in both plant and soil compartment and address the problem of climate change and food security.

The present book, *Climate Change and Agroforestry System: Adaptation and Mitigation Strategies,* addresses the burning issues of climate change impact, alteration of environmental quality, and its subsequent mitigation and adaptation strategies through various practices of agro-ecosystems. It also focuses on new insights related to updated research, development, and extension activities for combating climate change.

The book comprises of 12 chapters. Chapter 1 describes the scenario of climate change, its issues, causes, impacts on agro-ecosystems, and provides an in-depth analysis of the potential of various modern, improved, and scientific farming practices like climate-smart agriculture and agroforestry system towards climate change mitigation and adaptation. Chapter 2 addresses the role of agroforestry and potential in natural resource management under changing climate and global warming. Food, nutritional security, and people's livelihood security, along with environmental health, are also discussed in this chapter.

Chapter 3 discusses the greening environment and landscape management through the adoption and promotion of urban forestry and agroforestry systems along with its socioeconomic and environmental benefits. Chapter 4 emphasizes the climate change impacts on biodiversity and their mitigation through forestry and agroforestry technology along with other various technologies, such as carbon sequestration in agricultural soil and other activities. Chapter 5 emphasizes climate change mitigation through carbon sequestration of the agroforestry system and its role in soil health and quality management. Chapter 6 provides detailed insight about a nexus between soil sustainability and food security under changing the climate, which is linked and managed through better farming practices. Chapter 7 discusses a linking concept between social dimensions of climate change, its impact, mitigation, and adaptation through integrating social dimensions with climate change policies and programs. Chapter 8 focuses on recent climate fluctuations and related invasion of major fungal diseases in crop plants and forest trees. Chapter 9 reflects on the research findings of the utilization of boiling water of rice for irrigation purposes as a sustainable water management strategy at a laboratory scale. Chapter 10 addresses the potential role of forests and agriculture activities in climate change mitigation. Chapter 11 looks at the research findings related to the solid waste scenario and its subsequent management from the natural resource management perspective. Chapter 12 provides an introduction and discussion of the importance of the non-formal system towards imparting environmental education.

This book will be a standard reference work for disciplines such as forestry, agriculture, ecology, and environmental science, as well as will be a way forward towards strategy formulation for combating climate change.

The editors would appreciate receiving comments from readers that may assist in the development of future editions.

—Abhishek Raj, PhD
Manoj Kumar Jhariya, PhD
Dhiraj Kumar Yadav, PhD
Arnab Banerjee, PhD

Acknowledgments

Life is a continuous journey full of challenges and hurdles that need to be overcome by the blessings of the Almighty. Since pre-historic times it was found that individuals with a hardworking nature, tenacity, and enthusiasm usually reach the peak of success. Above all, spiritual bliss also helps to progress through the life. The present book is the result of untiring, continuous effort from the team of editors who have been actively engaged in the compilation of the book. Further, such book title of applied science addressing current issues and scientific findings are no doubt the need of the hour. Blessing from the parents, love, and best wishes from the friends have also supported this gesture.

It is our deep sense of gratitude towards Honorable Vice-Chancellor, IGKV, and Sarguja University, Ambikapur, for their continuous support, stimulation, and cooperation, which helped us to achieve the compilation of the book.

We are also grateful to Mr. Ashish Kumar, President, Apple Academic Press, to accomplish our dream of publishing this book. Further, we acknowledge Dr. Mohammed Wasim Siddiqui, Assistant Professor and Scientist, Bihar Agriculture University, India, and Prof. M.N. Hoda, Director Bharati Vidyapeeth's Institute of Computer Application & Management, New Delhi, and Dr. Ritika Wason (BVICAM) for their continuous help and support for execution of the project.

CHAPTER 1

Impact of Climate Change on Agroecosystems and Mitigation Strategies

ABHISHEK RAJ,[1] MANOJ KUMAR JHARIYA,[2] DHIRAJ KUMAR YADAV,[2] and ARNAB BANERJEE[3]

[1]*Department of Forestry, College of Agriculture, Indira Gandhi Krishi Vishwavidyalaya, Raipur–492012, Chhattisgarh, India, Mobile: +00-91-8269718066, E-mail: ranger0392@gmail.com*

[2]*Assistant Professor, University Teaching Department, Department of Farm Forestry, Sarguja Vishwavidyalaya, Ambikapur–497001, Chhattisgarh, India, Mobile: +00-91-9407004814, E-mail: manu9589@gmail.com (M. K. Jhariya); Mobile: +00-91-9926615061, E-mail: dheeraj_forestry@yahoo.com (D. K. Yadav)*

[3]*Assistant Professor, University Teaching Department, Department of Environmental Science, Sarguja Vishwavidyalaya, Ambikapur–497001, Chhattisgarh, India, Mobile: +00-91-9926470656, E-mail: arnabenvsc@yahoo.co.in*

ABSTRACT

Natural and anthropogenic factors (agricultural activities) accelerate the emissions of greenhouse gases (GHGs) into the atmosphere. Although since the green revolution some practices of intensive agricultural, i.e., huge application of chemical fertilizers, biomass burning, tillage practices, faulty land-use conversion, deforestation, and other human activities results several GHGs (CO_2, CH_4, N_2O, etc.) leads warming effects of global earth's ecosystem that affects all living treasure (loss of biodiversity, emergence of infectious disease and insect pest, poor quality of timbers and overall biomass and soil productivity) through extreme weather events. Climate change (CC) affects agroecosystem structure and functions, which in turn affects overall production and ecosystem services, which are prerequisites for sustaining humans and animal life. However, the agroecosystem itself contributes in GHGs emissions through the application of chemical fertilizers, application of mechanized technology, a huge amount of pesticides and herbicides, etc. that give higher and ample production along with releasing GHGs into the atmosphere. Therefore, climate-smart agriculture, location-specific agroforestry practices, conservation agriculture, non-tillage system, integrated farming systems (IFS), etc. are the viable practices that not only enhance overall biomass productivity but also mitigate the CC impact through enhancing the storage and sequestration capacity of carbon into the both vegetation and soils. Healthy agroecosystem provides healthy soils and plants which sustain all livestock and human population and maintains food and nutritional security (FNS) along with environmental security. Therefore, there is a linking concept among agroecosystems, food, and nutrition, and environmental security. Moreover, effective government policy, NGOs, private agencies, and educational institutions should take some prioritized thrust areas for the research and development for better agroecosystem structure and functions, which can be the betterment of all biodiversity and ecosystem health.

1.1 INTRODUCTION

Today, intensive agricultural practices emit several GHGs into the atmosphere and developing country contributes about three-quarters (3/4) of direct emissions of GHGs of the total emission in the world (Paustian et al., 2004; FAO, 2011; Hosonuma et al., 2012; Moreau et al., 2012). The process of enteric fermentation and flooded rice cultivation practices release methane (CH_4) in the atmosphere, microbial transformation of nitrogenous fertilizers and manures releases nitrous oxide (N_2O) whereas change in biogeochemical cycles, biomass burning, and organic residues decomposition affects our environment through emission of carbon dioxides (CO_2) (Janzen, 2004; Chirinda et al., 2018). Similarly, the process of deforestation has been intensifying the level of GHGs and promotes the burning issue of global warming leads to changing our climate (Oenema et al., 2005; Hosonuma et al., 2012; Scott et al., 2018).

Globally, agricultural activity like enteric fermentation, manure application on soil and pasture/grasses land, application of synthetic fertilizer, practices of rice cultivation, manure management practices and management of crop residues etc. contributes approximately 4354 $MtCO_2eq\ yr^{-1}$ emission of GHGs into the atmosphere of which Asian continent contributed maximum (45%) emission and least in Oceania (3%). Likewise, the anthropogenic activity as deforestation contributed approximately 3500 $MtCO_2eq\ yr^{-1}$ GHGs emission into our atmosphere and of which America contributed the highest emission (2100) whereas least in Oceania (105 $MtCO_2eq\ yr^{-1}$) and the GHGs emission from European country was nil in the period of one decade from 2000 to 2010 (Table 1.1) (Tubiello et al., 2013). Also, CH_4 emission from the rice cultivation practices by different countries in the world is depicted in Figure 1.1. It is very well known fact about nurturing significant of agriculture sector to burgeoning population but GHGs emission (13.5% of the total emitted GHGs) from agricultural sector due to faulty land-use conversion, higher application of inorganic fertilizers and rice cultivation are one of the major issue world face today (Johnson et al., 2007). This cannot only affect the overall food production but promotes land degradation, enhancing desertification; deteriorate soil structure and its function due to poor soil quality, affects overall biodiversity along with food, health, and environmental security are under the question marks.

In this context, there is a need to adopt and modified natural ecosystems to more friendly and sustainable agricultural system, which can fulfill the human needs as direct and indirect services along with environmental health through minimizing the continuous emission of GHGs.

Therefore, climate-resilient agroecosystem practices are very good strategies for both enhancing the soil and plants productivity in the parallel of carbon sequestration (CS) capacity in the vegetation and soils (Jhariya et al., 2015; Singh and Jhariya, 2016). Practices like climate-smart agriculture, conservation farming, no-tillage practices, agroforestry systems (AFs), etc. enhance the vegetational and cropping diversity, improve landscape matrix through better structure and functions and improve the soil structure and functions which provide better services (Jhariya et al., 2018; Raj et al., 2018) (Figure 1.2). Moreover, various biotic and abiotic factors such as location-specific agricultural technology, its adoption and management practices, social, and cultural settings and environmental and ecological settings in the specific region affects the agroecosystem structure and functions (Figure 1.3). This chapter reviews the global climate situations, impacts on agroecosystem, GHGs emission from different practices of agroecosystem, mitigation strategies through the various climate-resilient agroecosystem practices viz., climate-smart farming system, agroforestry, conservation agriculture, non-tillage practices, etc. This chapter also represents the role of agroecosystem in soil organic carbon (SOC) and fertility maintenance which will be

TABLE 1.1 GHGs Emission from Agricultural Activity and Deforestation in the World Over the Period 2000–2010

Continent	Agricultural Activity (Value in $MtCO_2eq\ yr^{-1}$)								Deforestation Activity (Value in $MtCO_2eq\ yr^{-1}$)
	Practices of Enteric Fermentation	Manure Left Over on Pasture Land	Application of Synthetic Fertilizer	Flooded Rice Cultivation	Manure Management Practices	Crop Residues Management	Addition of Manure on Soil Pools	Total Agricultural Practices	
Asia	698.04	224.75	366	436.1	125.64	64.86	30.8	45%	140
America	659.26	239.25	120	24.5	94.23	39.48	30.8	27%	2100
Europe	232.68	36.25	84	4.9	104.7	25.38	44	13%	0
Africa	271.46	181.25	18	19.6	13.96	8.46	3.3	12%	1120
Oceania	77.56	43.5	6	4.9	10.47	2.82	1.1	3%	105
World Total	1939	725	600	490	349	141	110	4354	3500

(Adapted from Tubiello et al., 2013).

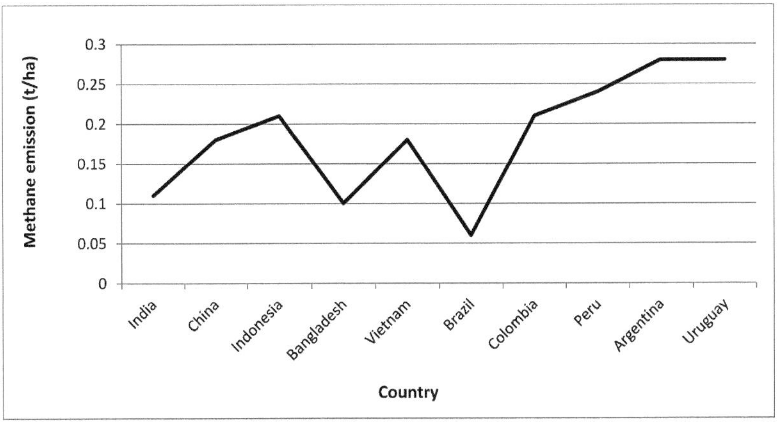

FIGURE 1.1 Methane emission from the rice cultivation practices by different countries in world (FAOSTAT, 2016).

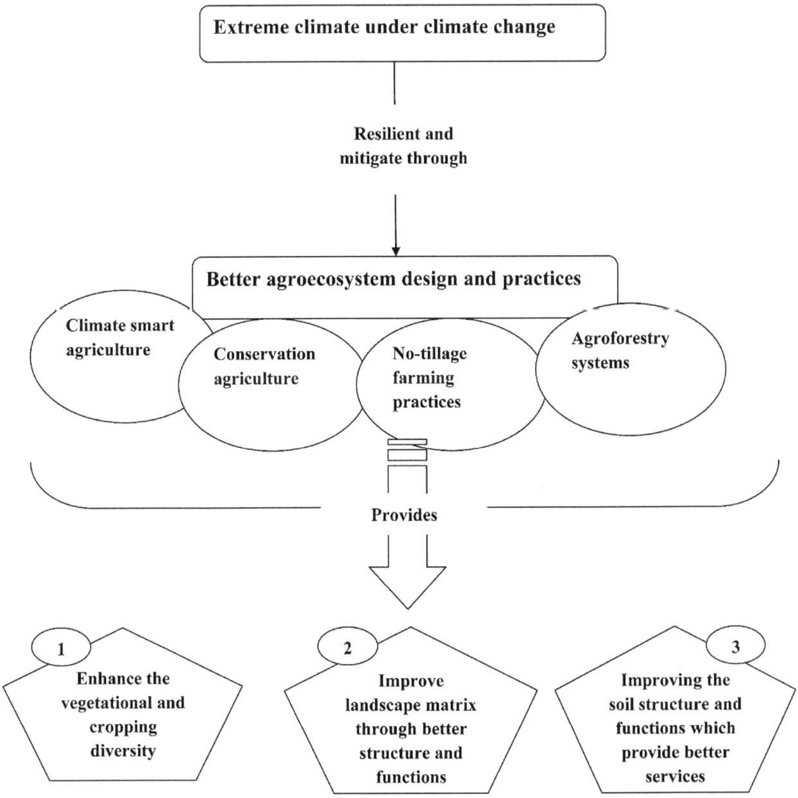

FIGURE 1.2 Extreme climate-resilient through the better agroecosystem practices.

helpful for achieving the food and nutritional security (FNS), maintaining soil ecosystem services and helps in enhancing storage and sequestration capacity of soil for mitigating climate change (CC) and global warming issues.

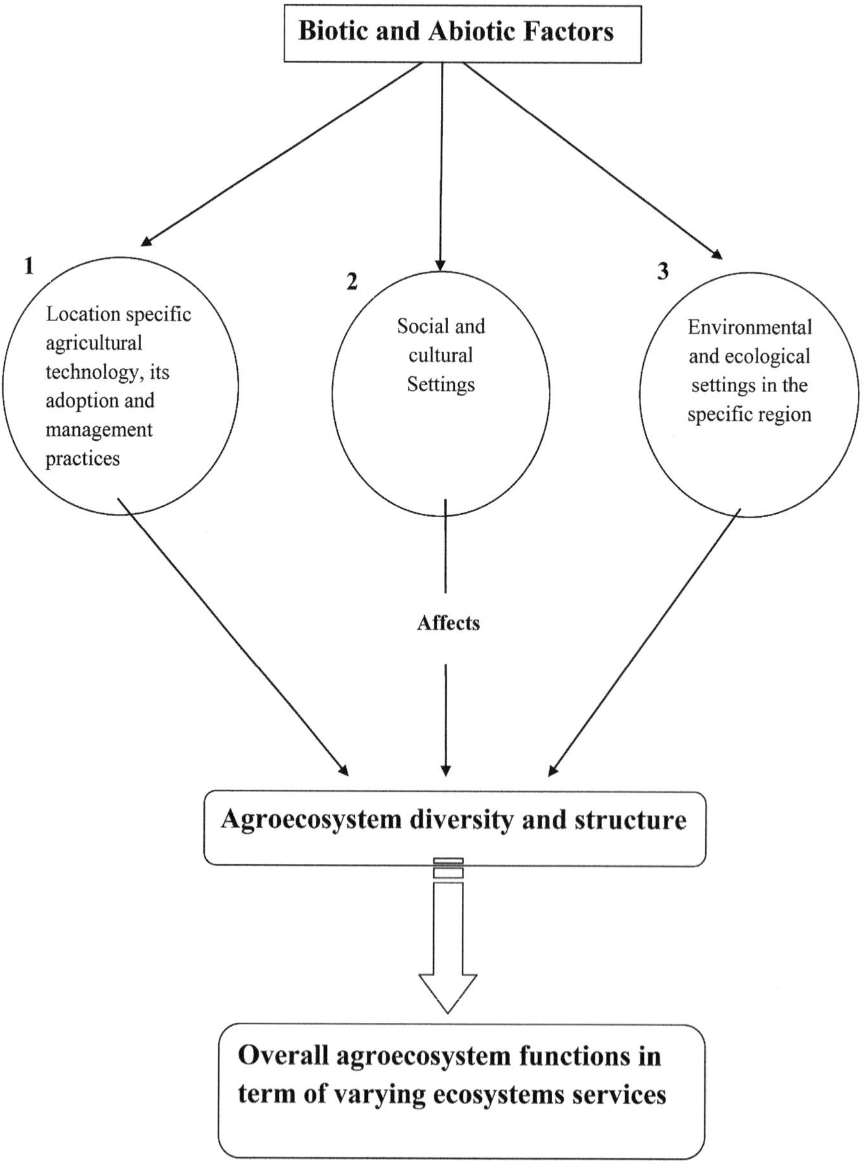

FIGURE 1.3 Factors affecting agroecosystem practices.

1.2 A GENERAL INTRODUCTION ON AGROECOSYSTEM

Agroecosystem represents modified and manipulated form of natural ecosystems through ecological intensification to enhance more productive and profitable farming system for people and surrounding environment and provides various essential ecosystem services (Zhang et al., 2007). Conservation agriculture, climate-smart agriculture, agroforestry, integrated farming system, organic farming, integration of microbial technology can helps in enhancing carbon absorption and retention in both soil and vegetation which in turn gives higher biomass along with reduction in continuous emission of GHGs (carbon emission) that represents more sustainable agroecosystem in contrary to conventional and intensive agricultural practices (Raj and Jhariya, 2017).

In general, scientific management of input elements (land, capital, and labor) are interlinked among them which provides better health of agroecosystem and extent of interactions determine the functions as ecosystem services through agroecosystem (Ilbery, 1985; Rodríguez et al., 2005; Felipe-Lucia et al., 2014). Also study of agroecosystem based on three essential dimensions viz., environment (ecosystem) level, economic, and human dimensions and these dimensions are needed to very well study for analyzing the complexity of agroecosystem on the both temporal and spatial scale (Smit and Smithers, 1994; Bernués et al., 2014). Therefore, the study of agroecosystem is largely depended on each and every components' and interaction among different dimensions.

1.3 GREENHOUSE GAS (GHG) EMISSION AND CLIMATE CHANGE (CC): A GLOBAL SITUATION

Climate is the long-term phenomenon that has changed and changing rapidly due to continuous emission of several GHGs into the atmosphere through natural and anthropogenic factors. Industrial development, mining, deforestation, application of a huge amount of chemical fertilizer in the agricultural field produce CO_2, which is the most potent gas cause changing climate and global warming due to extreme weather phenomenon. Livestock's also contributed somewhat in GHGs emissions by rigorous production manure by the production of pork, beef, and poultry (Leip et al., 2010).

There is a comparative study of GHGs emissions between developed and developing countries from various anthropogenic and natural activities. For example, CH_4 production from rice cultivation and manure management released 97% and 48% of GHGs in developing countries than developed countries having 52% of GHGs emissions. Similarly, developing countries alone release 92% of world GHGs emissions through biomass burning. Moreover, 82% of total CH_4 emission is contributed by South and East Asia through rice cultivation practices, whereas African and Latin American countries released approximate 74% of total emission through biomass burning process (US-EPA, 2006a).

1.4 FUTURE TRENDS OF GHGS EMISSIONS

The continuous emission of GHGs since the pre-industrial era to till date represents changing climate and global warming. As per one estimate, CH_4 and N_2O contributed enormous emissions which increased by 17% in the duration of 15 years (1990–2005) of which 88% of emission contributed by the practices of biomass burning, event of N_2O emission from the soil pools and enteric fermentation together (US-EPA, 2006a). On increasing demand for food and nutrition, the emission of N_2O will increase by 50% up to the year of 2020 due to intensive agricultural practices (Mosier and Kroeze, 2000). Similarly, due to increasing number of livestock's and cattle population will promote overall CH_4 production by 60% up to the year 2030 (FAO, 2003), and this can be happened due to changing manure management and feeding practices.

As per FAO (2003) and US-EPA (2006), the increasing area of rice cultivation (4.5%) by 2030 will promote 16% in the emission of CH_4 in the duration of 15 years (2005–2020). This will surely affect our ecosystem and biodiversity, and this can be mitigated by adopting conservation tillage practice with another scientific and eco-friendly farming practices along with minimizing deforestation activity (FAO, 2001).

1.5 GHGS FROM DIFFERENT AGROECOSYSTEM

Agricultural practices itself contributes in the emission of GHGs (approximately 10–12%, IPCC, 2007; Smith et al., 2007, 2008) through various unsustainable way of practices viz., heavy application of chemical fertilizers instead of biofertilizers and flooding of field under rice cultivation practices that create anaerobic conditions which can enhance the CH_4 emissions up to 90% in the atmosphere and this emissions are regulated by the both morphological and physiological characters of rice crops (Butterbach-Bahl et al., 1997; Das and Buruah, 2008; Baruah et al., 2010; Boateng et al., 2017; Wang et al., 2017). However, the application of mechanization, chemical, and inorganic fertilizers, pesticides, herbicides, etc. along with the adoption of that crops having the characteristics of high yields in agroecosystem release GHGs into the atmosphere.

Due to agricultural intensifications from the era of green revolutions through high inputs of chemical fertilizers has surely intensified the production but at the cost of environmental degradation by releasing GHGs into the atmosphere (Ajmal et al., 2018). Out of all GHGs, CO_2 is leading potent gas release from the soil through the transformations of the biogeochemical cycle. Similarly, soil releases the largest amount of N_2O through the synthesis and transformation of nitrogen-based fertilizers through the microbial works in the soil. Similarly, arid, and semi-arid agroecosystem have been experiencing intensive type of practices and management (due to extreme environment and water shortage event) such as higher disturbance of soil through tillage practices (emit 21% more CO_2 than untilled soil), application of higher dose of fertilizers, which affects the soil properties and environment and influence microbial activity which potentially leads to higher emission of GHGs (Abdalla et al., 2016; O'Dell et al., 2018; Tang et al., 2018; Behnke et al., 2018). Similarly, it has been seen about decreasing emission value (4.126 Tg yr^{-1} > 0.265 Tg yr^{-1} > 52.6 Tg yr^{-1}) of CH_4, $N_2O$and CO_2, respectively from intensively agroecosystem practices in the different land-use system of Indus Basin of Pakistan. Practice of rice cultivation and enteric fermentation are the major region for CH_4 emission, mineralization of OM and microbial process of denitrifications are the cause of emission of N_2O in the atmosphere whereas soil respiration, crop residue burning and OM decomposition leads CO_2 emission in the atmosphere (Iqbal and Goheer, 2008). Thus, the agroecosystem has two faces; either it can work as carbon absorption or carbon emission, which is based on the nature of practices and management.

1.6 EXTREME CLIMATIC IMPACTS ON AGROECOSYSTEM

The climate impacts are well understood for all the agriculturalists, industrialists, policymakers, developers, etc. The emissions of GHGs by both natural and anthropogenic activity are the global concern today, which not only affects our environment through global warming but also affects both terrestrial and ocean ecosystems. Extreme weather can potentially affect the vegetational characteristics (species richness, diversity, morphology) and reproductive biology along with wild animals. Agroecosystems pertain desirable modification in natural ecosystems by ecological intensification for enhancing the services of various outputs for the human and livestock's welfare. Due to the changing climate (uncertain rainfall, rising temperature, and humidity) affects the overall structure and functions of the agroecosystem. For example, Ecosystem services as tangible and intangible benefits, productivity, and sustainability, Loss of biodiversity, soil nutrient cycling and health, Fluctuation in ground and surface water results drought and flood, Affects FNS and human health's and land degradation and desertification's are the major impacts due to changing climate and global warming (Figure 1.4).

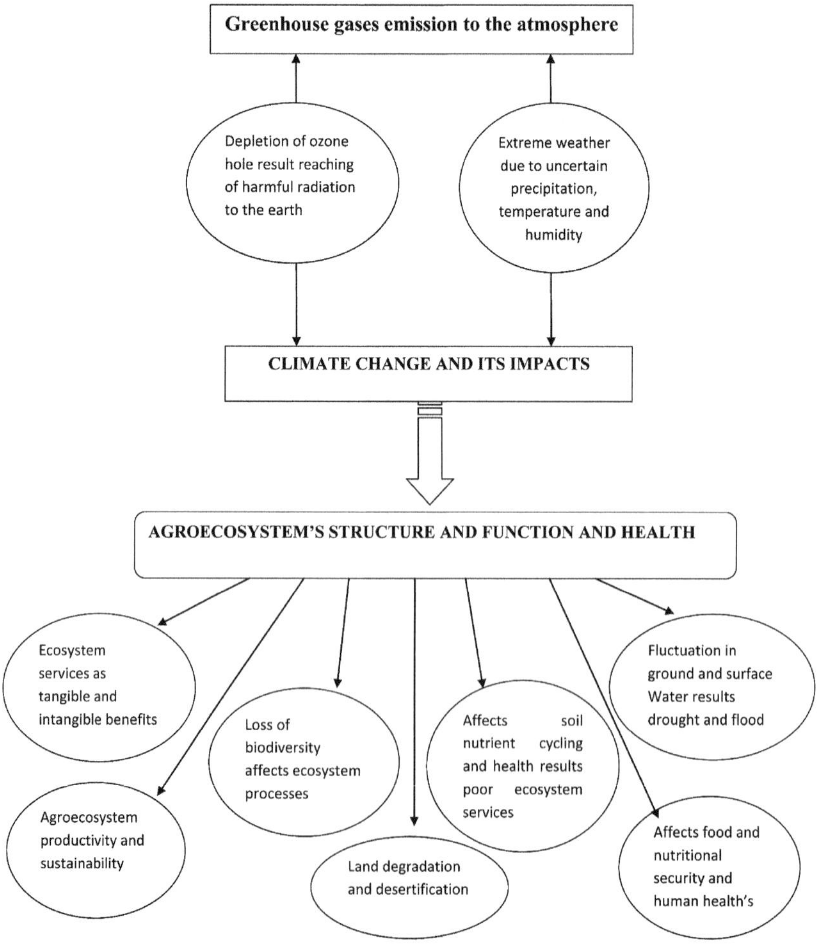

FIGURE 1.4 Climate change impacts on agroecosystem.

1.7 CLIMATE RESILIENT AGROECOSYSTEMS

The different practices have the capacity to build up carbon in the soil and vegetation system along with diversifying production and ecosystems. Therefore, climate-smart scientific farming practices of agroecosystem along with desirable ecological intensification and modification in natural ecosystem for satisfying the human needs in term of tangible and intangible benefits are the very authentic steps towards adaptation and mitigation of extreme climatic events along with sustainable environment and ecosystems (Jhariya et al., 2015, 2018).

Better management and practices of agroecosystem provides better landscape structure and management through various intermediate benefits such as enhancement in the overall biomass and productivity which satisfy the human's food and nutrition, promotes agroecosystem to provides both tangible (direct) and intangible benefits to the people and surrounding environment, provides efficient ecosystem services to the environment, Maintain soil structure and essential nutrient availability which is prerequisites for both plants and livestock's health and productivity, enhance the biodiversity which is very essential for better landscape, enhance carbon storage and sequestration capacity to both vegetation and soil pools that mitigate global CC issues, enhance the pollination an breeding program which helps in maintain progeny and overall population and reducing the land degradation and desertification problems (Figure 1.5) The major question today "How climate change can be minimized/checked through the practices of sustainable agroecosystem?" and "What are the different tools and practices to accomplish this goal?."

1.7.1 CLIMATE-SMART AGRICULTURE

Climate-smart agriculture (which is ecofriendly and in favor of socioeconomic development) is a very good strategy for mitigating extreme weather events (CC) by maximizing the potential of storage and sequestration of excess carbon from the atmosphere into the woody vegetation and soils that can enhance biomass productivity and strengthen farmer's income and livelihood. The practices of agriculture in a climate-smart way are not only helps in mitigating GHGs emission from the agricultural sector but also maintain soil, health, and environmental security. Thus, climate-smart agriculture meet out three objectives as reducing GHGs emission form several agricultural activity, enhancing the productivity and incomes that helps in FNS at national level and helps in making resilience to CC and global warming phenomenon (FAO, 2013, 2015; Lipper et al., 2014; Akrofi-Atitianti et al., 2018).

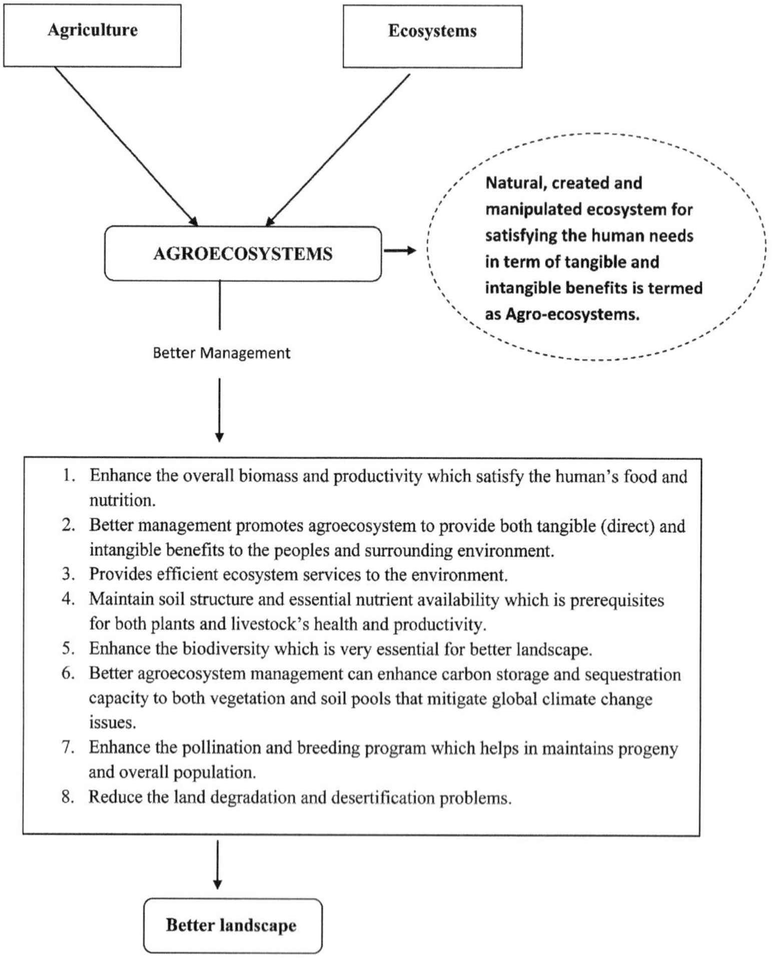

FIGURE 1.5 Agroecosystem management for better landscape.

1.7.2 AGROFORESTRY

CC event can be adapted through the scientific practices of various location-specific agroforestry models which is good strategy for minimizing the deleterious impacts changing climate through diversifying the structure and productivity which helps in atmospheric carbon fixation in the vegetation parts, promotes close and efficient nutrient cycling along with enhancement in SOC pools and enhance the production of goods and services in the sustainable ways along with carbon offset credits through better practices and management of AFs (Jhariya et al., 2015; Singh and Jhariya, 2016; Raj and Jhariya, 2017) (Figure 1.6).

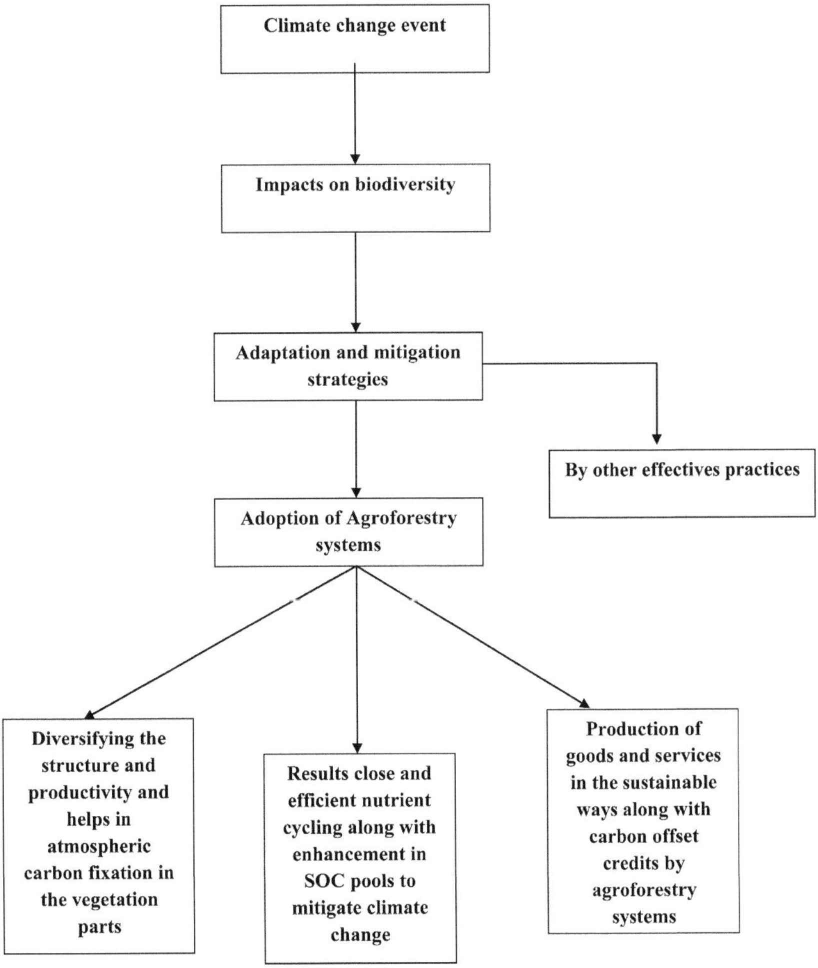

FIGURE 1.6 Model of agroforestry for climate mitigation.

Agroforestry (in contrary to monoculture system) is well known, location-specific, and resource conservation-based farming systems in the tropics that can diversify the yield and productivity along with the provision of various direct and indirect benefits to the rural people. Enhancement in the overall biomass productivity, improvement in the soil health and quality, soil, and water conservation, close, and efficient nutrient cycling, production of bio-energy and fuel woods, enhancement in storage and sequestration of carbon in soil and vegetation along with employment generation is the most peculiar significance which can be achieved through the adoption sustainable land use farming systems i.e., AFs (Dhyani et al., 2003; Buchman, 2008; Fanish and Priya, 2013; Jhariya et al., 2015; Raj et al., 2016).

Moreover, proper and scientific management of agroforestry can improve social and economic settings along with better landscape matrix. However, Agroforestry as the best substitute to fossils fuel, can mitigate the ongoing issues of extreme weather due to CC through the process of CS and its storage in both vegetation parts and soil (stores as organic carbon pools) (Mutuo et al., 2005; Sudha et al., 2007). Also, the pruning (tending operation) materials and liter falls add organic matter to the soils, which can soil carbon through the microbial decomposition of these organic residues. Several studies have been conducted in lieu of carbon storage and sequestration capacity of agroforestry models, and integration of trees in the cropping situations can impacts on carbon footprint in a positive way. For example, integration of *Cordia alliodora* in the AFs sequestered 14.2 Mg CO_2/hectare/year than the sole coffee plantation, which having less sequestration value (Andrade et al., 2008, 2014; Soto-Pinto et al., 2010).

1.7.3 CONSERVATION AGRICULTURE

Conservation agriculture is the great strategy to combat CC problem through its peculiar significance in soil and water conservation, maintaining the plant and human health through nutrient availability to the plants and FNS, efficient natural resource utilization and its management, enhancing livestock productivity and health, improving the overall farm productivity and quality, minimizing surplus emission of GHGs in the atmosphere along with the maintaining environmental sustainability and ecological security.

The main characteristics of conservation agriculture is tillage minimization, diversifying the yield and maintain vegetation covers throughout the years, addition of plant residues to the soil which helps in organic carbon building and application of several organic amendments to soils which keeps healthy and potential soil-inhabiting microorganism (Kassam et al., 2009; Hobbs and Govaerts, 2010; NASS, 2012; FAO, 2013a). Thus the good practices of conservation agriculture can enhance the storage and sequestration capacity of atmospheric carbon in the both vegetation and soils (as SOC) and which is the better option for CC mitigation (Govaerts et al., 2009; Milder et al., 2011; Sharma, 2011; Patle et al., 2013; Srinivasarao et al., 2015).

1.7.4 NO-TILLAGE PRACTICES

It is the well understanding subject of no-tillage, and zero tillage practices (contrary to tillage practices), which has very significant effects on the soil as these practices promotes the availability of organic carbon into the soils and helps in minimizing CO_2emission into the atmosphere resulting CC phenomenon. As per one estimate, around 50 million hectares (3.5%) areas of total arable land covered by zero or no-tillage practices (FAO, 2001). Disturbance of soils through various anthropogenic activities and tillage practices disturbs the soil and promotes an emission of CO_2 by loss of SOC through its decomposition and erosion activities (Madari et al., 2005). The key concern of today: Is the practices of zero/non-tillage always helpful in carbon-storing activity in the soils? There are various literatures that are available on the support of tillage practices and its significant effects on enhancing carbon gaining capacity of soil. Of course, it is often seen, but not always, and reported by various researchers (Gregorich et al., 2005; Alvarez, 2005). Similarly, non-tillage practices are not always supports reduction in N_2O emissions, but sometimes it promotes the emissions which depend on the soil characteristics along with prevailing climatic situations in that areas (Marland et al., 2001).

As per Chirinda et al., (2018), CH_4 emission can be reduced (up to 21 and 25%) by the adoption of zero tillage practices in contrary to tillage and soil disturbing practices. Similarly, a reduction in CH_4 emission (by 21 and 25%) through the adoption of no-tillage practice instead of conventional tillage method was observed in the loamy soil of the Southern Brazil region (Bayer et al., 2014, 2015). The above studies are also supported by Hanaki et al., (2002) and Ahmad et al., (2009) (21–60% CH_4 reduction), Ali et al., (2009) (54% CH_4 reduction), Bayer et al., (2014), and Zhang et al., (2014).

1.7.5 LIVESTOCK'S MANAGEMENT

Livestock's and cattle populations are also one of the contributors of GHGs emission into the atmosphere and emit CH_4 as 1/3 of world anthropogenic emissions (US-EPA, 2006). CH_4 and N_2O are produced by enteric fermentation and manures (Ji, 2012; Chhabra et al., 2013; Bhatta et al., 2015). Improved nutritional and feeding practices, effective dietary additives, efficient manure management practices, increased livestock's productivity through better breeding technology, etc. are the some effective mitigation strategies for reducing the higher emission of CH_4 and other harmful gases in the atmosphere (Burman et al., 2001; Soliva et al., 2006; Montes et al., 2013).

Moreover, the improvement of pasture quality and addition of oilseeds crops to their feeding materials can help to reduce the CH_4 emissions along with enhancement in the livestock's productivity and health (Jordan et al., 2006; Alcock and Hegarty, 2006; Toprak, 2015).

1.8 AGROECOSYSTEM DESIGN AND PRACTICES UNDER CHANGING CLIMATE

As we know, agroecosystem represents the changing/modification of natural ecosystem comprising plant (woody perennial trees, herbaceous crops, climbers, liana, etc.) and animals (livestock's) communities and their interaction with surrounding environments for various tangible and intangible benefits to the people and promotes the socio-economic upliftments, livelihood security along with environmental sustainability. But this sustainability can be achieved through a well-maintained scientific oriented design on the location-specific region. Also, the ecological approach and intensification are needed for studying the structure and function in agroecosystem design and management. Therefore, the design of farming systems should be in the view of maximizing the production of both crop and soil along with maximizing environmental health. Better agroecosystem design, according to location-specific results better soil health, overall productivity (Raj et al., 2018; Jhariya et al., 2015, 2018). The design should be in the eye of enhancing the productivity, diversifying yield, improving soil fertility along with enhancing the carbon storage and sequestration capability of both vegetation and soils (Figure 1.7). Although, good design can fulfill all the ecological, social, and economic goals.

1.9 AGROECOSYSTEM FOR TERRESTRIAL CARBON STOCKS AND FERTILITY OF SOILS

Soil is the largest terrestrial pools, and precious natural resource holds all living organism, provides various ecosystem services and stores approx. 1500 gigaton organic carbon which is two times more than both sum of vegetational biomass and atmospheric carbon (Lal, 2004; Davidson and Janssens, 2006; Tarnocai et al., 2009). As per one estimate, faulty land-use conversion accelerate CC phenomenon which contributes decline in total SOC at the rate of 10% of total fossil fuel SOC were estimated to contribute to a loss in SOC at a rate equivalent to 10% emission of the total fossil fuel in European country (Janssens et al., 2005). Intensive agriculture and agroecosystem farming practices affects the total organic carbon stocks in the soils as loss of 30–50% in the US due to adoption of conventional method of farming systems (Kucharik et al., 2001; Reicosky, 2003). This loss can be minimized through either reduction in tillage and SOC mineralization along with the enhancement/addition of organic matter and their proper decomposition in the soil (Lal, 2003).

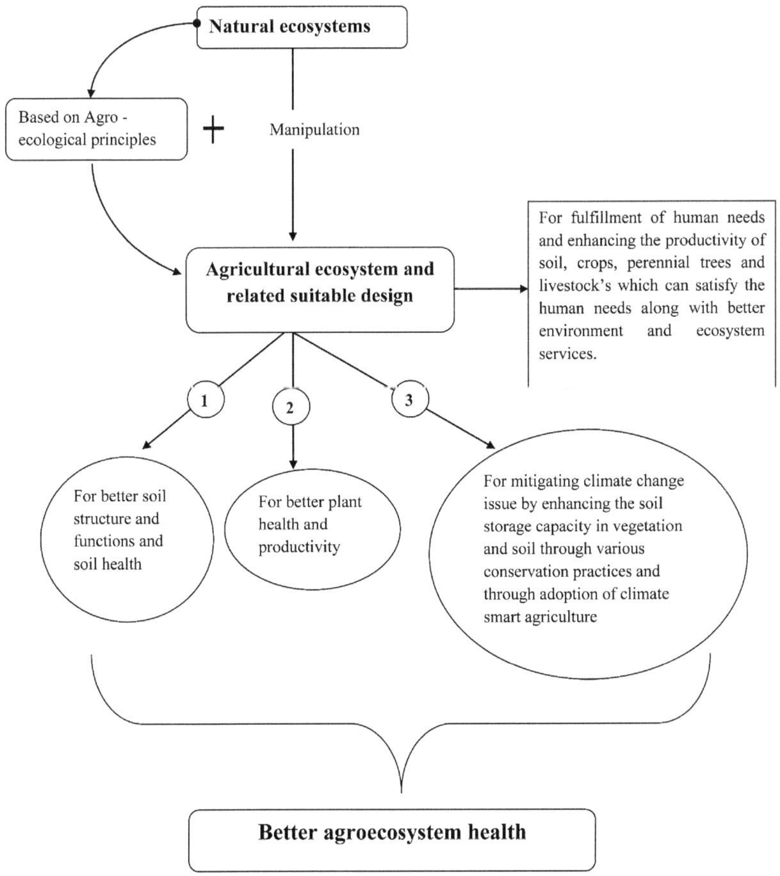

FIGURE 1.7 Agroecosystem design to agroecosystem health.

Soil has elastic in properties which can regain the lost SOC with the greater potential through the adoption of conservation agriculture practices and has various advantages as non-disturbance of soil in the tillage practices, provides plant residues for soil cover undercover crops systems and enhancing the plant diversity to maintain great structure and function of the ecosystem. Therefore, this practice can improve the soil structure and quality which is helpful in enhancing the sequestration capacity of carbon in the soils for mitigating CC impacts along with increasing the SOC pools by 0.2 to 0.7 ton/hectare/year (Lal et al., 1998; Pacala and Socolow, 2004; Basch et al., 2012).

1.10 RESEARCH AND DEVELOPMENT FOR CLIMATE CHANGE (CC) MITIGATION AND ADAPTATION

Today, a major challenge of all academicians, researchers, and scientist are to enhancing the greater productivity and profitability of agroecosystem practices without further emission of GHGs. This can be achieved through various ecofriendly farming practices and technology. Therefore, adoption of improved agronomical practices, proper rice management, developing ecofriendly farming models, location specific agroforestry model, sustainable land use practices, fire management, nutrient management practices, grazing, and livestock's management and research for pasture and grasses management are the prioritized thrust areas for the research and development for better agroecosystem structure and functions by mitigating continuous emissions of GHGs through enhancing CS and mitigating the impacts of CC and global warming (Follett, 2001; Aulakh et al., 2001; Yan et al., 2003; Lovett et al., 2003; Korontzi et al., 2003; Oelbermann et al., 2004; Paustian et al., 2004; Van Wilgen et al., 2004; Jordan et al., 2004; Clark et al., 2005; Pattey et al., 2005). Also, climate resilient agroecosystem practices are needed to maintain FNS, which in turn maintains socioeconomic and environmental security.

1.11 CONCLUSION

Adopting scientific and eco-friendly farming practices and technology are the potential options for enhancing the productivity of agroecosystem and maintenance of its structure and function along with better ecosystem services under the ongoing issues of extreme weather due to changing climatic events. Agroecosystem provides ecosystem services to all living treasure on the mother earth in direct (tangible) and indirect way (intangible). The better management of cropping system, reducing deforestation activity, adopting location specific farming practices, climate-smart agriculture, integration of leguminous tree in AFs, zero tillage practices, management of manure and livestock's population and its productivity are a good steps ahead for protection of our environment through minimizing the GHGs emission and its deleterious impacts on our ecosystem.

KEYWORDS

- **agroecosystem**
- **biodiversity**
- **climate-smart agriculture**
- **conservation agriculture**
- **ecosystem health**
- **greenhouse gases (GHGs)**
- **integrated farming systems (IFS)**

REFERENCES

Abdalla, K., Chivenge, P., Ciais, P., & Chaplot, V., (2016). No-tillage lessens soil CO_2 emissions the most under arid and sandy soil conditions: Results from a meta-analysis. *Biogeosciences, 13*, 3619–3633.

Ahmad, S., Li, C., Dai, G., Zhan, M., Wang, J., Pan, S., & Cao, C., (2009). Greenhouse gas emission from direct seeding paddy field under different rice tillage systems in central China. *Soil Tillage Res., 106*, 54–61.

Ajmal, M., Ali, H. I., Saeed, R., Akhtar, A., Tahir, M., Mehboob, M. Z., & Ayub, A., (2018). Biofertilizer as an alternative for chemical fertilizers. *Research and Reviews: Journal of Agriculture and Allied Sciences, 7*(1), 1–7.

Alcock, D., & Hegarty, R. S., (2006). Effects of pasture improvement on productivity, gross margin and methane emissions of a grazing sheep enterprise. In: Soliva, C. R., Takahashi, J., & Kreuzer, M., (eds.), *Greenhouse Gases and Animal Agriculture: An Update* (pp. 103–106). International Congress Series No. 1293, Elsevier, The Netherlands.

Ali, M. A., Lee, C. H., Lee, Y. B., & Kim, P. J., (2009). Silicate fertilization in no-tillage rice farming for mitigation of methane emission and increasing rice productivity. *Agric. Ecosyst. Environ., 132*, 16–22.

Alvarez, R., (2005). A review of nitrogen fertilizer and conservative tillage effects on soil organic storage. *Soil Use and Management, 21*, 38–52.

Andrade, H. J., Segura, M. A., Canal, D. S., Feria, M., Alvarado, J. J., Marín, L. M., Pachón, D., & Gómez, M. J., (2014). *The Carbon Footprint of Coffee Production Chains in Tolima, Colombia*. Wageningen Academic Publishers. doi: 10.3920/978–90–8686–788–2_3.

Andrade, H. J., Segura, M., Somarriba, E., & Villalobos, M., (2008). Valoraciónbiofísica y financiera de la fijación de carbono por uso del sueloenfincascacaoterasindígenas de Talamanca, Costa Rica. RevistaAgroforesteríaen las Américas *46*, 45–50.

Aulakh, M. S., Wassmann, R., Bueno, C., & Rennenberg, H., (2001). Impact of root exudates of different cultivars and plant development stages of rice (*Oryza sativa* L.) on methane production in a paddy soil. *Plant Soil, 230*, 77–86.

Baruah, K. K., Gogoi, B., & Gogoi, P., (2010). Plant physiological and soil characteristics associated with methane and nitrous oxide emission from rice paddy. *Physiology and Molecular Biology of Plants : An International Journal of Functional Plant Biology*, *16*(1), 79–91. http://doi.org/10.1007/s12298–010–0010–1 (Accessed on 9 October 2019).

Basch, G., Kassam, A., Gonzalez-Sanchez, E. J., & Streit, B., (2012). *Making Sustainable Agriculture real. CAP 2020*. Life + Agricarbon project. Available at: http://www.ecaf.org/docs/ecaf/ca%20and%20cap%202020.pdf (Accessed on 9 October 2019).

Bayer, C., Costa, F. S., Pedroso, G. M., Zschornack, T., Camargo, E. S., Lima, M. A., Frigheto, R. T. S., Gomes, J., Marcolin, E., & Macedo, V. R. M., (2014). Yield-scaled greenhouse gas emissions from flood irrigated rice under long-term conventional tillage and no-till systems in a Humid Subtropical climate. *Field Crop Res.*, *162*, 60–69.

Bayer, C., Zschornack, T., Pedroso, G. M., Da Rosa, C. M., Camargo, E. S., Boeni, M., Marcolin, E., Cecilia Dos Reis, C. E. S., & Dos Santos, D. C., (2015). A seven-year study on the effects of fall soil tillage on yield-scaled greenhouse gas emission from flood irrigated rice in a humid subtropical climate. *Soil Tillage Res.*, *145*, 118–125.

Behnke, G. D., Zuber, S. M., Pittelkow, C. M., Nafziger, E. D., & Villamil, M. B., (2018). Long-term crop rotation and tillage effects on soil greenhouse gas emissions and crop production in Illinois, USA. *Agriculture, Ecosystems and Environment*, *261*, 62–70.

Bernués, A., Rodríguez-Ortega, T., Ripoll-Bosch, R., & Alfnes, F., (2014). Socio-cultural and economic valuation of ecosystem services provided by Mediterranean mountain agroecosystems. *Plos One*, *9*(7), e102479.

Bhatta, R., Malik, P., Prasad, C., & Bhatta, R., (2015). Enteric methane emission: Status, mitigation and future challenges: An Indian perspective. *Livest. Prod. Clim. Chang.*, *6*, 229–244.

Boateng, K. K., Obeng, G. Y., & Mensah, E., (2017). Rice cultivation and greenhouse gas emissions: A review and conceptual framework with reference to Ghana. *Agriculture*, *7*(1), 7.

Buchman, N., (2008). *Agroforestry for Carbon Sequestration to Improve Small Farmers' Livelihoods*. From the North-South Centre Research for development.

Burman, K., Mohini, M., & Singhal, K. K., (2001). Effect of supplementation of rumens in and level of roughages on methane production. *Ind. J. Anim. Nutri.*, *18*, 325–329.

Butterbach-Bahl, K., Papen, H., & Rennenberg, H., (1997). Impact of gas transport through rice cultivars on methane emission from rice paddy fields. *Plant Cell Environmental*, *20*, 1175–1183.

Chhabra, A., Manjunath, K. R., Panigrahy, S., & Parihar, J. S., (2013). Greenhouse gas emissions from Indian livestock. *Clim. Change*, *117*, 329–344.

Chirinda, N., Arenas, L., Katto, M., Loaiza, S., Correa, F., Isthitani, M., et al., (2018). Sustainable and low greenhouse gas emitting rice production in Latin America and the Caribbean: A review on the transition from ideality to reality. *Sustainability*, *10*(671), 1–16.

Clark, H., Pinares, C., & De Klein, C., (2005). Methane and nitrous oxide emissions from grazed grasslands. In: McGilloway, D., (ed.), *Grassland: A Global Resource* (pp. 279–293). Wageningen Academic Publishers; Wageningen, The Netherlands.

Das, K., & Baruah, K. K., (2008). Methane emission associated with anatomical and morphophysiological characteristics of rice (*Oryza sativa*) plant. *Physiologia Plantarum, 134*, 303–312.

Davidson, E. A., & Janssens, I. A., (2006). Temperature sensitivity of soil carbon decomposition and feedbacks to climate change. *Nature, 440*, 165–173.

Dhyani, S. K., Sharda, V. N., & Sharma, A. R., (2003). Agroforestry for water resources conservation: Issues, challenges and strategies. In: Pathak, P. S., & Ram, N., (eds.). *Agroforestry: Potentials and Opportunities*. Agribios (India) and ISA.

Fanish, S. A., & Priya, R. S., (2013). Review on benefits of agro forestry system. *International Journal of Education and Research, 1*(1), 1–12.

FAO, (2001). *Soil Carbon Sequestration for Improved Land Management* (p. 58). World Soil Resources Reports No. 96. FAO, Rome.

FAO, (2003). *World Agriculture: Towards 2015/2030* (p. 97). An FAO Perspective. FAO, Rome.

FAO, (2011). *Linking Sustainability and Climate Financing: Implications for Agriculture*. FAO, Rome.

FAO, (2013). *Climate Smart Agriculture Source Book*. http://www.fao.org/docrep/018/i3325e/i3325e.pdf (Accessed on 9 October 2019).

FAO, (2013a). *Basic Principles of Conservation Agriculture*. www.fao.org/ag/ca/1a.html (Accessed on 9 October 2019).

FAO, (2015). *Final Report for the International Symposium on Agroecology for Food Security and Nutrition*. Rome, Italy.

FAO, (2016). *FAOSTAT Emissions Database, Agriculture, Rice Cultivation*. Available online: http://www.fao.org/faostat/en/#data/GR (Accessed on 9 October 2019).

Felipe-Lucia, M. R., Comín, F. A., & Bennett, E. M., (2014). Interactions among ecosystem services across land uses in a floodplain agroecosystem. *Ecology and Society, 19*(1), 20.

Follett, R. F., (2001). Organic carbon pools in grazing land soils. In: Follett, R. F., Kimble, J. M., & Lal, R., (eds.), *The Potential of U. S. Grazing Lands to Sequester Carbon and Mitigate the Greenhouse Effect* (pp. 65–86). Lewis; Boca Raton, FL.

Govaerts, B., Verhulst, N., Navarrete, C., Sayre, A., Dixon, K. D., & Dendooven, J. L., (2009b). Conservation agriculture and soil carbon sequestration: Between myth and farmer reality. *Crit. Rev. Plant Sci., 28*, 97–122.

Gregorich, E. G., Rochette, P., Van Den Bygaart, A. J., & Angers, D. A., (2005). Greenhouse gas contributions of agricultural soils and potential mitigation practices in Eastern Canada. *Soil and Tillage Research, 83*, 53–72.

Hanaki, M., Toyoaki, I., & Saigysa, M., (2002). Effect of no-tillage rice (*Oryza sativa* L.) cultivation on methane emission in three paddy fields of different soil types with rice straw application. *Jpn. Soc. Soil Sci. Plant Nutr., 73*, 135–143.

Hobbs, P. R., & Govaerts. B., (2010). How conservation agriculture can contribute to buffering climate change. In: Reynolds, M., (ed.) *Climate Change and Crop Production*. CABI, Wallingford, UK. Chapter 10.

Ilbery, B. W., (1985). *Agricultural Geography: A Social and Economic Analysis*. Oxford University Press, New York.

IPCC, (2007). Climate change 2007: Impacts, adaptation and vulnerability. In: Parry, M. L., Canziani, O. F., Palutikof, J. P., Van Der Linden, P. J., & Hanson, C. E., (eds.), *Contribution of Working Group II to the Fourth Assessment Report of the Intergovernmental Panel on*

Climate Change. Cambridge University Press, Cambridge, United Kingdom and New York, NY, USA.

Iqbal, M. M., & Goheer, M. A., (2008). Greenhouse gas emissions from agro-ecosystems and their contribution to environmental change in the Indus Basin of Pakistan. *Adv. Atmos. Sci., 25*, 1043.

Janssens, I. A., Freibauer, A., Schlamadinger, B., Ceulemans, R., Ciais, P., Dolman, A. J., Heimann, M., Nabuurs, G. J., Smith, P., Valentini, R., & Schulze, E. D., (2005). The carbon budget of terrestrial ecosystems at country scale – a European case study. *Biogeosciences, 2*, 15–26.

Janzen, H. H., (2004). Carbon cycling in earth systems–a soil science perspective. *Agriculture, Ecosystems and Environment, 104*, 399–417.

Jhariya, M. K., Banerjee, A., Yadav, D. K., & Raj, A., (2018). Leguminous trees an innovative tool for soil sustainability. In: Meena, R. S., Das, A., Yadav, G. S., & Lal, R., (eds.), *Legumes for Soil Health and Sustainable Management* (pp. 315–345). Springer, ISBN 978–981–13–0253–4 (eBook), ISBN: 978-981-13-0252-7 (Hardcover). https://doi.org/10.1007/978–981–13–0253–4_10 (Accessed on 9 October 2019).

Jhariya, M. K., Bargali, S. S., & Raj, A., (2015). Possibilities and perspectives of agroforestry. In: Miodrag, Z., (Ed.) *Chhattisgarh, Precious Forests – Precious Earth* (pp. 237–257). ISBN: 978–953–51–2175–6, In- Tech. doi: 10.5772/60841.

Ji, E. S., & Park, K. H., (2012). Methane and nitrous oxide emissions from livestock agriculture in 16 local administrative districts of Korea. *Asian-Australasian Journal of Animal Sciences, 25*(12), 1768–1774. http://doi.org/10.5713/ajas.2012.12418 (Accessed on 9 October 2019).

Johnson, J., Franzluebbers, A. J., Weyers, S. L., & Reicosky, D. C., (2007). Agricultural opportunities to mitigate greenhouse gas emissions. *Environmental Pollution, 150*, 107–124.

Jordan, E., Lovett, D. K., Hawkins, M., & O'Mara, F. P., (2004). The effect of varying levels of coconut oil on methane output from continental cross beef heifers. In: Weiske, A., (ed.), *Proc. Int. Conf. on Greenhouse Gas Emissions from Agriculture—Mitigation Options and Strategies* (pp. 124–130). Leipzig, Germany: Institute for Energy and Environment.

Jordan, E., Lovett, D. K., Hawkins, M., Callan, J., & O'Mara, F. P., (2006). The effect of varying levels of coconut oil on intake, digestibility and methane output from continental cross beef heifers. *Animal Science, 82*, 859–865.

Kassam, A., Friedrich, T., Shaxson, F., & Pretty, J., (2009). The spread of conservation agriculture: Justification, sustainability, and uptake. *International Journal of Agricultural Sustainability, 7*, 292–320.

Korontzi, S., Justice, C. O., & Scholes, R. J., (2003). Influence of timing and spatial extent of savanna fires in southern Africa on atmospheric emissions. *J. Arid. Environ., 54*, 395–404.

Krofi-Atitianti, F., Speranza, C. I., Bockel, L., & Asare, R., (2018). Assessing climate smart agriculture and its determinants of practice in Ghana: A case of the cocoa production system. *Land, 7*(30), 1–21.

Kucharik, C. J., Brye, K. R., Norman, J. M., Foley, J. A., Gower, S. T., & Brundy, L. G., (2001). Measurements and modeling of carbon and nitrogen cycling in agroecosystems of southern Wisconsin: Potential for SOC sequestration during the next 50 years. *Ecosystems, 4*, 237–258.

Lal, R., (2003). Global potential of soil carbon sequestration to mitigate the greenhouse effect. *Critical Reviews in Plant Sciences, 22*, 157–184.

Lal, R., (2004). Soil carbon sequestration to mitigate climate change. *Geoderma, 123*, 1–22.

Lal, R., Kimble, J. M., Follett, R. F., & Cole, C. V., (1998). *The Potential of US Croplands to Sequester Carbon and Mitigate the Greenhouse Effect.* Ann Arbor Press, Ann Arbor, MI, USA.

Leip, A., Weiss, F., Wassenaar, T., Perez, I., Fellmann, T., Loudjani, P., Tubiello, F., Grandgirard, D., Monni, S., & Biala, K., (2010). *Evaluation of the Livestock Sector's Contribution to the EU Greenhouse Gas Emissions (GGELS) –Final Report.* European Commission, Joint Research Centre.

Lipper, L., Thornton, P., Campbell, B. M., Baedeker, T., Braimoh, A., Bwalya, M., et al., (2014). Climate-smart agriculture for food security. *Nat. Clim. Change, 4,* 1068–1072.

Lovett, D., Lovell, S., Stack, L., Callan, J., Finlay, M., Connolly, J., & O'Mara, F. P., (2003). Effect of forage/concentrate ratio and dietary coconut oil level on methane output and performance of finishing beef heifers. *Livest. Prod. Sci., 84,* 135–146.

Madari, B., Machado, P. L. O. A., Torres, E., Andrade, A. G., & Valencia, L. I. O., (2005). No tillage and crop rotation effects on soil aggregation and organic carbon in a Fhodic Ferralsol from southern Brazil. *Soil and Tillage Research, 80,* 185–200.

Marland, G., McCarl, B. A., & Schneider, U. A., (2001). Soil carbon: policy and economics. *Climatic Change, 51,* 101–117.

Milder, J. C., Majanen, T., & Scherr, S. J., (2011). *Performance and Potential of Conservation Agriculture for Climate Change Adaptation and Mitigation in Sub-Saharan Africa.* WWF-CARE Alliance's Rural Futures Initiative. CARE: Atlanta, GA. (www.careclimatechange.org).

Montes, F., Meinen, R., Dell, C., Rotz, A., Hristov, A. N., Oh, J., Waghorn, G., Gerber, P. J., Henderson, B., Makkar, H. P., & Dijkstra, J., (2013). Mitigation of methane and nitrous oxide emissions from animal operations: II. A review of manure management mitigation options. *Journal of Animal Science, 91,* 5070–5094.

Moreau, T. L., Moore, J., & Mullinix, K., (2012). Mitigating agriculture greenhouse gas emissions: A review of scientific information for food system planning. *Journal of Agriculture, Food Systems, and Community Development, 2*(2), 237–246.

Mosier, A., & Kroeze, C., (2000). Potential impact on the global atmospheric N2O budget of the increased nitrogen input required to meet future global food demands. *Chemosphere-Global Change Science, 2,* 465–473.

Mutuo, P. K., Cadisch, G., Albrecht, A., Palm, C. A., & Verchot, L., (2005). Potential of agroforestry for carbon sequestration and mitigation of greenhouse gas emissions from soils in the tropics. *Nutr. Cycl. Agroecosyst., 71,* 43–54.

NASS, (2012). *Management of Crop Residues in the Context of Conservation Agriculture* (p. 12). National Academy of Agricultural Sciences, New Delhi; Policy paper no. 58.

O'Dell, D., Eash, N. S., Hicks, B. B., Oetting, N. N., Sauer, T. J., Lambert, D. M., Logan, J., Wright, W. C., & Zahn, J. A., (2018). Reducing CO_2 flux by decreasing tillage in Ohio: Overcoming conjecture with data. *Journal of Agricultural Science, 10*(3), 1–15.

Oelbermann, M., Voroney, R. P., & Gordon, A. M., (2004). Carbon sequestration in tropical and temperate agroforestry systems: A review with examples from Costa Rica and southern Canada. *Agric. Ecosyst. Environ., 104,* 359–377.

Oenema, O., Wrage, N., Velthof, G. L., Van Groenigen, J. W., Dolfing, J., & Kuikman, P. J., (2005). Trends in global nitrous oxide emissions from animal production systems. *Nutrient Cycling in Agroecosystems, 72,* 51–65.

Olander, L., Wollenberg, E., Tubiello, F., & Herold, M., (2013). Advancing agricultural greenhouse gas quantification. *Environmental Research Letters, 8*(1), 1–7.

Pacala, S., & Socolow, R., (2004). Stabilization wedges: Solving the climate problem for the next 50 years with current technologies. *Science*, *305*, 968–972.
Patle, G. T., Bandyopadhyay, K. K., & Singh, D. K., (2013). Impact of conservation agriculture and resource conservation technologies on carbon sequestration-a review. *Indian J. Agr. Sci.*, *83*(1), 3–13.
Pattey, E., Trzcinski, M. K., & Desjardins, R. L., (2005). Quantifying the reduction of greenhouse gas emissions as a result of composting dairy and beef cattle manure. *Nutr. Cycl. Agroecosyst.*, *72*, 173–187.
Paustian, K., Babcock, B. A., Hatfield, J., Lal, R., McCarl, B. A., McLaughlin, S., Mosier, A., Rice, C., Robertson, G. P., Rosenberg, N. J., Rosenzweig, C., Schlesinger, W. H., & Zilberman, D., (2004). *Agricultural Mitigation of Greenhouse Gases: Science and Policy Options* (p. 120). CAST (Council on Agricultural Science and Technology) Report, R141 2004, ISBN 1–887383–26–3.
Raj, A., & Jhariya, M. K., (2017). Sustainable agriculture with agroforestry: Adoption to climate change. In: Kumar, P. S., Kanwat, M., Meena, P. D., Kumar, V., & Alone, R. A., (ed.), *Climate Change and Sustainable Agriculture* (pp. 287–293). ISBN: 9789–3855–1672–6. New India Publishing Agency (NIPA), New Delhi, India.
Raj, A., Jhariya, M. K., & Bargali, S. S., (2016). Bund based agroforestry using *Eucalyptus* Species: A review. *Current Agriculture Research Journal, 4*(2), 148–158.
Raj, A., Jhariya, M. K., & Bargali, S. S., (2018). Climate smart agriculture and carbon sequestration. In: Pandey, C. B., Gaur, M. K., & Goyal, R. K., (ed.), *Climate Change and Agroforestry: Adaptation Mitigation and Livelihood Security* (pp. 1–19). ISBN: 9789–386546067. New India Publishing Agency (NIPA), New Delhi, India.
Reicosky, D. C., (2003). Tillage-induced CO_2 emissions and carbon sequestration: Effect of secondary tillage and compaction. In: Garcia-Torres, L., Benites, J., Martinez-Vilela, A., & Holgado-Cabrera, A., (eds.), *Conservation Agriculture* (pp. 291–300). Kluwer Academic Publishers, Dordrecht, the Netherlands.
Rodríguez, J. P., Beard, T. D. Jr., Agard, J. R. B., Bennett, E., Cork, S., Cumming, G., et al., (2005). Interactions among ecosystem services. In: Carpenter, S. R., Pingali, P. L., Bennett, E. M., & Zurek, M. B., (eds.), *Ecosystems and Human Well-Being: Scenarios* (Vol. 2, pp. 431–448). Findings of the Scenarios Working Group. Island, Washington, D. C., USA.
Scott, C. E., Monks, S. A., Spracklen, D. V., Arnold, S. R., Forster, P. M., Rap, A., et al., (2018). Impact on short-lived climate forcers increases projected warming due to deforestation. *Nature Communications*, *9*(1), doi: 10.1038/s41467–017–02412–4.
Sharma, K. L., (2011). Conservation agriculture for soil carbon build up, mitigation of climate change. In: Srinivasarao, C. H., Venkateswarlu, B., Srinivas, K., Kundu, S., & Singh, A. K., (eds.), *Soil Carbon Sequestration for Climate Change Mitigation, Food Security* (p. 322). Central Research Institute for Dryland Agriculture, Hyderabad.
Singh, N. R., & Jhariya, M. K., (2016). Agroforestry and agrihorticulture for higher income and resource conservation. In: Narain, S., & Rawat, S. K., (ed.), *Innovative Technology for Sustainable Agriculture Development* (pp. 125–145). ISBN: 978–81–7622–375–1. Biotech Books, New Delhi, India.
Smit, B., & Smithers, J., (1994). Sustainable agriculture and agroecosystem health. In: Nielson, N. O., (ed.), *Proceedings of an International Workshop on Agroecosystem Health* (pp. 31–38). University of Guelph, Guelph, Ontario.
Smith, P., Martino, D., Cai, Z., Gwary, D., Janzen, H., Kumar, P., McCarl, B., et al., (2007). Agriculture. In: Metz, B., Davidson, O. R., Bosch, P. R., Dave, R., & Meyer, L. A., (eds.), *Climate Change 2007: Mitigation. Contribution of Working Group III to the Fourth Assessment*

Report of the Intergovernmental Panel on Climate Change. Cambridge University Press, Cambridge, United Kingdom and New York, NY, USA.

Smith, P., Martino, D., Cai, Z., Gwary, D., Janzen, H., Kumar, P., & Smith, J., (2008). Greenhouse gas mitigation in agriculture. *Philosophical Transactions of the Royal Society B: Biological Sciences, 363*(1492), 789–813. http://doi.org/10.1098/rstb.2007.2184 (Accessed on 9 October 2019).

Soliva, C. R., Takahashi, J., & Kreuzer, M., (2006). *Greenhouse Gases and Animal Agriculture: An Update* (p. 377). International Congress Series No. 1293, Elsevier, The Netherlands.

Soto-Pinto, L. M., Anzueto, J., Mendoza, G., & Jimenez-Ferrer, B., (2010). Carbon sequestration through agroforestry in indigenous communities of Chiapas, Mexico. *Agroforestry Systems, 78*, 39–51.

Srinivasarao, C., Lal, R., Kundu, S., & Thakur, P., (2015) Conservation Agriculture and Soil Carbon Sequestration. In: Farooq, M., & Siddique, K., (eds), *Conservation Agriculture* (pp. 479–524). Springer, Cham.

Sudha, P., Ramprasad, V., Nagendra, M. D. V., Kulkarni, H. D., & Ravindranath, N. H., (2007). Development of an agroforestry carbon sequestration project in Khammam district, India. *Mitigat. Adapt. Strat. Climate Change, 12*, 1131–1152.

Tang, J., Wang, J., Li, Z., Wang, S., & Qu, Y., (2018). Effects of irrigation regime and nitrogen fertilizer management on CH_4, N_2O and CO_2 emissions from saline–alkaline paddy fields in Northeast China. *Sustainability, 10*(475), 1–15.

Tarnocai, C., Canadell, J. G., Schuur, E. A. G., Kuhry, P., Mazhitova, G., & Zimov, S., (2009). Soil organic carbon pools in the northern circumpolar permafrost region. *Global Biogeochemical Cycles, 23*, 1029–1040.

Toprak, N. N., (2015). Do fats reduce methane emission by ruminants? – a review. *Animal Science Papers and Reports, 33*(4), 305–321.

Tubiello, F. N., Salvatore, M., Rossi, S., Ferrara, A., Fitton, N., & Smith, P., (2013). The FAOSTAT database of greenhouse gas emissions from agriculture. *Environ. Res. Lett., 8*(015009), p. 10.

US-EPA, (2006). *Global Anthropogenic Non-CO_2 Greenhouse Gas Emissions: 1990–2020.* United States Environmental Protection Agency, EPA 430-R-06–003. Washington, D.C. http://www.epa.gov/nonco2/econ-inv/downloads/GlobalAnthroEmissionsReport.pdf (Accessed on 9 October 2019).

Van Wilgen, B. W., Govender, N., Biggs, H. C., Ntsala, D., & Funda, X. N., (2004). Response of savanna fire regimes to changing fire-management policies in a large African National Park. *Conserv. Biol., 18*, 1533–1540.

Wang, C., Lai, D. Y. F., Sardans, J., Wang, W., Zeng, C., & Peñuelas, J., (2017). Factors related with CH_4 and N_2O emissions from a paddy field: Clues for management implications. *PLoS One, 12*(1), e0169254. http://doi.org/10.1371/journal.pone.0169254 (Accessed on 9 October 2019).

Yan, X., Ohara, T., & Akimoto, H., (2003). Development of region-specific emission factors and estimation of methane emission from rice field in East, Southeast and South Asian countries. *Global Change Biol., 9*, 237–254.

Zhang, W., Ricketts, T. H., Kremen, C., Carney, K., & Swinton, S. M., (2007). Ecosystem services and dis-services to agriculture. *Ecological Economics, 64*, 253–260.

Zhang, X. X., Yin, S., Li, Y. S., Zhuang, H. L., Li, C. S., & Liu, C. J., (2014). Comparison of greenhouse gas emissions from rice paddy fields under different nitrogen fertilization loads in Chongming Island, Eastern China. *Sci. Total Environ., 472*, 381–388.

CHAPTER 2

Agroforestry for Climate Change Mitigation, Natural Resource Management, and Livelihood Security

ABHISHEK RAJ,[1] MANOJ KUMAR JHARIYA,[2] DHIRAJ KUMAR YADAV,[2] ARNAB BANERJEE,[3] and PRATAP TOPPO[1]

[1]Department of Forestry, College of Agriculture, Indira Gandhi Krishi Vishwavidyalaya, Raipur–492012, Chhattisgarh, India, Mobile: +00-91-8269718066, E-mail: ranger0392@gmail.com (A. Raj)

[2]Assistant Professor, University Teaching Department, Department of Farm Forestry, Sarguja Vishwavidyalaya, Ambikapur–497001, Chhattisgarh, India, Mobile: +00-91-9407004814, E-mail: manu9589@gmail.com (M. K. Jhariya); Mobile: +00-91-9926615061, E-mail: dheeraj_forestry@yahoo.com (D. K. Yadav)

[3]Assistant Professor, University Teaching Department, Department of Environmental Science, Sarguja Vishwavidyalaya, Ambikapur–497001, Chhattisgarh, India, Mobile: +00-91-9926470656, E-mail: arnabenvsc@yahoo.co.in

ABSTRACT

Agroforestry–a science-based sustainable land use farming system, commonly location-specific, and also a type of integrated farming system–is due to the integration of three elements (woody trees, crops/herbaceous annual plants and livestock's animals) in the unit piece of land of either marginal or large landholding farming community. This system gains wider recognition due to its wider adoptability and applicability in the tropics and multifunctional role in resource conservation, production potential as timber, fuelwood, biofuel, and non-timber forest products (NTFPs) (including several nutritious and healthy fruits), protection potential i.e., minimizing the curse of climate change (CC) and global warming through buffer maximum temperature, improving livelihood generation through enhancing socioeconomic status of farmers. Carbon (C) storage and sequestration are another potential of agroforestry system (AFs) (also known as C farming system) in which excessive atmospheric can be fixed, absorb, and stabilize into vegetation (woody part) and soils. Thus, there is need to adopt location specific agroforestry models which is managed in such a way that will be benefited in term of high production, provide more protection to crops, provide highest net income, improve soil health and quality, enhance socioeconomic and living standard of people, minimize CC and global warming and utilize natural resource in more efficient way.

2.1 INTRODUCTION

Today, the word agroforestry is well understood among all farming practitioners, experts, and policymakers, etc. due to its easy adoptability, accessibility, and practices in different agro-ecological regions. Different types of agroforestry models are exist due to its location-specific adoptability of combination of three components such as woody tree species, herbaceous crop, and livestock's/pastures (Figure 2.1). Three components viz., trees (leguminous preferable), annual crops, and livestock's/grasses are the prerequisites for any type of agroforestry model which can be adopted on a single piece of land, which are socio-economically and environmentally acceptable (Nair, 1979).

Agroforestry contributes conservation and management of natural resources, helps in meeting out national demands for timber, fuelwood, and non-timber forest products (NTFPs) along with neutralizing climate change (CC) and global warming (Verchot et al., 2007; Raj et al., 2018). However, the practices of agroforestry can help in meeting people's needs and reduce the dependency on natural forest and maintain forest covers (33% recommended by national forest policy), which are affected by deforestation and other illegal practices. Large and progressive farmers have remarkable adaptations and strategies for mitigating CC phenomenon by adopting AFs and other scientific land use management practices, whereas small and marginal farmers could not be able to do this due to lack of scientific knowledge, skill, and natural resources.

As per Verchot et al., (2007) and Nair and Nair (2014), agroforestry practices could help in mitigating extreme weather events and reducing greenhouse gases (GHGs) through the process of carbon (C) sequestration, i.e., storing of atmospheric C into the vegetation and soils. Thus the intimate mixture of agroforestry elements (trees, crops, and animals) are the basis of sustainable land use system, which is socioeconomically viable and ecologically acceptable (Nair, 1979). Therefore, agroforestry has wider recognition in the biomass production enhancement, soil fertility improvement, storage, and sequestration of C, microclimate improvement, and the production of timber, fuelwood, NTFPs, and biofuel (Fanish and Priya, 2013). Although, incorporation of nitrogen-fixing trees (multipurpose in nature) in agricultural farm and their interaction with crops affects structure, adoptability, and overall productivity of the models along with soil health and quality and inhabiting organisms functions which can maintain the ecosystem (Raj et al., 2014a; Jhariya et al., 2015, 2018).

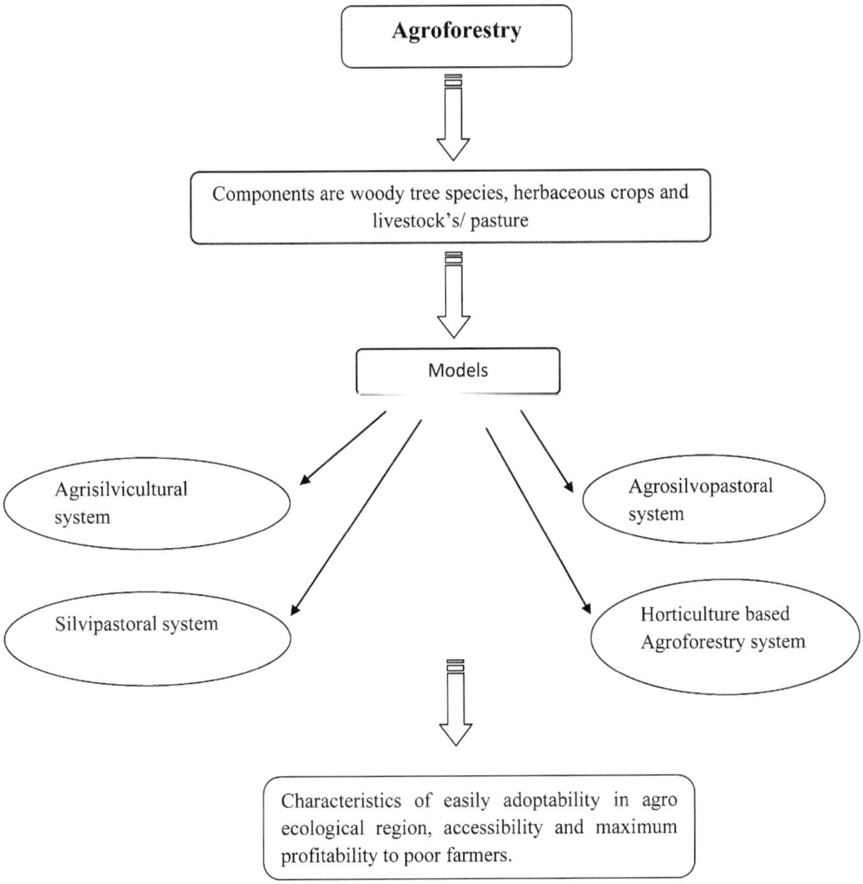

FIGURE 2.1 Agroforestry classification.

The biological properties of soil, i.e., soil microbial activity and its functions can determine the soil quality through its linking concept with soil physical properties (Raj et al., 2014b). Of all functional potential of agroforestry, C storage and sequestration are gaining wide recognition due to its properties of reducing GHGs in the atmosphere and mitigating CC and global warming phenomenon by fixation of atmospheric C into vegetation and soils. This C fixation has two benefits, i.e., first represent lowering the excessive heat in our environment whereas seconds represents stored C represents biomass production in term of timber, fuelwood, biofuel, and NTFPs which is basic materials utilized by the poor farmers and their source of income to strengthening livelihood security (Alavalapati and Nair, 2001; Sudha et al., 2007; Parihaar, 2016). As per Dhyani et al., (2013), biomass production through AFs in rainfed, arid, and semiarid areas were varied from 2–10 ton/ha/yr. Thus, the decision to adopting agroforestry and its scientific management never goes vain, and its contribution to resource management and its utilization along with CC mitigation are the appreciable.

In this context, this chapter review about possibilities and potential of AFs and its multifunctional role in the productivity enhancement, socio-economic upliftments, natural resource management (NRM) (soil and others), and CC mitigation.

2.2 AGROFORESTRY: AN AREA COVERAGE

Agroforestry is adopted throughout the tropics due to its diverse nature of tree-crop combinations and wide adoptability in agro-ecological region (location specific), which entails its characteristics of sustainable land use systems along with economically profitability. Various authors have been reported areas of agroforestry in India (25.32 Mha by Dhyani et al., 2013) and the world (1,023 Mha by Nair et al., 2009) (Table 2.1). In world review, agroforestry practices such as silvopastoral system, alley-cropping, riparian buffers, and wind and shelterbelt covered an area of 235.2 million hectares in the United States of America (Nair and Nair, 2003). Similarly, the area of the home garden has been reported as 8.0 million hectares in S.E. Asia (Kumar, 2006).

2.3 POSSIBILITIES AND POTENTIAL OF AGROFORESTRY

Diversifying production, biodiversity management, CC mitigation, water, and nutrient management, enhancing soil fertility, etc. are the foremost potential of AFs (Singh and Jhariya, 2016). Agroforestry can be potentially adopted in all tropic regions whenever and wherever possible. A small and large area is available in the plain and hilly region (modified by land topography, slope, and aspects) for adopting different agroforestry models as sustainable land-use practices. Both leguminous (nitrogen-fixing) and multipurpose tree species (MPTs) are the basic elements for practicing AFs which entails the structure, management, and interaction among various components of trees (perennial trees and other timber trees) and crops and delivers both tangible (timber, fuelwood, biofuel, and NTFPs) and intangible benefits (climate and environment improvement, soil management, efficient nutrient cycling, food and nutritional security (FNS) and livelihood improvement etc.) to ecosystem (Jhariya et al., 2015).

TABLE 2.1 Agroforestry Area in India and World

World	India	Area	Divisible Into
1,023 Mha by Nair et al., (2009)	25.32 Mha by Dhyani et al., (2013)	20 M ha	Irrigated area covers 7.0 Mha
			Rainfed area covers 13.0 Mha
		5.32 Mha	Shifting cultivation covers 2.28 Mha
			Home gardens covers 2.93 Mha

Thus, the practices of agroforestry are the foremost option to ameliorate the waste and degraded land through its wider adoptability, extreme climate solving strategies, diversifying food productivity, maintaining soil fertility, reducing soil erosion, combating desertification, enhancing employment opportunities, improving economic and maintaining livelihood security (Raj and Jhariya, 2017).

2.4 AGROFORESTRY FOREST FOR MULTIFARIOUS BENEFITS

Agroforestry provides diverse ecosystem services and has gained importance due to characteristics of multifarious benefits such as provides multiple products (timber and NTFPs) (Rahman et al., 2016) which is economically benefited to farmers and helps in strengthening livelihood security (Kassie, 2018); provides diverse types of food and fruits which guarantees health and nutritional security; protect and provides shelter to all organism which helps in management and conservation of biodiversity and absorbs atmospheric C and stores into both vegetation and soils which shows CC mitigation and adaptation strategies along with environmental security (Figure 2.2).

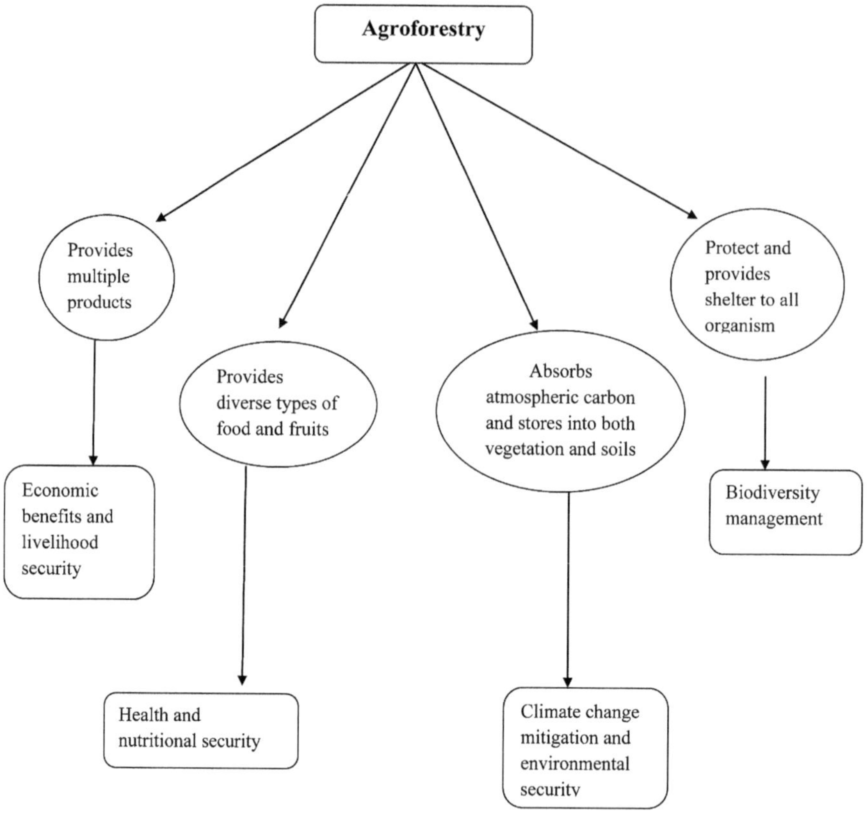

FIGURE 2.2 Agroforestry for multifarious benefits.

2.5 AGROFORESTRY FOR NATURAL RESOURCE MANAGEMENT (NRM)

Natural resources are the treasure of our motherland, which sustains our lives and maintains the structure and functions of ecosystem. For the perspective of its management, a scientific management and location-specific farming practices are the necessary for the proper utilization of resources like soil, water, nutrients, etc. Therefore, agroforestry is the best strategy for the resources conservation and their efficient utilization. In agroforestry and other integrated types of farming system, all components are interdependent to each other and waste from one component are efficiently utilized by others, i.e., the chances of losses are minimized along with the provision of ecosystem services.

Leaching losses of essential elements and ions can be minimized through the incorporation of long taped leguminous and MPTs, i.e., it follows the close type of nutrient cycling rather than the open type of nutrient cycling in sole based cropping system. Similarly, better choice to use some efficient nitrogen-fixing trees (MPTs) in an agricultural farm under AFs, which not only improve the crop and soil productivity but also helps in proper utilization of natural resources (Jhariya et al., 2018). Therefore, agroforestry plays a big role in NRM and its conservation, such as soil health and fertility management, biodiversity management, nutrient management, and efficient nutrient cycling, food, and nutritional management and climate and environment management (Figure 2.3). Thus, agroforestry performs remarkable functions and gaining wider scope in terms of resource conservation and productivity than the sole based farming system. However, scientific management and a combination of suitable species (tree-crop combination) are the prerequisite for long-term benefits (direct and indirect) form any agroforestry models along with conservation of biodiversity and soil and water resources.

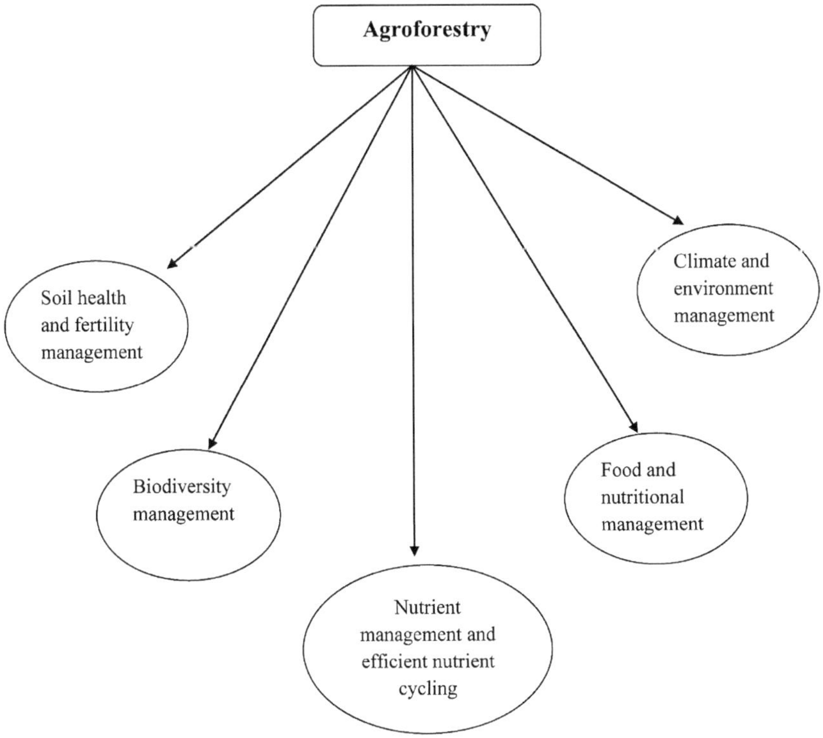

FIGURE 2.3 Agroforestry for natural resource management.

We should not overlook tree-crop interaction because it plays a major and potential role in overall productivity enhancement. A good and managed farming system can reduce competition among crops for light, space, soil, nutrition, and water and helps in resource distribution in tree and crop systems in efficiently and properly. Of all, water is very important resource and gaining utmost recognition in whole plant metabolic function, growth, and development. As per Ong et al., (2006), good, and suitable combination of species can maximize the water use efficiency and its distribution among plants because it is major source of translocation and distribution of ion, minerals, and essential nutrients. Several questions have been arises in the context of agroforestry and its role in NRM along with yield enhancement. But one question is "how can we enhance both yield productivity of plants along with improving resource captures and its distributions among the different elements/components of AFs?"

2.6 BIODIVERSITY MANAGEMENT THROUGH AFS

AFs provide shelter to different organisms such as different trees, crops, and livestock animals, which are integrated itself and maintain biodiversity that plays a crucial role in structure and diversity maintenance. Similarly soil parts stores variety of organism such as beneficial microorganism, bacteria, fungi, protozoan's, earthworm, and other macro organism which not only helps in decaying of organic matter and other residues but also maintain biodiversity along with nurturing plants and animals along with healthy ecosystem and environment under changing climate and global warming (Korn et al., 2003; Schroth et al., 2004; McNeely and Schroth, 2006; Vallejo-Ramos et al., 2016). Incorporating woody perennial trees (having MPTs and leguminous in characteristics) into agricultural farm (either as scattered or on boundary) of small landholding farmers (having 0.01–5.00 hectare land) in the tropics are the good examples of AFs which provides ecosystem services, guarantees socio-ecological adaptation and maintains biodiversity conservation in any circumstances (Albrecht and Kandji, 2003; Montagnini and Nair, 2004; Scales and Marsden, 2008; Nair et al., 2009; Perfecto and Vandermeer, 2010; Zimmerer, 2015).

Moreover, AFs has greatest potential in biodiversity conservation and higher biomass production in addition to promising natural resources management and its utilization than other conventional agricultural practices. However, subsistence farmers in the tropics face the constraint of low productive soil having characteristics of low availability of essential nutrients, acidic, and highly leached soils and these farmers generally cannot access mineral fertilizers for their farms (Glaser et al., 2002). Similarly, soil fertility and quality enhancement, C input, organic matter addition and nutrient status maintenance through litter inputs other than biodiversity conservation are another function of agroforestry ecosystems relative to sole cropping or other pasture production system (Kirby and Potvin, 2007; Lal, 2004; Jose, 2009; Wright et al., 2012; Tscharntke et al., 2012).

2.7 SOIL MANAGEMENT THROUGH AFS

Without soil, the imagination of lives does not exist. Surely, soils are the largest pools of the terrestrial ecosystem, can deliver several ecosystem services which plays as basic substratum for flora and fauna, stores millions and billions of micro and macro organism, fungi, protozoan's, earthworms, stores essential elements which plays major role in growth and development of plants and maintaining ecosystem structure and functions. As we discussed earlier, agroforestry deserves close and efficient nutrient cycling rather than an open type in the single cropping system. In close types, nutrients are not leached and recycle and mobilize from earth to plants and reverse in very efficiently, whereas chances of nutrient losses are prominent in open type nutrient cycling (Jhariya et al., 2018).

Although, provision of ecosystem service through soil is gaining wider recognition in terms of soil structure maintenance, fertility enhancement, maintaining nutrient loads, nurturing inhabiting organism, efficient nutrient cycling, storage, and sequestration of C, etc. and these all services are important for maintaining soil health. The litterfall and other debris input from trees in the AFs can add organic matter in soil, which disintegrate (with the help of microorganism) into simpler form and can be utilized by tree crop system which can maintain the productivity status of whole system along with healthy ecosystem. Similarly, disintegration, and decaying of belowground root system can add essential nutrient and enhance fertility of soils and these nutrients are the primary source of tree crop system.

Similarly, agroforestry can improve the soil biological systems i.e., population, and health of soil-inhabiting organism, root nodule Rhizobium, fungi, protozoan's, earthworm, and several others organism which is the basic organism for both disintegration of organic matter and its proper distribution and translocation in plants which shows healthy sign of ecosystem structure and functions (Bertin et al., 2003). For example, earthworm (farmer-friendly) gain wide recognition in term disintegration and decaying of coarse organic substance and its proper distribution in lower to upper soil layer (also in reverse), which maintains status of organic and mineral soil layers. Moreover, disintegration of organic matter can add C input in soil and accumulation of C in soils can depends on tree species types, their nature, litterfall types, its amount, rate of disintegration, species involves in decaying of organic residues, etc. This C input can help in addition to more C into soils, which plays a basic functions in C storage and sequestration and helps in mitigating CC and global warming phenomenon of today. Thus, AFs add litter and other debris on agroforestry floor and its decomposition release essential nutrients which improve the soil fertility and nutrient status that results in close and efficient nutrient cycling and healthy ecosystem and environment (Figure 2.4).

2.8 NUTRIENT CYCLING IN AFS

Integrating woody trees with herbaceous crops in agricultural farms arbitrate an efficient close nutrient cycling and enhance availability, supplying, and distribution of nutrients in the root zone of crops and minimize the nutrient losses. As we aware about AFs and its potential in both environment and economic benefits whereas nutrient cycling and its management are the new paradigm which is highly linked with productivity and profitability enhancement (Nair et al., 2008; Gold and Garrett, 2009). Although, the practices of AFs guarantees the soil health improvement through fertility enhancement in varying soil-plant systems and related ecological functions. Similarly, biological parts of soil under AFs and other forest system are playing an inevitable role in decaying and decomposition of organic residues that plays a major role in nutrient cycling and related fertility management (Kaur et al., 2000; Salamanca et al., 2002; Sharma et al., 2004).

Similarly, the study has been conducted to identify linking concept of soil microbial biomass, essential nutrients like C, nitrogen, and phosphorus with organic matter and fertility of soil in various AFs (Kaur et al., 2000; Amatya et al., 2002; Lee and Jose, 2003; Zaia et al., 2008). However, soil-inhabiting essential microbes have subtle importance in both decomposition of organic matter, and it is cycling as nutrient cycling process. Although, soil microbes are a good indicator to understand changing soil quality and health status, which varies from soil types and its management practices under varying environmental conditions in various agroecosystems (Dinesh et al., 2003; Ndaw et al., 2009). Therefore, the productivity of AFs and its soil nutrient availability generally depend on population and functional activity of inhabiting microbes and its role in decaying and decomposing of organic matter along with nutrient releasing and its availability to plants (Moore et al., 2000). Similarly, leguminous trees having profound potential to fix atmospheric nitrogen with the help of Rhizobium bacteria in root nodules that impact positively in both soil improvement and productivity of AFs. Through litterfall and its decomposition along with root decaying can release this fixed nitrogen and cycling of nutrient occurs (Khanna, 1998; Mafongoya et al., 2004).

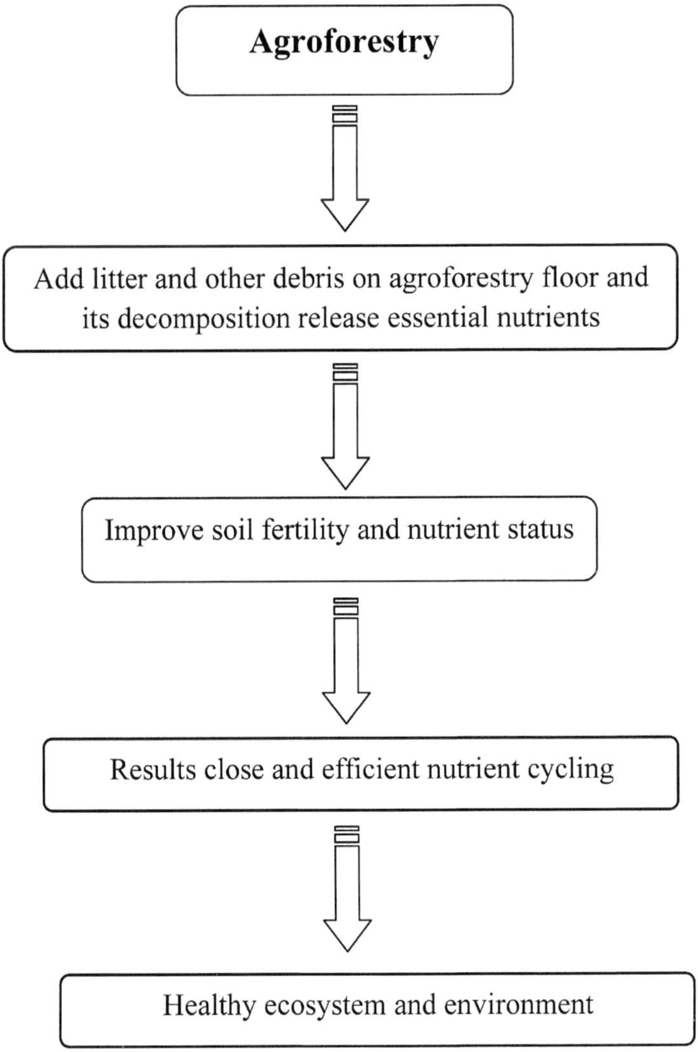

FIGURE 2.4 Agroforestry for healthy soil and efficient nutrient cycling.

2.9 CLIMATE CHANGE (CC) MITIGATION THROUGH AGROFORESTRY SYSTEM (AFS)

CC is the burning topic today which is the major curse of our environment affects growth and development of all organisms including plant community in term of morphological, anatomical, physiological, and phonological changes along with reduction in health and productivity. Developmental projects, industrialization, and other anthropogenic and natural activities release the major potent gas C dioxide in the atmosphere along with other GHGs which warms and raise the earth atmospheric temperature results loss of health of organism and wealth of national properties (Forster et al., 2007).

In this context, forest and AFs plays undeniable role in absorption and fixation of atmospheric C and other gases into woody perennial trees and its components (foliage, bole, branch, and underground root system) along with soil (through soil C sequestration). Although, integration of woody perennial trees with some pastures/grasses under silvopastoral systems has greatest potential in both land reclamation and CC mitigation through C sequestration process (Hoosbeek et al., 2018).

Moreover, the fixation and distribution of C is directly linked with the species types, their absorption capacity, agroforestry model types, and its absorption potential and prevailing climatic situation. Similarly, the practices of agroforestry, including biofuels and other fuel production system, are a good option to reduce dependency on fossil fuel, which is the major source of GHGs emissions in the atmosphere. It is a good choice to incorporate a fast-growing tree in the agricultural farming system, which helps in C absorption and its retention in the woody parts of tree and whole system based on C farming system. As per one estimates, AFs in humid tropic retain high C which varies from 0.3–15.2 MC per ha per year (Nair et al., 2011). Similarly, integrating MPTs into agricultural farm under AFs can reduces N_2O emissions by through the uptake of nitrogen and its storage by woody perennial tree (Allen et al., 2004, 2009, 2010; Bergeron et al., 2011).

2.10 FOOD AND NUTRITIONAL SECURITY (FNS) THROUGH AF

Indeed, food is the prime and foremost commodity for all living organism on the earth and without it lives does not exist. Quality and nutritious food are the basis of healthy and quality life. In this context, agroforestry can diversify the food system along with the provision of nutritious food and fruits which maintain the food and nutrition security. Therefore, agroforestry is linked with healthy and nutritious food and fruits which is the basis of healthy and quality life results FNS which is the pillar of national security and socioeconomic development and livelihood security (Figure 2.5).

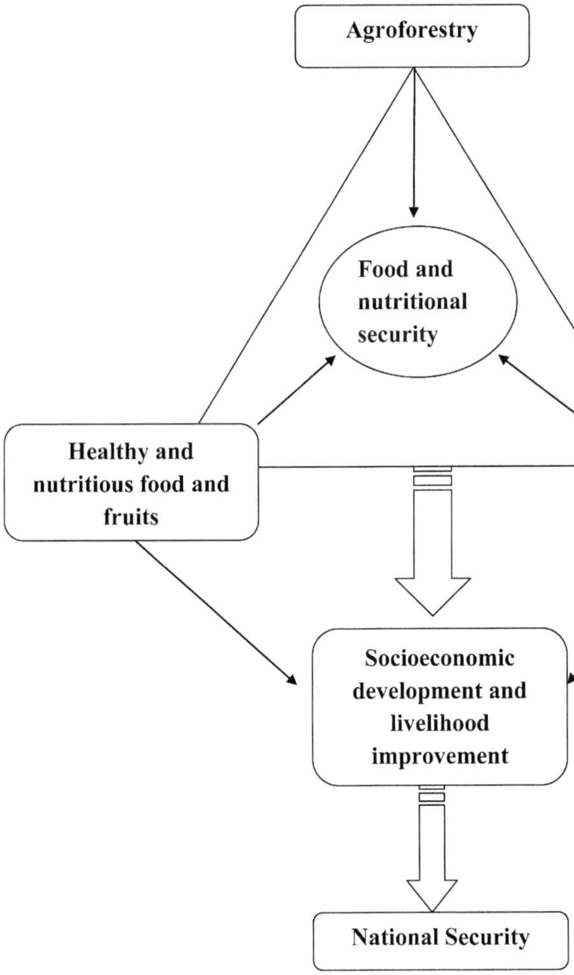

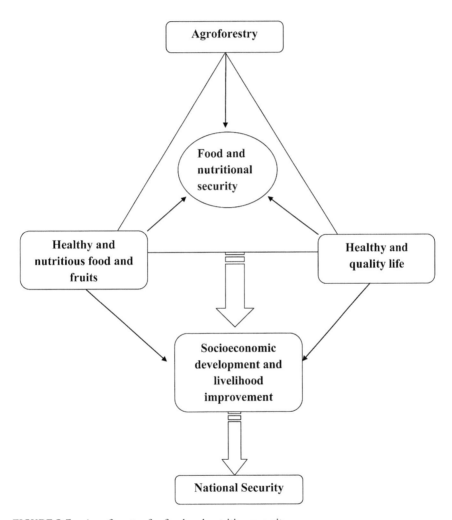

FIGURE 2.5 Agroforestry for food and nutrition security.

Also, integration of fruit trees in horticultural based AFs could potentially provide various nutritious fruits and other food materials to the farmers which help in maintaining health and nutrition along with enhancement in economic profitability through the sale of surplus production results socioeconomic and livelihood improvement. The yield potential of fruits, vegetables, medicinal plants, tubers, and some fiber products are the basis characteristics of horti-based farming systems which can utilized all resources efficiently and properly. This system provides essential nutrients such as protein, vitamins, and others minerals to people and helps in accomplishing the daily diet of fruits (85 grams) and vegetables (220 grams) (Roy, 2011). Mango, papaya, Aonla, and citrus fruits based AFs are very prominent in tropics, which provide all nutritious and healthy substance to people. The scientific management and proper care of these systems provide healthy nutritious fruits along with timber, fuelwood, etc. which provide cash income to poor farmers and helps in augmenting employment and maintaining FNS system along with CC mitigation (Samara, 2010).

2.11 LIVELIHOOD SUSTENANCE THROUGH AF

Indeed, there is cent percent agreement between tree-based farming systems and livelihood upliftments of poor people. Besides nutritional and environmental security, AFs assures incomes to farmers through provision of diversifying products such as timber, fuelwood, NTFPs such as edible fruits, medicinal plants, gum, resin, tendupatta, katha, etc. on which rural poor are depends for their sustenance (WAC, 2010; Jhariya and Raj, 2014). As per Dhyani (2012), agroforestry practices comprises sericulture, Lac culture, apiculture, and gum exudation techniques through modern and scientific way are the another important practices which can add additional incomes to the farmers and maintaining socioeconomic status of farmers along with biodiversity conservation (Dhyani, 2012; Das et al., 2014; Raj et al., 2017).

Moreover, the practices of gum and resin production under AFs results 14 to 23% augmentation in farmers income in Ethiopia (African region) (Abtew, 2014). Similarly, as per one estimates fruit tree like Guava based AFs contributed higher net economic return as 2.96 times higher than single cropping system in Meghalaya (Bhattacharya and Mishra, 2003). It is already discussed that AFs including various location-specific models are gaining wider recognition in term of higher productivity and socioeconomic upliftments under extreme climate event (buffer higher temperature) whereas employment generation is another important significant under AFs and according to NRCAF (2007) the agroforestry area of 25.4 mha contributed 943 million people annually which shows remarkable contribution through the adoption of AFs. Similarly, agroforestry of the Himalaya region potentially generates employment by the figure of 5.76 million people/day/year, which is very helpful in rural development (Dhyani et al., 2005).

2.12 RESEARCH AND DEVELOPMENT IN AFS

Today, AFs are gaining wide importance among farmers, planners, policymakers over time more to more emphasis has been given on location-specific models, and related research has been prioritized. The research and development activity of agroforestry have been changed from the past decade and research on agroforestry models has become more rigorous and scientific. Scientific oriented research and development should be undertaken for solving the land-use problem, farmer's socioeconomic issues, soil health and quality management, food, and nutritional constraints under changing climate and extreme weather events.

The research agenda has been started with a high priority on soil fertility, health, and quality improvement, tree-crop interaction, socioeconomic issues on a small scale to large landscape, and more to more consideration on maintaining food and environmental security. C storage and sequestration, biodiversity conservation, wasteland development, and rehabilitation of degraded lands are other areas where scientific research is needed. Policymakers, government, and non-governmental institutions, NGO's, private organization, etc. should work together in the direction of encouragement of extending the agroforestry areas under changing climate situations. People and farmers should be aware the multifarious benefits of agroforestry and its ecosystem services, which not only sustain the human being but also nurture our nature and mother earth. Thus, the role of agroforestry in maintaining FNS, resources conservation, soil health maintenance, storage, and sequestration of C in both vegetations and soils, socioeconomic improvement of resource-poor people, provision of several important ecosystem services have gained a great attention and need to work in this direction of better ecosystem and environment with strengthening national security (Lal, 2010; van Noordwijk et al., 2012)

2.13 CONCLUSION

The practice of agroforestry is not shrink and confine, but it has wide scope, tremendous potential and higher adoptability in agro-ecological regions of the tropics and now becoming heart of global farming community for solving the hunger problem, land degradation, maintaining health and food security, resource management and its proper utilization, maintenance of soil fertility through enhancing soil quality, biodiversity management, promoting efficient nutrient cycling, guarantees farmers income security under changing climate and mitigating global warming. A better management and scientific-based research are needed for understanding tree-crop interaction and resource management and conservation, which will be helpful in augmenting both crop and soil productivity along with higher profitability to farmers as the source of income in the era of changing the climate. Thanks to all agroforestry practitioners, government, and non-government institutions, NGO's, private organization, educational universities, policymakers and planners who work in the direction of sustainable land management and banish hunger and poverty and rebuild resilient rural environments at regional, national, and international level.

KEYWORDS

- **agroforestry**
- **biofuel**
- **C farming**
- **climate change**
- **natural resource management**

REFERENCES

Abtew, A. A., Pretzsch, J., Secco, L., & Mohamod, T. E., (2014). Contribution of small-scale gum and resin commercialization to local livelihood and rural economic development in the dry-lands of Eastern Africa. *Forests*, *5*, 952–977.

Alavalapati, J. R. R., & Nair, P. K. R., (2001). Socioeconomic and institutional perspectives of agroforestry. In: Palao, M., & Uusivuori, J., (eds.), *World Forest Society and Environment-Markets and Policies* (pp. 71–81). Kulwer Academic Publishers, Dordrecht.

Albrecht, A., & Kandji, S. T., (2003). Carbon sequestration in tropical agroforestry systems. *Agric. Ecosyst. Environ.*, *99*, 15–27.

Allen, C. D., Macalady, A. K., Chenchouni, H., Bachelet, D., McDowell, N., Vennetier, M., et al., (2010). A global overview of drought and heat-induced tree mortality reveals emerging climate change risks for forests. *Forest Ecology and Management, 259*, 660–684.

Allen, D. E., Mendham, D. S., Singh, B., Cowie, A., Wang, W., Dalal, R. C., & Raison, R. J., (2009). Nitrous oxide and methane emissions from soils are reduced following afforestation of pasture lands in three contrasting climatic zones. *Australian Journal of Soil Research, 47*, 443–458.

Allen, S. A., Jose, S., Nair, P. K. R., Brecke, B. J., NkediKizza, P., & Ramsey, C. L., (2004). Safety-net role of tree roots: Evidence from a pecan (*Caryaillinoensis* K. Koch)-cotton

(*Gossypium hirsutum* L.) alley cropping system in the southern United States. *Forest Ecology and Management*, *192*, 395–407.
Amatya, G., Chang, S. X., Beare, M. H., & Mead, D. J., (2002). Soil properties under a pinus radiate—ryegrass silvopastoral system in New Zealand. Part II. C and N of soil microbial biomass, and soil N dynamics. *Agrofor. Syst.*, *54*, 149–160.
Bergeron, M. S., Lacombe, R. L., Bradley, J., Whalen, A., Cogliastro, M. F. J., & Arp, P., (2011). Reduced soil nutrient leaching following the establishment of tree-based intercropping systems in eastern Canada. *Agroforestry Systems*, *83*, 321–330.
Bertin, C., Yang, X., & Weston, L. A., (2003). The role of root exudates and allelochemicals in the rhizosphere. *Plant and Soil*, *256*(1), 67–83.
Bhattacharya, B. P., & Misra, L. K., (2003). Production potential and cost-benefit analysis of agrihorticulture agroforestry systems in Northeast India. *Journal of Sustainable Agriculture*, *22*, 99–108.
Das, I., Katiyar, P., & Raj, A., (2014). Effects of temperature and relative humidity on ethephon induced gum exudation in *Acacia nilotica*. *Asian Journal of Multidisciplinary Studies*, *2*(10), 114–116.
Dhyani, S. K., (2012). Agroforestry interventions in India: Focus on environmental services and livelihood security. *Indian Journal of Agroforestry*, *13*(2), 1–9.
Dhyani, S. K., Handa, A. K., & Uma, (2013). Area under agroforestry in India: An assessment for present status and future perspective. *Indian Journal of Agroforestry*, *15*(1), 1–11.
Dhyani, S. K., Sharda, V. N., & Sharma, A. R., (2005). Agroforestry for sustainable management of soil, water and environmental quality: Looking back to think ahead. *Range Management and Agrofores*try, *26*(1), 71–83.
Dinesh, R., Chaudhuri, S. G., Ganeshamurthy, A. N., & Dey, C., (2003). Changes in soil microbial indices and their relationships following deforestation and cultivation in wet tropical forest. *Appl. Soil Ecol.*, *24*, 17–26.
Fanish, S. A., & Priya, R. S., (2013). Review on benefits of agro forestry system. *International Journal of Education and Research*, *1*(1), 1–12.
Forster, P., Ramaswamy, V., Artaxo, P., Berntsen, T., Betts, R., Fahey, D. W., et al., (2007). Changes in atmospheric constituents and in radiative forcing. In: Solomon, S., Qin, D., Manning, M., Chen, Z., Marquis, M., Averyt, K. B., Tignor, M., & Miller, H. L., (eds.), *Climate Change 2007: The Physical Science Basis. Contribution of Working Group I to the Fourth Assessment Report of the Intergovernmental Panel on Climate Change*. Cambridge University Press, Cambridge, United Kingdom and New York, NY, USA.
Glaser, B., Lehmann, J., & Zech, W., (2002). Ameliorating physical and chemical properties of highly weathered soils in the tropics with charcoal—A review. *Biol. Fertil. Soils*, *35*, 219–230.
Gold, M. A., & Garrett, H. E., (2009). Agroforestry nomenclature, concepts, and practices. In: Garrett, H. E., (ed.), *North American Agroforestry: An Integrated Science and Practice* (pp. 45–56). Agronomy Society of America, Madison.
Hoosbeek, M. R., Remme, R. P., & Rusch, G. M., (2018). Trees enhance soil carbon sequestration and nutrient cycling in a silvopastoral system in south-western Nicaragua, *Agroforest. Syst.*, *92*(2), 263–273.
Jhariya, M. K., & Raj, A., (2014). Human welfare from biodiversity. *Agrobios. Newsletter*, *12*(9), 89–91.
Jhariya, M. K., Banerjee, A., Yadav, D. K., & Raj, A., (2018). Leguminous trees an innovative tool for soil sustainability. In: Meena, R. S., Das, A., Yadav, G. S., & Lal R., (eds.), *Legumes for Soil Health and Sustainable Management* (pp. 315–345). Springer,

ISBN 978–981–13–0253–4 (eBook), ISBN: 978–981–13–0252–7 (Hardcover). https://doi.org/10.1007/978–981–13–0253–4_10 (Accessed on 9 October 2019).

Jhariya, M. K., Bargali, S. S., & Raj, A., (2015). Possibilities and perspectives of agroforestry in Chhattisgarh. In: Miodrag, Z., (ed.), *Precious Forests - Precious Earth* (pp. 238–257). InTech E- Publishing Inc.

Jose, S., (2009). Agroforestry for ecosystem services and environmental benefits: An overview. *Agrofor. Syst., 76*, 1–10.

Kassie, G. W., (2018). Agroforestry and farm income diversification: Synergy or trade-off? The case of Ethiopia. *Environ. Syst. Res., 6*, p. 8. https://doi.org/10.1186/s40068–017–0085–6 (Accessed on 9 October 2019).

Kaur, B., Gupta, S. R., & Singh, G., (2000). Soil carbon, microbial activity and nitrogen availability in agroforestry systems on moderately alkaline soils in northern India. *Appl. Soil Ecol., 15*, 283–294.

Khanna, P. K., (1998). Nutrient cycling under mixed-species tree systems in Southeast Asia. *Agrofor. Syst., 38*, 99–120.

Kirby, K. R., & Potvin, C., (2007). Variation in carbon storage among tree species: Implications for the management of a small-scale carbon sink project. *For Ecol. Manage., 246*, 208–221.

Korn, H., Ntayombya, P., Berghall, O., Cotter, J., Lamb, R., Ruark, G., & Thompson, I., (2003). Climate change mitigation and adaptation options: Links to, and impacts, on biodiversity. In: *Interlinkages Between Biological Diversity and Climate Change: Advice on the Integration of Biodiversity Considerations Into the Implementation of the United Nations Framework Convention on Climate Change and its Kyoto protocol* (pp. 48–87). CBD Technical Series no. 10. Montreal, Canada: Secretariat of the Convention on Biological Diversity. http://www.biodiv.org/doc/publications/cbd-ts-10.pdf (Accessed on 9 October 2019).

Kumar, B. M., (2006). Carbon sequestration potential of tropical home gardens. In: Kumar, B. M., & Nair, P. K. R. (eds.), *Tropical Home Gardens: A Time Tested Example of Sustainable Agroforestry* (pp. 185–204). Advance in Agroforestry 3. Springer, Dordrecht, the Netherlands.

Lal, R., (2004). Soil carbon sequestration impacts on global climate change and food security. *Science, 304*, 1623–1627.

Lal, R., (2010). Beyond Copenhagen: Mitigating climate change and achieving food security through soil carbon sequestration. *Food Secur., 2*, 169–177.

Lee, K. H., & Jose, S., (2003). Soil respiration and microbial biomass in a pecan-cotton alley cropping system in southern USA. *Agrofor. Syst., 58*, 45–54.

Mafongoya, P. L., Giller, K. E., Odee, D., Gathumbi, S., Ndufa, S. K., & Sitompul, S. M., (2004). Benefiting from N2-fixation and managing rhizobia. In: Noordwijk, M., Van Cadisch, G., & Ong, C. K., (eds.), *Below-Ground Interactions in Tropical Agroecosystems: Concepts and Models with Multiple Plant Components* (pp. 227–242). CABI International, Wallingford.

McNeely, J. A., & Schroth, G., (2006). Agroforestry and biodiversity conservation–traditional practices, present dynamics, and lessons for the future. *Biodivers. Conserv., 15*(2), 549–554. doi: 10.1007/s10531–005–2087–3.

Montagnini, F., & Nair, P. K. R., (2004). Carbon sequestration: An underexploited environmental benefit of agroforestry systems. *Agrofor. Ecosyst., 61*, 281–295.

Moore, J. M., Klose, S., & Tabatabai, M. A., (2000). Soil microbial biomass carbon and nitrogen as affected by cropping systems. *Biol. Fertil. Soils, 31*, 200–210.

Nair, P. K. R., & Nair, V. D., (2003). Carbon storage in North American agroforestry systems. In: Kimble, J., Heath, L. S., Birdsey, R. A., & Lal, R., (eds.), *The Potential of U. S. Forest Soils to Sequester Carbon and Mitigate the Green House Effect, Boca Raton* (pp. 333–346). *FL*. CRC Press LLC.
Nair, P. K. R., & Nair, V. D., (2014). Solid-fluid-gas: The state of knowledge on carbon sequestration potential of agroforestry systems in Africa. *Current Opinion in Environ. Sustain.*, *6*, 22–27.
Nair, P. K. R., (1979). Agro-forestry research: A retrospective and prospective appraisal. *Proc. Int. Conf. International Cooperation in Agro-Forestry* (pp.275–296). ICRAF, Nairobi.
Nair, P. K. R., Gordon, A. M., & Mosquera-Losada, M. R., (2008). Agroforestry. In: Jorgensen, S. E., & Fath, B. D., (eds.), *Encyclopedia of Ecology* (Vol. 1, pp. 101–110). Elsevier, Oxford.
Nair, P. K. R., Kumar, B. M., & Nair, V. D., (2009). Agroforestry as a strategy for carbon sequestration. *J. Plant Nutr. Soil Sci.*, *172*, 10–23.
Nair, P. K. R., Nair, V. D., Kumar, B. M., & Haile, S. G., (2009). Soil carbon sequestration in tropical agroforestry systems: A feasibility appraisal. *Environ. Sci. Policy.*, *12*, 1099–1111.
Nair, P. K. R., Vimala, D. N., Kumar, B. M., & Showalter, J. M., (2011). Carbon sequestration in agroforestry systems. *Advances in Agronomy*, *108*, 237–307.
Ndaw, S. M., Gama-Rodrigues, A. C., Gama Rodrigues, E. F., Sales, K. R. N., & Rosado, A. S., (2009). Relationships between bacterial diversity, microbial biomass and litter quality in soils under different plant covers in northern—Rio de Janeiro State, Brazil. *Can. J. Microbiol.*, *55*, 1089–1095.
Ong, C. K., Black, C. R., & Muthuri, C. W., (2006). Modifying forests and agroforestry for improved water productivity in the semi-arid tropics. *CAB Reviews: Perspectives in Agriculture, Veterinary Science, Nutrition and Natural Resources*, *65*, 1–19.
Parihaar, R. S., (2016). Carbon stock and carbon sequestration potential of different land-use systems in hills and bhabhar belt of Kumaun Himalaya. *PhD Thesis* (p. 350). Kumaun University, Nainital.
Perfecto, I., & Vandermeer, J., (2010). The agroecological matrix as alternative to the land-sparing/agriculture intensification model. *Proc. Natl. Acad. Sci.*, *107*, 5786–5791.
Rahman, S. A., Sunderland, T., Roshetko, J. M., Basuki, I., & Healey, J. R., (2016). Tree culture of smallholder farmers practicing agroforestry in Gunung Salak Valley, West Java, Indonesia. *Small Scale For.*, *15*(4), 433–442.
Raj, A., & Jhariya, M. K., (2017). Sustainable agriculture with agroforestry: Adoption to climate change. In: Kumar, P. S., Kanwat, M., Meena, P. D., Kumar, V., & Alone, R. A., (ed.), *Climate Change and Sustainable Agriculture* (pp. 287–293). ISBN: 9789–3855–1672–6. New India Publishing Agency (NIPA), New Delhi, India.
Raj, A., & Singh, L., (2017). Effects of girth class, injury and seasons on ethephon induced gum exudation in *Acacia nilotica* in Chhattisgarh. *Indian Journal of Agroforestry*, *19*(1), 36–41.
Raj, A., Jhariya, M. K. & Toppo, P., (2014b). Cow dung for ecofriendly and sustainable productive farming. *International Journal of Scientific Research, 3*(10), 42–43.
Raj, A., Jhariya, M. K., & Bargali, S. S., (2018). Climate smart agriculture and carbon sequestration. In: Pandey, C. B., Gaur, M. K., & Goyal, R. K., (ed.), *Climate Change and Agroforestry: Adaptation Mitigation and Livelihood Security* (pp. 1–19). ISBN: 9789–386546067. New India Publishing Agency (NIPA), New Delhi, India.
Raj, A., Jhariya, M. K., & Pithoura, F., (2014a). Need of agroforestry and Impact on ecosystem. *Journal of Plant Development Sciences, 6*(4), 577–581.
Roy, A., (2011). *"Requirement of Vegetables and Fruit" The Daily Star (A English Newspaper).*

Salamanca, E. F., Raubuch, M., & Joergensen, R. G., (2002). Relationships between soil microbial indices in secondary tropical forest soils. *Appl Soil Ecol*, 21, 211–219.
Samara, J. S., (2010). Horticulture opportunities in rainfed areas. *Indian J. Hort.*, *67*(1), 1–7.
Scales, B. R., & Marsden, S., (2008). Biodiversity in small-scale tropical agroforests: A review of species richness and abundance shifts and the factors influencing them. *Environ. Conserv.*, *35*, 160–172.
Schroth, G. G., Da Fonseca, A. B., Harvey, C. A., Gascon, C., Vasconcelos, H. L., & Izac, A. M. N., (2004). *Agroforestry and Biodiversity Conservation in Tropical Landscapes*. USA: Island Press.
Sharma, P., Rai, S. C., Sharma, R., & Sharma, E., (2004). Effects of land use change on soil microbial C, N and P in a Himalayan watershed. *Pedobiologia*, *48*, 83–92.
Singh, N. R., & Jhariya, M. K., (2016). Agroforestry and Agrihorticulture for Higher Income and Resource Conservation. In: Narain, S., & Rawat, S. K. (ed.), *Innovative Technology for Sustainable Agriculture Development* (pp. 125–145). ISBN: 978–81–7622–375–1. Biotech Books, New Delhi, India.
Sudha, P., Ramprasad, V., Nagendra, M. D. V., Kulkarni, H. D., & Ravindranath, N. H., (2007). Development of an agroforestry carbon sequestration project in Khammam district, India. *Mitigation and Adaptation Strategies for Climate Change*, *12*, 1131–1152.
Tscharntke, T., Clough, Y., Wanger, T. C., Jackson, L., Motzke, I., & Perfecto, I., (2012). Global food security, biodiversity conservation and the future of agricultural intensification. *Biol. Conserv.*, *151*, 53–59.
Vallejo-Ramos, M., Moreno-Calles, A. I., & Casas, A., (2016). TEK and biodiversity management in agroforestry systems of different socio-ecological contexts of the Tehuacán Valley. *Journal of Ethnobiology and Ethnomedicine*, *12*, 31. http://doi.org/10.1186/s13002–016–0102–2 (Accessed on 9 October 2019).
Van Noordwijk, M., Tata, H. L., Xu, J., Dewi, S., & Minang, P., (2012). Segregate or integrate for multifunctionality and sustained change through landscape agroforestry involving rubber in Indonesia and China. In: Nair, P. K. R., & Garrity, D., (eds.), *Agroforestry: The Future of Global Land Use* (pp. 69–104). Springer, Dordrecht.
Verchot, L. V., Noordwijk, M. V., Kandji, S., Tomich, T., Ong, C., Albrecht, A., Mackensen, J., Bantilan, C., Anupama, K. V., & Palm, C., (2007). Climate change: Linking adaptation and mitigation through agroforestry. *Mitg. Adapt. Strat. Glob. Change, 12*, 901–918
World Agro-forestry Centre, (2010). *Transforming Lives and Landscapes* (pp. 1–5).
Wright, H. L., Lake, I. R., & Dolman, P. M., (2012). Agriculture-a key element for conservation in the developing world. *Conserv. Lett.*, *5*, 11–19.
Zaia, F. C., Gama-Rodrigues, A. C., Gama-Rodrigues, E. F., & Machado, R. C. R., (2008). Fosforo organic oem solos sob agrossistemas de cacau. *Rev. Bras Ci. Solo, 32*, 1987–1995.
Zimmerer, K. S., (2015). Understanding agrobiodiversity and the rise of resilience: Analytic category, conceptual boundary object or meta-level transition? *Resilience, 3293*, 1–16.

CHAPTER 3

Potential of Agroforestry and Environmental Greening for Climate Change Minimization

A. O. AKANWA,[1] H. C. MBA,[2] E. B. OGBUENE,[3] M. U. NWACHUKWU,[2] and C. C. ANUKWONKE[1]

[1]*Chukwuemeka Odumegwu Ojukwu University, (COOU) Uli Campus, Anambra State, Nigeria, Mobile: +234-8065813596, E-mail: angela.akanwa1@gmail.com (A. O. Akanwa)*

[2]*University of Nigeria Nsukka (UNN), Department of Urban and Regional Planning, Faculty of Environmental Sciences, Enugu Campus, Enugu State, Nigeria*

[3]*Centre for Environmental Management and Control (CEMAC), University of Nigeria Nsukka, Enugu State, Nigeria*

ABSTRACT

Suggestions have been rife and profuse about measures necessary for minimizing climate change (CC). Most of the suggestions have commonly been geared towards the reduction of effluent discharges from industrial establishments; thus implying the controversial issue of restrictions on the growth of the manufacturing industry. However, environmental researchers are currently calling the attention to the potential contribution of intensive agroforestry and environmental greening in the minimization of CC. This chapter orchestrates the salient information about components of agroforestry and environmental greening as adaptation measures for the minimization of CC. Agroforestry measures elaborated on include urban agriculture, agroforestry, all-season farming, intensive forestry activities in rural areas, and conservation of existing forest reserves in urban areas as well as legislative measures against tree felling. Environmental greening measures discussed include land surface greening and landscaping as well as gardening, horticulture, and legislative measures for minimization of land surface exposure. These measures are highlighted as an efficient means of minimizing CC. Given that the extent to which greenhouse gases (GHG) deplete the ozone layer protecting the earth's surface from direct rays of the sun is greatly reduced. In effect, agroforestry, and environmental greening are capable of reducing the excessive imbalance in the earth's heat budget that results in CC.

3.1 INTRODUCTION

Climate change (CC) happens to be a true titan above all environmental crisis facing humanity today. Apparently, it is no longer just a scientific phenomenon, but the glaring global sporadic reports and incessant weather-related crises have made it an unarguable reality and overwhelming evidence to an unaware person. Globally, the increasing trends in global warming have exacerbated climatologically changes causing (90%) of disasters like floods, storms, rising sea levels, heatwaves, and other weather-related events. This has resulted in massive destruction of properties, human lives, spread of diseases, food shortage, poverty, and disruption of socio-economic activities and environmental resources and ultimately made millions of people homeless across the globe (UNEP, 2007).

Regrettably, the global situation can be said to be an environmental state of emergency. However, unsustainable anthropogenic efforts such as the burning of fossil fuels, agriculture, exploration of natural resources, industrialization, urbanization, and many more has released hydrocarbons which contribute to the greenhouse effect and global warming and consequently CC. Given that the world is a whole ecological system that is grossly inter-dependent, this makes the persistent climate-related problems and unprecedented incidents obviously connected. The melting ice in the arctic increased aridity, heat waves, and widespread flooding events, hurricanes, varying temperatures, rapid food shortages, water pollution, forest losses, and poverty in Africa can all be sourced from the varying CCs.

This present chapter deals with the potential roles played by agroforestry and environmental greening as quick and strategic actions required in safeguarding the varying impacts of CC from countries, regions, and continents. Further, it advocates agroforestry and environmental greening as effective tools that could be widely applicable, economically viable, and environmentally friendly to cushion these destructive incidents of CC, in order to make the built environment tenable, exuberant, and danger-free.

3.2 PROMOTION OF URBAN AGRICULTURE AS AN AGROFORESTRY SYSTEM (AFS) IN MINIMIZING CLIMATE CHANGE (CC)

Natural green areas adjoining urban areas (Figure 3.1) are heavily pressurized by increasing human activities that have destroyed forest areas, predominant in developing countries (Akanwa et al., 2017a). Urban agriculture promotion is a necessity as greening; forests and agriculture perform multifaceted roles in various spheres of global carbon cycling as well as helps in CC mitigation. Nowadays, climate variability has been considered as one of the recent environmental threats to the world, with studies showing that temperature change is the key factor of CC (Ogbuene, 2012; Shasur-Barand Hoffman, 2000). Urban areas are experiencing an exponential population increase whereby about 3% of earth's land surface, and presently more than half of the world's population lives in cities (Gossop, 2011; United Nations Environmental Program, 2009). This is as a result of the increased urban spread and preference due to the huge attractions, socio-economic facilities and driving opportunities that are readily available (Department of Economic and Social Affairs of the United Nations Population Division, 2012).

Urbanization trends have encouraged a large number of people to relocate to urban areas. In fact, in 2010, more than half of the world's population occupied urban areas. Unfortunately, the unprecedented increase in city expansion has placed pressure on food security resulting in an unusual demand for food cultivated, harvested, processed, and sold within these cities (Ericksen, 2007; Bryld, 2003; Bessie, 2005).

The rise in a large expanse of farmlands for food production in urban areas became a rescue approach to the hunger problems experienced in cities (Lang, 2010). However, urban settlements during ancient civilizations have always preferred to produce their own food. More than five thousand years ago, humans settled in long-term neighborhoods that have grown crops alongside with animals in cities (Taylor, 2010). Over the years, there has been a gradual combination of farming activities within urban areas as the demand for food continued to increase within cities (Dixon et al., 2009; CPRE, 2010).

Urban farming is referred to as an informal set of activities focusing on farm production (cultivation, processing, and distribution) in an urban environment (Brown and Carter, 2003). It is the interaction of socio-economic activities that combines farm crops and husbandry in large open spaces located in cities or in rural areas adjacent to city boundaries. It could also include fish, bee or animal farming, growing vegetables, fruits, flowers, and tree planting, and many more (Duchemin et al., 2009).

FIGURE 3.1 (See color insert.) Urban greening in Nigeria (Akanwa, 2018).

For purposes of this chapter, urban agriculture encompasses the practice of agricultural activities within the urban and peri-urban periphery. It also involves the application of sophisticated and technological farming innovations capable of supplying food for the larger economy (Dixon et al., 2009; Peters et al., 2008). This chapter reviews urban agriculture using the experiences of other countries practicing the activity.

City farming in Africa and developing countries is rather a potential socio-economic survival and livelihood enhancing strategy for those operating at the economic margin (Clemmitt, 2008; Godfray et al., 2010; Maxwell and Slater, 2003). As urban areas continue to harbor a large number of the human population especially in Africa, Asia, and Latin America, they are associated with many problems. These socio-economic problems include high unemployment rates, poverty, inadequate housing, and particularly food insecurity resulting in large number of poor city dwellers (Kortright and Wakefield, 2011; Morgan, 2009). Many urban dwellers are forced by their socio-economic disposition to farm and grow their own food within their location in order to survive abject poverty and hunger (McBeath and McBeath, 2009). Moreover, urban (vertical) farming in developing countries is encumbered by a number of factors especially electricity, water, and finance. This will entail that they maximize the available opportunities they have within them. For example, instead of depending on the LED lighting system for vertical farming, African and developing countries can utilize solar energy (Hu et al., 2011).

Entrepreneurs can grow small farms while applying low-technological skills that demand less finance, water, and light. Ugandans, for instance, are faced with lack of financial resources coupled with limited access to land and water to enable the building of modern vertical farms; however, the urban farmers have innovated wood-designed crates for the vegetables. Similarly, in Kenya, modern vertical farms have been built with several models which include; tower gardens, hanging gardens, A-Frame gardens and multifarious gardens, each was built with the ability to maximize water and protect the vegetables from insects (Zerbe, 2010; Taylor, 2010; Plyler, 2012).

Urban farming in developed countries provides food for the urban areas, and also involves a high technological application of sophisticated farming ventures in the urban environment (Steel, 2008). New scientific inventions have been applied in urban agriculture to facilitate water collection, distribution, and retention, protection, wetlands drainage, and slope terracing (Mougeout, 1994).

Urban agriculture in developed countries could be an unconventional activity that may include community gardens. As a conventional process, it involves networks of interested food-related groups or policymakers (Plyler, 2012; Cruz and Medina, 2001).

Developing countries like the United States, practice city agriculture, where people are involved in food farming as a communal or collective activity that may include gardening, tree planting on either private or public lands (Peters, 2011; Kaethler, 2006). In the Netherlands, a new innovation of large commercial indoor farms is practiced to furnish Europe's largest supermarkets with high quality, pesticide-free vegetables. These indoor farms grow certain crops by employing sophisticated lighting systems and human manipulated climate buildings that require less water and soil to flourish (Mougeout, 1994). It is an improved scientific approach whereby farms operate in a closed growing system. It helps to control evaporation from plants and hence, the farms tend to utilize less water than traditional farms (Jo and McPherson, 1995). Further, most vertical farms require that plants are grown in nutrient solutions with available air and water and not necessarily using soil as a medium, i.e., aeroponics or hydroponic systems (Plyler, 2012). However, the vertical farming technique is ideal since it is an indoor farm that requires less water, less soil, climate-controlled, and disease-free. Hence, it will meet the demands of the teeming population in cities and the rising demands for high-quality pesticide-free food (Reynolds, 2009).

A continuous increase in urban population without subsequent improved food supply/farming strategies is likely to be a threat to the food supply for urban residents. This can increase the potentials for food shortages in cities worldwide (Morgan, 2009; Sonnino, 2009). The need for a readily available urban food supply has led to an unconscious and an apparent marginalization of urban agriculture in cities within the past decades. It is unfortunate that the potentials of urban agriculture in cities are yet to be fully maximized. Considering that in 1993, only 15% of food consumed was grown in cities; in 2005, that number increased only to 30% (Morgan, 2014).

Urban agricultural activities perform various roles; firstly, it is capable of minimizing poverty and hunger of the urban dwellers, and secondly, through agricultural activities practiced, wealth is created through the sale of the products (Mayer and Knox, 2010). Russian cities generate huge income from the practice of city farming and gardening through commercial dealings, and also they provide food for their families (Lerner and akin, 2011). Urban farms generate wealth by selling food and generating profits (Chase, 2012; Hiranandani, 2010). Commercial farms can be small farms that make less than $250,000 in gross sales (Brown and Carter, 2003). Three groups of urban farms include *recreational farms* of less than 100 acres, *adaptive farms* of 100 to 200 acres, and *traditional farms* of more than 200 acres (Brown and Carter, 2003). Urban agriculture should be encouraged since it reduces unemployment by providing opportunities that generate income and thereby improve economic situation in urban areas (Van Rooyen et al., 1995).

Urban agriculture has benefits accrued to immediate areas. These benefits include job creation and promotion growth of vegetables and fruits, readily supply of healthy organic food, improved nutrition and mental alertness, environmental greening, CC mitigation, photosynthesis, and many more (Peters, 2011). It fosters food culture, traditions, and well-being (Harper et al., 2009; Taylor, 2010; Peters, 2011). Generally, urban agriculture plays a major role in sustainable city development by creating open green spaces, increasing the urban habitat diversity and thereby biodiversity in cities, thus reducing noise and pollution, closing the energy loops and making cities more habitable by mitigating CC Generally, urban agriculture has brought about huge benefits especially in providing food and vegetables (Steel, 2008; Drescher, 2010). However, there are certain problems that inhibit its full potentials in urban areas.

These problems include non-available landspaces, high start-up costs, traditional gardening skills, and the seasonality of the region as affected by CC (Brown and Carter, 2003). Possible solutions to these problems include strategic partnerships accompanied by intensive research studies (Angelo et al., 2001; Maxwell and Slater, 2003). Research is being carried out on urban food production with new prospects on waste streams, heating systems and well as improved organic agricultural methods (Bhatt and Kongshaug, 2005).

3.3 PROMOTION OF AGRO-FORESTRY IN RURAL AREAS

The rural traditional farming systems of shifting cultivation contribute to huge annual losses of forest cover that alter the distribution of species and results in biodiversity loss (Buresh and Cooper, 1999; Kumar, 2006). Agroforestry has been used as a major strategy to enjoin rural dwellers to become partners in rehabilitating degraded forestlands (FAO, 2013).

Agroforestry is a system that integrates trees in farms or rangelands. It sustains production of crops and livestock for increased sustainable advantages (Figure 3.2). Also, it is the application of diverse, sustainable, and high technological farming management systems (Leakey, 1996; Kareemulla et al., 2005; Saxena, 2000).

It plays an important role in the economic development of rural areas as well as poverty reduction. Likewise, it increases the yield and services of per unit agro-forest area (Ibrahim et al., 2010). Agroforestry reduces soil erosion, improves soil quality, vegetative cover, land productivity, and uplift

FIGURE 3.2 (See color insert.) Rural areas in Africa showing its rich agroforestry resources (Akanwa, 2016).

the local farmers' level of living through sustained farm productivity (FAO, 2013; Nair, 2012).

The high diversity of species in rural areas, its vegetal make-up, and richness encourages multiple and healthy ecological independency, which are synonymous with forest ecosystems (DeClerck and Negreros-Castillo, 2000). However, it is this rich quality of the species structure that generates higher levels of resilience to extreme climatic disasters. In such settings, there is usually the presence of high moisture content and higher trees/plants population densities creating a highly assorted agroforestry system (AFs). Obviously, this is responsible for the vast practice of improved AFs such tree fallows, taungya systems, alley cropping, plantation crop combinations (cacao, rubber, bananas), and home gardens in wet regions contributing to food security, food diversity and poverty reduction (Pezo and Ibrahim, 2001; Aktar et al., 1992; Samra et al., 2005; Kumar and Nair, 2004; Balooni, 2003).

The predominant agroforestry practices in rural areas include the production of fuelwood, plantation of crops/trees, animal husbandry, shelterbelts, which act as windbreaks to protect soils and crops, and many more (Macias, 2008). About 91% of rural areas represent the territory of the 27 Member States of the European Union (EU) that harbor more than 56% of its population. This makes rural development policy of extraordinary importance since agroforestry is an essential natural resource that helps in rural economic development (Anonymus, 2010; Akanwa et al., 2017b). Agroforestry practices such as farming, hunting, forestry, and fishery apparently remain the most vibrant economic activities operational in rural areas. These activities are highly practiced in different counties, and employ about 52 to 82% of the total rural population; hence, agroforestry promotion in rural areas is a means of rural diversification and sustainable development (Kedaitiene and Martinaviciene, 2005; Forestry/Fuel-wood Research and Development Project, 1992).

Agro-forestry promotion in rural areas has not received due attention and recognition probably due to its inability to generate significant economic returns. On the contrary, local communities have actively participated in management and conservation of wildlife through community-based natural resource management (CBNRM) initiatives, though, CBNRM initiatives are yet to receive adequate promotion for despite a large number of rural families that depend on its resources for survival (Slee and Snowdon, 1999; Ogbuene, 2012; Zerihum et al., 2014).

Globally, it is estimated that over 1 billion people depend on the numerous forest resources for their livelihood with close to 200 million indigenous communities fully dependent on forests (Chao, 2012). Moreover, 350 million people who live adjacent to dense forests depend on them for subsistence and income generation (Chao, 2012). Also, in developing countries, it has been estimated that more than half of the rural population income is sourced locally from environmental resources capable of supplying them during food shortage or crisis (Yoveva et al., 2000; Zakariah et al., 1998).

It has been proven that logging without the afforestation program is a major challenge to agroforestry promotion in rural areas (Ogbuene, 2010). Logging in developing nations is more intensive and can be quite destructive. Akanwa et al. (2017a and 2017b) also revealed in their study carried out in Southeastern, Nigeria, Ebonyi State to be precise that 402.855 hectares of vegetation cover was destroyed by intensive quarrying activities using the open-pit mining method. Quarrying operations involves the removal of the topsoils, trees, and vegetation, hence, destroying the landscape, massive loss of vegetation while endangering human, animal, and plant lives. The increase in the demand for natural resources has placed immense pressure on forest resources. Therefore, there are considerable negative environmental effects of quarrying activities (Figure 3.3) on trees and vegetation and generally escalating CCs problems.

Consequently, there is an urgent need for an increase in agroforestry education and awareness campaigns on the crucial benefit of agroforestry promotion or re-afforestation projects in rural areas (Okojia, 1993; Akanwa, 2016).

3.4 ALL-SEASON FARMING

Climate and weather conditions influence and limit all-season farming through changes in rainfall, temperature, moisture content, evaporation levels, and soil temperature at various depths among others (Bhalme, 1997; Ogbuene, 2010). All season farming will also include aggressive agriculture, which is capital and labor-intensive and requires the application of chemical fertilizers and pesticides. It has major advantages, which includes an increase in crop yield (Figure 3.4). Hence, farmers can easily monitor the land while protecting their livestock and producing large quantities of cheap food, vegetables, fruits, and poultry products.

Moreover, it meets the ever-increasing demand for food supplies and incorporates agricultural diversity in the growth of plants tree, livestock as well as contributes both to CC mitigation and adaptation (FAO, 2013).

FIGURE 3.3 (See color insert.) Unsustainable quarrying activities destroy lands, soils, trees, vegetation, landscapes, and displace animals from their habitat. The quarry sites are abandoned as artificial pools after closure without plans of re-afforestation (Akanwa, 2016).

FIGURE 3.4 (See color insert.) All-season farming involves the combination of agricultural activities with tree planting and growth. It also employs the application of improved seeds, chemical fertilizers, and pesticides to promote growth, disease-resistant, and pest-free crops (Ogbuene, 2012; Akanwa, 2016).

All season farming includes sustainable land-use strategies that use crop rotation and nutrient recycling (Taylor, 2010), and other improved or alternative practices as biological control, hydroponics, and organic farming. Trees are perennials and naturally endowed with the ability to survive for longer periods. Consequently, the positive results of all-season farming will last for a long time. Other types of farming systems maximize other by-products like the foliage and branches of trees capable to minimizing climate variability (Lowenfels and Lewis, 2010).

Trees are basic instruments that contribute both to crop enrichment and balancing the soil productivity levels for sustainable agroecosystems (Gliessman, 2015). However, it has negative impacts that can possibly lead to air, land, and water pollution. Due to the intensive nature of all season-farming it requires the constant application of pesticides that can equally disturb food chains thereby, affecting insect, bird, and mammal populations. However, a major limitation to all-season farming is water supply problem. Applying water harvesting technology by retaining water in the artificial lakes, dams, wetlands, soil, or tanks below ground could serve as a solution to recharge underground water aquifer for all-season farming (Stamets, 2005).

Another challenge facing all-season farming system is the appropriate integration of compatible crops and trees for optimum performance throughout the year providing the growth of crops, leaves, and the process of photosynthesis (Ong et al., 1996). Besides, excessive fertilization can cause N leaching into water systems and N losses to the atmosphere. However, sustainable management of the trees can provide the necessary nutrients for the durable production of the systems while, the nutrient contributions from the trees contribute to mitigate CC by reducing the needs for synthetic fertilizers (Muschler, 2001, 2004; Beer et al., 1998).

3.5 INTENSIVE FORESTRY ACTIVITIES IN RURAL AREAS

Intensive forestry requires the management of various forest resources and the application of sustainable best practices for maximum forest products (Figure 3.5). Forestry and forest management are major sources of farm products, biodiversity conservation and CC mitigation (Slee and Snowdon, 1999; Lutz et al., 2010).

Forests are major terrestrial ecosystems that are renewable, valuable, and provide sustainable products that occupy 30–43% of the world's land surface contributing to the national economy (Allen et al., 2010).

FIGURE 3.5 (See color insert.) Huge expanse of forest areas located in the rural areas (Akanwa, 2016).

Forests provide a catalog of services that include habitats for wildlife, clean water, fiber, and fuel, carbon storage, and climate mitigation. All these are of vital importance for the economy of many regions in the world (Lindquist et al., 2012; Dixon et al., 2009; Kupcak, 2011).

As regards climate mitigation, Allen et al., (2010) reported in their study that forest ecosystems happen to store large quantities of carbon, given that the total amount of carbon sequestered in forest vegetation is approximately 359 billion tons. Similarly, forest soil is another carbon pool that stores large amount of carbon. Consequently, the amount of carbon stored in forest ecosystems and soils happens to be twice that which is stored in the atmosphere (Lal, 2005).

There is a need for intensive forest activities, especially in rural areas for its functions in the global carbon cycling as long term – carbon repositories. Agroforestry can contribute both to CC mitigation and adaptation through different ways. In its contribution through mitigation it allows for C sequestration and retention in biomass and the soil, the substitution of synthetic inputs by biological mechanisms (this is of particular importance for N fertilizers which may release significant amounts of N_2O), and the reduction of enteric CH_4 emissions from ruminants by receiving improved feed and fodder. In its adaptation contribution, agroforestry can increase the resistance and resilience of the system to climate variability because the trees buffer against extreme climatic events, protect soils and watercourses, and diversify their production (FAO, 2013; Matocha et al., 2012; La Greca et al., 2011).

Moreover, the shade trees reduce heat stress on animals and crops; while fruits, timber, and fuelwood species provide additional products which buffer against price fluctuations of individual products; and fodder trees supply high-quality forage to reduce grazing pressure, land degradation, and methane emissions (Reid et al., 2004).

3.6 CONSERVATION OF EXISTING FOREST RESERVE IN URBAN AREAS

Numerous unguarded human activities such as agricultural activities, urban development, and resource extraction among others are responsible for the fast-shrinking, disappearance, and depleted world's forest resources/ reserves. It is given that forests have virtually disappeared in 25 countries; 18 have lost more than 95% of their forests, and another 11 have lost 90%. In developing countries, the deforestation rate is extremely high, and forest losses are equally alarming (World Commission on Forests and Sustainable Development, 1999; Akanwa et al., 2017a). About 15 million hectares of productive forests are being cleared each year, whereas only 10% area is brought back to forest vegetation on the global basis (World Commission on Forests and Sustainable Development, 1999).

According to National Research Council (1991), in declining or disappearing forests, trees are not the only lost resource; the many species of animal and plant life that depend on forests for survival are equally lost. Many of these species, their usefulness, intrinsic qualities, benefits, and biological relevance are yet to be established; hence, their genetic extinction is unavoidable (O'Neill et al., 2001).

The International Union for Conservation of Nature (IUCN) and UN defined protected areas as a clearly defined geographical space, recognized, dedicated, and managed, through legal or other effective means, to achieve the long-term conservation of nature with associated ecosystem services and cultural values." Conservation strategies are expedient to manage forest areas for sustainable forestry in urban areas. An urban vegetation or forestry plan has become necessary (Akanwa et al., 2017b).

IUCN protected areas are recognized as either Category II (national parks) or Category V (protected landscape or seascape). However, there are other urban protected areas that have received international recognition; these include marine areas, world heritage sites, UNESCO Geoparks, Ramsar sites, and biosphere reserves among others.

These urban protected areas can be better managed at the national level by governments, state or provincial governments in federal systems, local governments, non-governmental organizations, local community groups, or businesses. However, as a valuable asset, people must be seen as important stakeholders and hence be involved in the decisions that affect the sustainable management of the forests. The success of all the efforts aimed at the sustainable management of the forest resources depends on the combined efforts of the stakeholders.

An urban forest management plan is also important in protecting this remarkable resource. It involves a series of outlined duties drawn out for all vested authorities and stakeholders with an achievable goal and practical means of actualizing them (Wolf, 2004; The Tree Protection Legislation, 2006). The plan should be time-bound and yet, a long-term process that is practical, obtainable, and adjustable. Its ultimate aim is to promote a sustainable urban forest management system that will promote the development of an exuberant urban forest. A sustainable urban forest management system defines the essential elements of the urban forest that plays a protective role over the use of the forest resource. It covers the environmental, economic, social, and cultural aspects of the forest ecosystems (Clark et al., 1997; Dwyer et al., 1992; Coder, 1996; Kenney et al., 2011).

3.7 STRENGTHENING OF LEGISLATIVE MEASURES AGAINST TREE FELLING

Forests and trees through the process of photosynthesis sequester or capture carbon dioxide from the atmosphere to create energy for growth. The captured carbon is then stored in the tree's biomass over its lifetime as it performs important functions in the natural carbon cycle, and in the process, it helps in CC mitigation (Coder et al., 1996).

Trees store carbon in several parts of their system, such as their leaves, branches, trunks, stems, and roots, while their fallen leaves contribute biomass to the soil. In addition to the essential role played by trees in carbon storage, trees can also act as cooling agents when positioned effectively to protect residential buildings from the direct effect of sun rays (Dwyer et al., 1992).

The national average urban tree-canopy cover varies from city to city in developed countries; for instance, in Australian cities, the average urban tree-canopy is 39%; however, in Hobart, it varies from 59% to 13% in Melbourne. These results are similar to the average urban tree canopy cover in American cities, which ranges from 21% in New York to 42% in Pittsburgh (Mark, 2017).

In the next 20 years of the trees existence, it is expected that 39% of the urban tree population will gradually die off. This is based on the fact that the percentages of these present trees will decrease due to aging, pressures of drought, urban expansion/development, and water restrictions. The situation is not much different in the cities of other developed counties and their urban areas; clearly, there is need to develop and strengthen legislative measures for tree protection (The Tree Protection Legislation, 2006).

In developing countries, forests, and tree protection legislations are hardly enforced in contradistinction to developed areas where such measures are considered and implemented when necessary.

In developing countries, a comprehensive package of legislative measures for tree protection should be introduced; it should closely monitor and also formulate measures for further improvement. This will in protecting the valuable role that trees play in maintaining healthy environments, statutory legislations are thereby imperative. Some laws may be appropriate for individual trees, groups of trees or woodlands. Penalties can be imposed for unauthorized removal of such wood resources (The Tree Protection Legislation, 2006).

The government should be fully committed to protecting trees on government lands by ensuring that trees are not unnecessarily removed. Individuals should also be responsible for protecting trees within their premises in the urban space (see, A Guide to Tree Preservation Procedure; available from http://www.communitiesgov.uk/publications/planningand/building/protectedtrees). These regulatory measures should also be incorporated into regulatory conditions in public works contracts requiring removal of trees. These regulatory measures should prevent indiscriminate tree felling by the contractors, builders, and others with vested interest in development (The Tree Protection Legislation, 2006).

These measures should be further strengthened by involving the community and public participation in tree planting and greening. The public sector should cooperate with the private sector so as to facilitate urban greening schemes (Development Bureau, 2016). Publicity on greening can be effectively planned and implemented through mass media avenues such as the television, radio, newspapers, magazines, and social media with the participation of professional organizations. Properly organized educational talks in schools and local communities will also be useful in reinforcing the message of environmental greening towards minimization of CC challenge.

3.8 PROMOTION OF ENVIRONMENTAL/LAND SURFACE GREENING AS ESSENTIAL LANDSCAPING ACTIVITIES IN BOTH URBAN AND RURAL AREAS

Environmental greening includes all measures towards minimal disruption of the natural environment, which are known to be capable of retaining a sustainable and balanced climate. These measures are currently being highlighted for adaptation to the unpleasant phenomenon of CC. They include land surface greening with landscaping, gardening, horticulture, tree planting, and agroforestry.

Land surface greening is a popular concept that has gained prominence in recent times. The term evolved from Miller (1988), who defined it as a unified approach that covers all the aspects of the planting of vegetation through skillful management while harnessing all the multiple environmental and social benefits for users. It is concerned with all the aspects that involve the planting of vegetation, trees, grasses, flowers, and the enhancement of waterlogged areas whose design is intended to improve the environmental quality, economic value, and social opportunities associated with both city and rural landscape (Mohd-Yusof, 2012). It is a general term that embraces all vegetated land uses and activities aimed at providing, conserving, improving or extending vegetation in urban and rural areas.

Environmental greening is a potential tool for making cities and rural areas more sustainable (Sorensen et al., 1997). The examples of land surface greening projects include urban and rural forestry, small parks, botanical gardens, tree planting, parks, gardens, and creation of other kinds of green spaces (Mohd-Yusuf, 2012). Land surface greening is based on the recognition that green spaces in both urban and rural areas should be exploited in an integral form, further than the traditional recreation use and aesthetics. They should also be used for many other environmental and social benefits. These social benefits include improved air quality, sanitary conditions, water supply, controlling floodwater management, food security, moderating both macro and microclimates, and many more (Sorensen et al., 1997). The importance of green infrastructure has become paramount in environmental policy issues globally as a tool for CC minimization (Forest Research, 2010).

3.9 THE BENEFITS OF PROMOTING ENVIRONMENTAL/LAND SURFACE GREENING IN URBAN AND RURAL AREAS

There are benefits associated with the promotion of land surface greening, which helps in minimizing CC. They are classified into three categories, namely, environmental, social, and economic benefits.

3.9.1 *ENVIRONMENTAL BENEFITS*

a. **Air Quality Improvement:** Land surface greening is essential in mitigating the effects of air pollution generally. Trees and shrubs reduce air pollution by absorbing some gaseous pollutants as well as intercepting dust and smoke particles from the air. Plants also absorb poisonous gases such as carbon monoxide from automobile exhaust pipes. In Chicago (USA), trees were found to get rid of 6,190 tones air pollution per annum, thus resulting in 0.3% improvement in air quality (Nowak, 1994; Forest Report, 2010).

Several studies showed that trees can reduce air pollution by absorbing other dangerous and gaseous pollutants from the atmosphere (Tiwary et al., 2009; Broadmeadow and Free-Smith, 1996; Free-Smith et al., 2005; Jouraeva et al., 2002). It is estimated that increased tree cover improves air quality by 5 to 10% (Nowak, 1994).

b. **Improvement in Climate:** Land surface greening improves the climate in two distinct forms in urban and rural areas. Trees improve human comfort significantly with regard to moderation of energy budget by reducing the intensity of Urban heat island (UHI) in urban areas, aids air movement, humidity, and air temperature circulation in both urban and rural areas (Shashua-Bar and Hoffman, 2000). Urban heat island is caused by the absorption of direct solar radiation by buildings and paved areas, as well as absence of the cushioning effect of vegetation. This leads to increased energy budget of buildings and enhanced formation of pollutants and smog (Sorensen et al., 1997; Chem and Wong, 2006).

Green infrastructure acts as a cooling medium through the moderation the warming effects of CC and UHI (Dimoudi and Nikolopoulou, 2003; Forest Research, 2010). Vegetation helps in air pollution control by minimizing the effects of carbon dioxide, which is a major cause of air pollution. Moreover, carbon dioxide also destabilizes the ozone layer, resulting in greenhouse effect and exposure of the earth's surface to unpleasant solar radiation. However, green vegetation absorbs some of the carbon dioxide in the atmosphere and converts it to life-sustaining oxygen through the process of photosynthesis. Moreover, vegetation cover of tall trees acts as shade from direct heat rays of the sun as well as umbrella from the chill of heavy rainfall.

Several studies have shown that land surface greening has been effective in amelioration of high temperature in urban and rural areas. The Works of Potchter et al., (2006); Gill et al., (2007); Chang and Chang (2007) and Yu and Hien (2006) have also shown that green infrastructure has cooling effect on the environment.

c. **Improvement of Water Quality:** Provision of portable water is a major development challenge. It is critical for the health and well-being of human life. The Millennium Development Goal and Sustainable Development Goal have set means on providing the global population with safe drinking water and basic sanitation (UNDP, 2005). It was estimated by the World Health Organization (WHO) that 1.1 billion people across the globe do not have access to potable water (WHO/UNICEF, 2004). Water supply quality is reduced by high-speed run-offs, pollutants, sewerage, and debris (Forest Report, 2010). In developing countries, contaminated water surfaces are sources of water-borne diseases such as typhoid, dysentery, and diarrhea, which is responsible for deaths in infants and children (Sorensen et al., 1997).

Land surface greening is capable of improving water quality in both urban and rural areas. Several studies have shown that the growing of trees and vegetation improves water quality significantly. Stovin et al., (2008) posited that trees intercept rainfall and store them at source and also trees use their canopy to filter pollutants preventing them from infiltrating the groundwater. Similarly, Lawrence et al., (1984) and Jeffries et al., (2003) discovered that grasses and trees reduce the diffusion of pesticides and other pollution agents into the ground. Also, Dudley and Stolten (2003) discovered that majority of the world largest cities depend on forest protected areas for portable water supply.

d. **Flood Control:** Flooding is a major environmental disaster confronting the globe. It has a devastating effect on humans, infrastructure, and environment. An estimated one trillion USA dollars in damages is caused annually by floods. The causes of flooding have been attributed to alteration of the natural environment by man through the destruction of arable lands, economic trees and vegetation and the development of buildings, infrastructures, and other hard surfaces. This disrupts the movement of water and increases the runoff, thus resulting in flooding. Also intensive rainfall, due largely to CC, have resulted in the swelling of rivers and intense flood of stormwater, which causes severe flood across the globe.

Land surface greening can control flooding by increasing the permeable surface area in a watershed, which decreases runoff rates thereby minimizing water flow levels within urban areas (Sorensen et al., 1997). Several cities such as Durban (South Africa), Tulsa (Oklahoma, USA) and Curitiba (Brazil) have used urban greening to successfully minimize the devastating effect of flooding. Green belts have also been used as buffers against the menace of urban flooding (Nisbet et al., 2004).

e. **Noise Abatement:** Noise pollution is one of the environmental challenges facing large cities worldwide. It has reached unhealthy level in large cities such as Mexico City, Lagos, Santiago, Rio de Janeiro, and Mumbai. These cities have noise levels above 75 decibels which can cause damage to the ear (Sorensen et al., 1997). The sources of noise pollution have been attributed to heavy industries, commercial, and traffic corridors.

Land surface greening reduces noise pollution in five ways. First, trees and vegetation including grasses absorb sound from the atmosphere. Second, trees can also deflect sound away from the listeners. Large trees can reflect sound back to the source. Fourth, land surface greening refracts and dissipates noise that passes through and around it. Fifth, it can mask sound, thus enabling people to filter out unwanted noise. For example, people may choose to hear the natural sound like birds chirping and singing instead of the loud uncoordinated noise made by industrial, commercial, and traffic activities that emanate from the city (Miller, 1988; Sorensen et al., 1997). In general, plants are of immense advantage to humans because they effectively absorb high-frequency noise, which is stressful. Therefore, land surface greening projects that will minimize noise pollution should be a dense vegetable cover with a range of tall trees and plants.

f. **Erosion Control:** Erosion and landslides have become common occurrences in both developed and developing countries. This has been attributed to loss of vegetative cover and intense environmental rainfall as a result of climate variability. This has resulted in loss of lives and properties across the globe. Consequently, land surface greening has been adopted by many countries to solve erosion problems. For instance, reforestation program covering large expanse of watershed area of Bogota River was carried out by the Colombian Government to control erosion (Sorensen et al., 1997).

g. **Sustainable Urban Drainage:** A Sustainable Urban Drainage System (SUDS) is an improved approach developed to reduce the volume of floodwater runoff. It encourages the experimentation of every available means of green adaptation from trees and vegetation to botanical gardens to absorb water from drains, ponds, and wetlands, in order, to reduce stormwater runoff at point sources, and thus create an improved drainage system (Forest Research, 2010). It increases the permeable surface evadable for runoff, reduce flow rates, and eliminate damage to settlements (Sorensen et al., 1997).

Studies have shown that the SUDS incorporation with green infrastructure has improved drainage system in both urban and rural areas. Gill et al., (2007) experimented that developing vegetative areas would minimize runoff levels by 4.9%, 5.7%, and

11.8–14.1%, respectively. Similarly, Mentens et al., (2006) discovered that green roofs have higher rainfall retention capacity than the traditional roofing.

h. **Land Reclamation:** Land surface greening is an effective tool for reclaiming unused or degraded lands. It can minimize prevalent health problems associated with land degradation as well as improve the local environment (Forest Research, 2010). Several studies have shown the potentials of green infrastructure in the regeneration of previously developed land (PDL). Hutchings (2002) found out that vegetation uses the process of phyto-remediation and phyto-stabilization to enable the mobilization of pollutants from land. He opined that growing plants and trees on degraded lands can help in destabilizing pollutant sources and paths of movement within the environment. These processes, he argues, are cost-effective methods of remediation. Handley (1996) opined that planting of vegetation on oil contaminated lands can positively influence the soil, water, and ground water, thus resulting in the improvement of the local ecosystem services. Sorensen et al., (1997) and Akanwa et al., (2017) indicated that implementing reforestation practices on degraded lands such as abandoned mining sites, dumpsites, or other reclamation sites can improve the scenic view or urban aesthetics.

i. **Wildlife Habitat and Biodiversity:** Land surface greening influences wildlife habitat and biodiversity in both urban and rural areas (Heller and Zavaleta, 2008). Green space is a habitat for various species of plants and animals. Greening the built environment influences the growth and protection of various plants and animal species coupled with the conditions that make them thrive in order to avoid their relocation or even extinction (Forest Research, 2010). The habitat provided by the green infrastructure determines the range of species of plants and animals. This is because the vibrancy of the vegetation cover will eventually determine the quality, number, and type of species of plant and animals that is available for habitation.

Good quality land surface greening has been found by several studies to influence the wildlife habitat and biodiversity. Fernandez-Juric and Jokimaki (2001) found that the size of a habitat is accountable for the richness of the bio-species. They opined that parks of 10–35 hectares size can harbor a large species of birds any urban location or region. Hardy and Dennis (1999) disclosed that the presence of large quantity of nectar resources in urban areas would attract and increase the number of butterflies in urban areas than rural areas.

Angola et al., (2006) found that the range of species recorded in different groups and different types of urban spaces/sites reflected patch size, particularly for plants and habitat quality. Baker and Hanis (2007) discovered that the presence of green infrastructure increased the rapid urban growth of mammals.

Several studies have also shown that land surface greening increases the populations of some endangered species. Gibson (1998) found that the population of rare UK invertebrate species increased by 12–15% on Brownfield sites due to greening efforts. Similarly, Gibson (1998) found that greening efforts have resulted in the increase in population of four UK rare species especially the horehound long-horn moth, the streaked bombardier beetles, the red star-thistle and the oolite downy-back beetle.

Land surface greening gives opportunities for longer-distance movement of some animals. Rudd et al., (2002) indicated that urban gardens create opportunities for residents to connect and extend relationships with one another. Brown and Kodric-Brown (1977) forecasted that adequate habitat connectivity can enhance support from more productive species population to those in low-quality habitat. Morimoto and Katoh (2005) showed evidence that linear greenings increase the richness of birds' species in urban parks. Kupcak (2001) found that small green patches at roundabouts and road verges have a potential benefit for movement of biodiversity. Forest Research (2010) disclosed that some animal species are vulnerable to the changing climate conditions, hence possible ways of adaptation will be to move to a new habitat with a more suitable micro-climate.

j. **Environmental Quality and Aesthetics:** Land surface greening can improve the quality and aesthetics of both urban and rural environment. Vegetation reduces urban heat island while complementing the architectural features and the usual large expanses of concrete (Sorensen et al., 1997). Consequently, it improves the quality of life as well as opportunities for people living and working within the environment. The provision of high-quality greenery in an area is aesthetically pleasing and has positive effect on the residents, businesses, and investors (Land use Consultant, 2004; Sorensen et al., 1997). It also increases the property value and the aesthetics of private lands/buildings in the urban or suburban area.

Several studies have shown that land surface greening can help improve the environmental quality and aesthetics. Swanwick (2009) reported that green spaces improve the quality of urban lifeby creating varieties of avenues through its strategic locations attracting people with cultural differences to socialize. Tibbatts (2002) revealed that visual appearance and attractiveness of urban centers are strongly influenced by the provision and development of large expanse of green areas. Sorensen et al., (1997) found that the re-afforestation of lands with rich vegetation cover increases its value and attractiveness by far than development of buildings. Braatz (1993) disclosed that cities such as Singapore and Kuala Lumpur gained significant foreign investments due to the attention given to the development of green areas and aesthetics is one of the major factors that accelerated the cities rapid economic growth.

3.9.2 ECONOMIC BENEFITS

a. **Food and Agricultural Products:** Land surface greening projects, such as gardening and other forms of agriculture, produce food and cash crops for food supply to both urban and rural populations, enabling them to overcome the challenges posed by CC. In developing countries, 28% of households in urban areas practice urban agriculture, whereas it is the predominant occupation in the rural areas (Smith et al., 1996).

The importance of land surface greening cannot be over emphasized. It helped the urban poor in Haiti to provide from one-tenth to one-third of their annual family consumption, thus ensuring their families to have adequate food supply, better health when comparable to other low- income families (UNDP, 1996; Sorensen et al., 1997). Also, it enabled them to reduce pressure on their limited financial resource. In Arlington's Virginia of USA, farmers plant gardens at community lands located in city's highways. This exercise has helped in food production and maintenance of medians, which serves a case saving measures for the city. Urban agriculture is plays a vital role to the farmers, population, and the city in general since there is adequate food supplies as well as providing other amenities (Sorensen et al., 1997). In rural areas, the population depends on agriculture as the main source of livelihood.

Some studies have shown that land surface greening triggers economic growth through job provision and local investment by farmers and stakeholders, increase in land and property value and local economic growth. ONS (2010) discovered that the economy of UK improved tremendously as a result of investment in green infrastructure. It helped the UK companies succeed in international businesses that were able to attract foreign investments.

Venn and Niemela (2004) disclosed that land surface greening adds beauty and serenity to an area with propensity for socialization, relaxation, and business opportunities. This in turn attracts varieties of people infrastructures, goods, and services. Scottish Enterprise (2008) found that improved landscape greening at Riverside Park, Cydlebank, and Winsford yielded over 16% and 13% respectively of net growth in employment, and generated over 1 million pounds of private investment. CESR, (2004) opined that locally createdjobs increased by 41% after the creation of the national forest. Moreover, internal re-afforestation of green infrastructure attracted investments totaling to 96 million pounds.

CTLA (2003) showed that provision of trees and a well landscape environment increases property values by 15 to 25%. ECOTEC (2008) found that investment in green infrastructure is a good local economic regeneration measure. This is because it influences and stimulates business opportunities and industrial expansion at both the local and international levels to a region. This is achieved by designing, developing, and maintaining land surface greening that is good enough to produce a sensitive, living, and working environment that adds significant value to local economics.

b. **Forest Products and Fodders:** Land surface greening serves as a source of wood for electric wire support poles, firewood, and fodder in both urban and rural areas. Trees are also used for fences, agriculture, construction, furniture making, and crafts. Firewood is used significantly by urban poor and rural dwellers as source of cooking and heating fuel (Wolf, 2004). Majority of people in developing countries, especially in rural areas, rely on green areas for their source of wood. Many species of trees and grasses serve as a source of high-quality fodder for live stocks. Other economic products of value include fruits, nuts, and fiber that are cultivated in urban and rural green areas.

3.9.3 SOCIAL BENEFITS

a. **Health:** Land surface greening is highly beneficial to health and wellbeing of the society, which is germane for tackling the threats posed by CC. Vegetation improves air quality, thus impacting positively on health through reduction in respiratory diseases. It reduces stress and provides aesthetically pleasing and relaxing environment that are germane for good health (Sorensen et al., 1997; Nowak et al., 1996). Convalescing patients that are exposed to trees and outdoor settings have been found to recover faster than patients without such exposure (Ulrich, 1990). Also, shades from trees improve the health by minimizing exposure to the direct effect of sun rays, thereby reducing health risks (Heisler et al., 1995; Sorensen et al., 1997). In addition, surface greening reduces urban heat island effect as well as serves as a link between humans and the natural environment. This linkage is a germane for relaxation, increase in productivity levels and general wellbeing (Nowak et al., 1996; Sorensen et al., 1997).

Ulrich (1990) reported that the extent of green space provided in a residential area was positively related to their perceived levels general mental health and wellbeing. He opined that this relationship was strongest for the lower socio-economic group. Landscaping and green spaces obviously influenced the conditions which contribute significantly to a number of health problems. Buck (2016) showed that communities with more parks experienced significant higher levels of outdoor recreation, thus resulting in improved physical health.

Green spaces are pivotal to improved health conditions through outdoor recreational activities, relaxation, and social integration. Buck (2016) and Ulrich (1990) indicated that green spaces reduces blood pressure and less stress levels as outdoor events carried out in parks and open spaces has greater psychological and physiological benefits than in other settings. Hence, it helps people experience less stress and related illnesses with positive effect on mental health.

Children having attention deficit disorder (ADD) experienced minimized health risks when exposed to green-outdoor spaces for recreation; obviously, a greener and nature-friendlier setting lowers the severe tendencies of ADD symptoms. In addition, this confirmed the existence of relationship between green spaces and mental wholesomeness across a range of areas in Greenwich, UK (Tiwary et al., 2009).

b. **Recreation and Stronger Communities:** Generally, land surface greening projects such as parks and other open spaces are sources of recreation for all income groups. The presence of green space encourages residents to frequently use their outside spaces. This increases outdoor activities, improves community life, develops connectivity, discourages crime and juvenile delinquencies and hence, strengthens the partnership CC mitigation (Weldon et al., 2007). Cohen et al., (2008) discovered that there was a strong positive association between green environments and the ability of residents to interact positively and effectively. Furthermore, evidence from several studies have shown that green spaces have the potentials for enhancing social inclusion especially among the vulnerable groups with disabilities, minority groups, young, and old people and others with economic disadvantages.

Forest Research (2010) found that social ties were stronger in greener communities. The overall crime level was significantly lower in communities with high presence of green spaces. Jorgensen et al., (2007) disclosed that green space leads to community integration and increased security for residents. Sullivan et al., (2004) found that 83% of individuals engaged in social activity make use of parks and other green spaces.

c. **Education:** Promotion of land surface greening through parks and other green spaces provides educational/research opportunities for the population. These learning opportunities in form of botanical gardens, zoos, nature trails and others are source of education to the residents and visitors about the fauna and flora available in an area. This offers the people the opportunity to learn about their environment and natural processes, as well as the impact of the CC (Sorensen et al., 1997). It will raise the awareness of the public on the importance of land surface greening as adaption strategy for CC. Examples of tree planting exercises in cities include Sao Paulo (Brazil) and Via Del Mar (Chile) which Sorensen et al., (1997) found to have been successful school tree planting programmes.

In general, the benefits of promoting land surface greening in both urban and rural areas are enormous. They are significant in treating the problems associated with CC as well as in addressing other environmental, social, and economic challenges confronting the society today. There is the need to tackle the institutional, financial, land tenure, operational, legal, participatory, and ecological and other challenges facing land surface greening. This will result in the creation of a high quality environment that promotes aesthetics, attraction, and well-being of its residents.

3.10 PROMOTION OF GARDENING AND HORTICULTURAL ACTIVITIES

Gardening is the practice of growing and cultivating plants around homes. It involves growing food, flowers, vegetables, fruits herbs, plants, shrubs, orchards, and trees (Buck, 2016). Horticulture, which is often known as gardening, is the art and science of growing ornamental plants, flowers, fruits, and vegetables (Reif, 1992; Jim and Chen, 2003). Based on the definitions, gardening, and horticulture are similar activities.

Promoting gardening and horticultural activities have environmental benefits, especially as planned mitigation strategy for CC challenges, which will result in improved environmental quality, climate amelioration and conservation of biodiversity (Shan, 2009; Officha et al., 2012).

Gardening and horticultural activities can improve the quality of the environment through minimization of CC causal agents from the environment. Plants reduce carbon dioxide known for ozone layer depletion. Plants also reduce smug from industrial, commercial, and trade activities. Jo and McPherson (1995) estimated that plants and trees remove 5,575 metric tons of air pollutant from the atmosphere in Chicago, USA. Similarly, Nowak et al., (2006) found that 71,000 metrics tons of air pollutants were removed by plants and trees per annum in 55 cities across the United States of America. In China, Yang et al., (2005) indicated that plants and trees removed over 1000 tons of air pollutants in Beijing. The works of Bealey et al., (2007), McDonald et al., (2007), and Fowler et al. (2004) have found that plants and trees were effective in reducing the concentration of PM_{10} and capturing of aerosols and air-borne metal particles from the atmosphere.

Also, gardening and horticultural activities could reduce noise level and enhance infiltration of rainwater, thereby minimizing stormwater runoff. This reduces the risk of flooding, non-point pollution, and pathogen contamination associated with CC (Shan, 2009).

Global warming, which is attributed to greenhouse gases (GHGs), is a major cause of CC. It has resulted in severe environmental challenges such as flooding, desertification, drought, erosion, urban heat island effect, and other hash climatic conditions worldwide. Gardening and horticultural activities are effective adaptation strategies for minimizing the effects of CC. Plants, trees, orchards, and other green spaces can reduce GHGs by taking carbon from the atmosphere and storing them in the ground. A study by Jo and McPherson (1995) showed that gardens extracted 20kg/m^2 of carbon from the atmosphere in two residential blocks in Chicago, USA, and stored permanently in the soil. Similarly, Akbari (2002) found that plants and trees reduce carbon emission annually by about 10–11kg.

Also, gardens and horticulture can regulate micro-climate and thus improve the thermal comfort in both urban and rural areas. Trees and plants provide cooling effects that can reduce high temperature in urban areas thereby mitigating the effects of urban heat island (Shan, 2009). This contributes to energy saving by reducing air-conditioning energy use in hot condition. A study by Wong et al., (2007) showed that buildings surrounded by or near gardening and horticultural activities have lower ambient temperature than buildings far away from them. Furthermore, gardens, and horticulture positively ensure urban humidity and water balance. Chen and Wong (2006) found that average relative humidity was higher in gardening and horticultural activities than their surrounding environment.

CC is a threat to biodiversity across the globe. The habitats of many animal and plants species have been destroyed by flooding, human development, wildfire, harsh climate condition, and other phenomena associated with CC. This negatively impacted the existence of such species of animals and plants. Gardening and horticultural activities can enhance biodiversity conservation (Officha et al., 2017a). The works of Godefroud and Koedam (2003) and Alvey (2006) have shown that there was increase in level of biodiversity in urban areas with high degree of gardening and horticultural activities. Cornelis and Hermy (2004) found that gardens and horticulture accommodate a very high proportion of wild plant species, breeding birds, butterflies, and amphibians in Flanders, Belgium. Stewart et al., (2004) found that Christchurch, the second-largest city in New Zealand, had higher floral diversity than its surroundings because of its prevalence of gardening and horticultural activities.

In general, CC phenomenon poses serious threat to mankind. The threat has become enormous even as many more people are now living in urban areas, thus abandoning the rural areas. There is need to encourage gardening and horticultural activities especially in the urban environment (Officha et al., 2017b). This is capable of enhancing measures towards mitigation of the threat posed by CC.

3.11 LEGISLATIVE MEASURES FOR MINIMIZING LAND SURFACE EXPOSURE

There are several legislative measures across the globe that has been targeted at protection of the land surface from human activities. The measures are found in the environmental laws and policies of various countries which were influenced by various international conventions and treaties especially the Earth Summit of 1992. For instance, in Nigeria, the environmental laws on protecting the land surface were aimed at minimizing land pollution from industrial activities, preventing, and reversing desertification, managing forest, and loss of wildlife resources and combating floods and erosion (Muhammed, 2012). In Europe, the USA, Australia, and South-east Asia, there are strong regulations protecting the land surface against the negative impact of human activities.

Despite the availability of comprehensive policy and legal framework, land surface degradation has continued to persist. This has been attributed to weakness in policy implementation and enforcement of environmental laws (Olarinde and Orecho, 2015). There is the need to strengthen the existing legislation measures aimed at minimizing the exposure of the land surface to environmental degradation. Consequently, legislative measures should be introduced to address the weaknesses of the existing regulations specifically, and take into account new and emerging environmental threats on the land surface. Furthermore, new developments in knowledge and best practices, as well as the outcomes of new international conventions should be integrated into the existing environmental laws and policies by all countries in order to make them more responsive to the challenges of CC.

3.12 CONCLUSION

This book chapter reviewed strategies that are sustainably appropriate to mitigate the global consequences of CC. In recent times, the industrial processes and activities have become intense releasing hydrocarbons which contribute to the greenhouse effect and, hence, global warming and CC has become unavoidable.

The study suggested widely applicable, economically viable and environmentally friendly measures such as agroforestry and environmental greening. The practical applications include; intensive and all-season agriculture, tree planting with attention to tree species and forestry practices that are less vulnerable to storms and fires; effective mitigation policies, utilizing new technologies, new agricultural practices among others as suggested in this chapter. This chapter also affirmed that conservation of forest reserves, environmental management policies and legislations protecting tree felling and greening should be applied in all aspects of development, urban design and in rural areas to make human settlements sustainably green and CC-resilient.

KEYWORDS

- **agriculture**
- **agroforestry**
- **climate change**
- **conservation**
- **greenhouse gases**
- **greening**

REFERENCES

Akanwa, A. O., (2016). *Effect of Quarrying Activities on Local Vegetation Cover in Ebonyi State, Nigeria.* A PhD Thesis from the center for environmental management and control (CEMAC), University of Nigeria Nsukka (UNN), Nigeria.

Akanwa, A. O., (2018). *Aerial Photographs Showing Urban Greening/Farming in the Central District Area.* Maitama, Abuja, Nigeria.

Akanwa, A. O., Okeke, F. I., Nnodu, V. C., & Iortyom, E. T., (2017a). Quarrying and its effect on vegetation cover for a sustainable development using high resolution satellite image and GIS. *Journal of Environmental Earth Sciences, 76*(4), 1–12.

Akanwa, A. O., Onwuemesi, F. E., Chukwurah, G. O., & Officha, M. C., (2017b). Effect of open cast quarrying technique on vegetation cover and the environment in south eastern, Nigeria. *American Scientific Research Journal for Engineering, Technology and Sciences (ASRJETS), 21*(1), 227–240.

Akbari, H., (2002). Shade trees induce building energy use and CO_2 emissions from power plants. *Environmental Pollution, 116*, 119–126.

Aktar, M. S., Abedin, M. Z., & Quddus, M. A., (1992). Trees in crop field under agroforestry system in Bangladesh. *Journal of Training and Development*, *5*(2), 115–119.

Allen, C. D., Macalady, A. K., & Chenchouni, H., (2010). A global overview of drought and heat induced tree mortality reveals emerging climate change risks for forests. *Ecology Management*, *259*, 660–684.

Alvey, A. A., (2006). Promoting and preserving biodiversity in the urban forest. *Urban Forest and Urban Greening, 5*, 195–201.

Angelo, M. J., Timbers, A., Walker, M. J., Donabedia, J. B., & Van Noble, D., (2001). Essays on building a more sustainable and local food system: Small, slow, and local. *Vermont Journal of Environmental Law, 12*(2), 353–378.

Angola, R. G., Gadler, J. P., Hill, M. O., Pullin, A., Rushton, S., Austin, K., Small, E., Wood, B., Wadsworth, R., Sanderson, R., & Thompson, K., (2006). Biodiversity in urban habitat patches. *Science of the Total Environmental*, *360*(1–3), 196–204.

Anonymous, (2010). *Rural Development Policy from the Period 2007–2013.* http://ec.europa.eu/agriculture/rurdev/index_cs.htm (Accessed on 9 October 2019).

Baker, P. J., & Hanis, S., (2007). Urban mammals: What does the future hold? An analysis of the factors affecting patterns of use of residential gardens in Great Britain. *Mammal Review, 37*(4), 297–315.

Balooni, K., (2003). Economics of wasteland afforestation in India: A review. *New Forum, 26*, 101–136.

Bealey, W. J., McDonald, A. G., Nemitz, E., Donavan, R., Dragosits, U., Duffy, T. R., & Fowler, D. (2007). Estimating the reduction of urban PM_{10} concentrations by trees within an environmental information system. *Journal of Environmental Management, 85*, 44–58.

Beer, J. W., Muschler, R. G., Somarriba, E., & Kass, D.m (1998). Shade management in Coffee and Cocoa plantations–A review. *Agroforestry System, 38*, 139–164.

Bessie, D., (2005). *Food Policy: Urban Farming as a Supplementary Food Source*. Walden University Doctoral Studies.

Bhalme, H. N., (1997). *The Climate of Arid Zone and Crops, World Climate Programme, Application and Service*. Indian Institute of Tropical Meteorological, Geneva.

Bhatt, V., & Kongshaug, R., (2005). *Making the Edible Landscape: A Study of Urban Agriculture in Montreal*. MacGill University, Montreal.

Braatz, S. M., (1993). *Urban Forestry in Developing Countries: Status and Issues, in Proceedings of Sixth National Urban Forest Conference*, Minnesota: American Forests.

Broadmeadow, M. S. J., & Freer-Smith, P. H., (1996). *The Improvement of Urban Air Quality by Trees*. Arboriculture research and information note. AALS, Farham.

Brown, H. K., & Carter, A., (2003). *Urban Agriculture and Community Food Security in the United States: Farming from the City Center to the Urban Fringe* (pp. 1–32). Amherst, MA: Community food security coalition. Non-profit. Retrieved from: http://www.aerofarms.com/wordpress/wpcontent/filesf/1265604354urbanagricultureandcommunity food security.pdf (Accessed on 9 October 2019).

Brown, J., & Kodric-Brown, A., (1977). Turnover rates in insular biogeography: Effects of immigration on extinction. *Ecology, 58*, 445–449.

Bryld, E., (2003). Potentials, problems, and policy. Implications for urban agriculture in developing countries. *Agriculture and Human Values, 20*(1), 79–86.

Buck, D., (2016). *Gardens and Health Implication for Policy and Practice*. A report submitted to the national gardens scheme.

Buresh, R. J., & Cooper, P. J. M., (1999). The science and practice of improved fallows. *Agroforest. System, 47*, 13–58.

Campaign to Protect Rural England (CPRE), Natural England, (2010). *Green Belts: A Greener Future*. London.

CESR, (2004). *Much More Than Trees: Measuring the Social and Economic Impact of the National Forest- Staffordshire University Centre for Economic and Social Regeneration.*

Chang, C. R., & Chang, C., (2007). A preliminary study on the local cool-island intensity of Taipei city parks. *Energy and Buildings, 38*, 105–120.

Chao, S., (2012). *Forest People: Numbers Across the World, Forest Peoples Program*. Morton-in- marsh, UK.

Chase, K. L., (2012). *From Hometown to Grow Town: A Study of Permaculture-Based Neighborhood Revitalization Strategies for Muncie, Indiana, Thesis*. Ball State University, Muncie, India.

Clark, J. R., Matheny, N. P., Cross, G., & Wake, V., (1997). A model of urban forest sustainability. *Journal of Arboriculture, 23*(1), 17–30.

Clemmitt, M., (2008). Global food crisis: What's causing the rising prices? Online publication. *CQ Researcher, 18*(24), 555–575.

Coder, K. D., (1996). *Identified Benefits of Community Trees and Forests: University of Georgia Cooperative Extension Service Forest Resources Publication FOR96–39*.
Cohen, O. A., Inagami, S., & Finch, B., (2008). The built environment and collective efficacy. *Health and Place, 14*, 198–208.
Cornelis, J., & Hermy, M., (2004). Biodiversity relationships in urban and suburban parks in Flanders. *Landscape and Urban Planning, 69*, 385–401.
Cruz, M. C., & Medina, R. S., (2001). *Agriculture in the City: A Key to Sustainability in Havana*. Cuba, IDRC, Ottawa, Canada.
CTLA, (2003). *Summary of Tree Valuation Based on CTLA Approach Council of Tree and Landscape Appraisers*.
De Clerck, F. A. J., & Negreros-Castillo, P., (2000). Plant species of traditional Mayan Home gardens of Mexico as an analog for multistrata-agroforests. *Agroforest. System, 48*, 303–317.
Department of Economic and Social Affairs of the United Nations Population Division, (2012). *World Urbanization Prospects: The 2011 Revision [Place Unknown]*. Retrieved from: http://esa.un.org/unpd/wpp/ppt/CSIS/WUP_2011_CSIS_4.pdf (Accessed on 9 October 2019).
Development Bureau, (2016). *Greening, Landscape and Tree Management*. The government of the Hong Kong special administrative region of the People's Republic of China.
Dimoudi, A., & Nikolopoulou, M., (2003). Vegetation in the urban environments: Microclimatic analysis and benefits. *Energy and Buildings, 35*, 69–76.
Dixon, J. M., Donati, K. J., Pike, L. L., & Hattersley, L., (2009). Functional foods and urban agriculture: Two responses to climate change-related food insecurity. *New South Wales Public Health Bulletin, 20*(1/2), 14–18.
Drescher, A., (2001). *The Integration of Urban Agriculture Into Urban Planning–An Analysis of the Current Status and Constraints*. University of Freiburg, Germany.
Duchemin, E., Wegmuller, F., & Legault, A. M., (2009). Urban agriculture: Multi-dimensional tools for social development in poor neighborhoods. *Field Actions Science Reports, 2*(1), 43–52.
Dwyer, J. F., McPherson, E. G., Schroeder, H. W., & Rowntree, R. A., (1992). Assessing the benefits and costs of the urban forest. *Journal of Arboriculture, 18*(5), 227–234.
ECOTEC, (2008). *The Economic Benefits of Green Infrastructures: The Public and Business Case for Investing in Green Infrastructure and a Review of the Underpinning Evidence*. Report for natural economy Northwest. Available at: http://www.naturaleconomynorthwest.co.uk/resources+reports.php (Accessed on 9 October 2019).
Ericksen, P. J., (2007). Conceptualizing food systems for global environmental change research. *Global Environmental Change, 18*(1), 234–245.
FAO, (2013). *Climate-Smart Agriculture Sourcebook* (p. 570). Food and agriculture organization, Rome.
Fernandez-Juric, E., & Jokimaki, J., (2001). A habitat island approach to conserving birds in urban landscapes: Case studies from Southern and Northern Europe. *Biodiversity and Conservation, 10*, 2023–2043.
Forest Research, (2010). *Benefits of Green Infrastructure*. Report by forest research. Forest research, Fernham.
Forestry/Fuel-wood Research and Development Project, (1992). *Growing Multipurpose Trees on Small Farms*. Bangkok, Thailand: Winrock International. 195 + ix pp. (including 41 species fact cards). To order in the USA, call: 703/351–4006 and request book order no. PNABR667.

Fowler, D., Skiba, U., Nemitz, E., Choubedar, F., Branford, D., Donovan, R., & Rowland, P., (2004). Measuring aerosol and heavy metal disposition on urban woodland and grass using inventories of ^{210}Pb and metal concentrations in soil, water, air and soil pollution. *Focus, 4*(2–5), 483–499.

Gibson, C. W. D., (1998). *Brownfield: Red Data the Values Artificial Habitats Have for Uncommon Invertebrates*. ENRR 273 English Nature, Peterborough.

Gill, S. E., Handley, J. F., Ennos, A. R., & Pauliet, S., (2007). Adapting cities for climate change: The role of the green infrastructure. *Built Environment, 33*, 115–133.

Gliessman, S. R., (2015). Agroecology. *The Ecology of Sustainable Food Systems* (3rd edn., p. 371). CRC Press, Boca Raton.

Godefroud, S., & Koedam, N., (2003). Distribution pattern of the flora in a peri-urban forest: An effect of the city-forest ecotone. *Landscape and Urban Planning, 65*(4), 169–185.

Godfray, H. C. J., Beddington, J. R., Crute, I. R., Haddad, L., Lawrence, D., Muir, J. F., & Toulmin, C., (2010). *Food Security: The Challenge of Feeding 9 Billion People.*

Gossop, C., (2011). Low carbon cities: An introduction to the special issue. *Cities (London, England), 28*, 495–497.

Handley, J. F., (1996). *The Post-Industrial Landscape: A Groundwork Status Report.*

Hardy, P. B., & Dennis, R. L. H., (1999). The impact of urban development on butterflies within a city region. *Biodiversity and Conservation, 8*(9), 1261–1279.

Harper, A., Shattuck, A., Holt-Gimenez, E., Alkon, A. H., & Limbrick, G., (2009). Food policy councils: Lessons learned &food first + institute for food and development, policy development report no. 1. Retrieved from: http://www.baylor.edu/content/services/document.php/104981.pdf (Accessed on 9 October 2019).

Heisler, G. M., Grant, R. H., Grumond, S., & South, C., (1995). Urban forest-cooling our communities. In: Kollen, C., & Bawatt, M., (eds.), *Proceeding of the Seventh National Urban Forest Conference* (pp. 31–34). Washington D.C. American Forest.

Heller, N. E., & Zavaleta, E. S., (2008). Biodiversity management in the face of climate change: A review of 22 years of recommendations. *Biology Conservation, 124*, 14–32.

Hiranandani, V., (2010). Sustainable agriculture in Canada and Cuba: A comparison. *Environmental Development and Sustainability, 12*(5), 763–775.

Hu, A., Acosta, A., McDaniel, A., & Gittelsohn, J., (2011). Community perspectives on barriers and strategies for promoting locally grown produce from an urban agriculture farm. *Health Promotion Practice, 14*(69), 1–7.

Hutchings, T., (2002). *The Opportunities for Woodland on Contaminated Land, Information Note, 44*, Forestry Commission, Edinburgh.

Ibrahim, M., Casasola, F., Villanueva, C., Murgueitio, E., Ramırez, E., Sáenz, J., & Sepulveda, C., (2010). Payment for environmental services as a tool to encourage the adoption of silvo-pastoral systems and restoration of agricultural landscapes dominated by cattle in Latin America. In: Montagnini, F., & Finney, C., (eds.), *Restoring Degraded Landscapes with Native Species in Latin America* (p. 244). Nova Science, New York.

Jeffries, R., Darby, S. E., & Sear, D. A., (2003). The influence of vegetation and organic debris on flood-plain sediments dynamics: A case study of a low-order stream in the New Forest, England. *Geomorphology, 51*, 61–80.

Jim, C. Y., & Chen, S. S., (2003). Comprehensive green space planning based on landscape ecology principles in compact Nanjing City, China. *Landscape and Urban Planning, 65*, 95–116.

Jo, H. K., & McPherson, E. G., (1995). Carbon storage and flux in urban residential greenspace. *Journal of Environmental Management, 45*, 109–133.

Jorgensen, A., Hitchmough, J., & Danneth, N., (2007). Woodland as a setting for housing appreciation and fear and the contribution to residential satisfaction and place identity in Warrington New Town, UK. *Landscape and Urban Planning, 79*, 273–287.

Jouraeva, V. A., Johnson, D. L., Hassett, J. P., & Nowak, D. L., (2002). Differences in accumulation of PAHS and metals on leaves of Tilia x euchlora and pyruscalleryana. *Environmental Pollution, 120*, 331–338.

Kaethler, T. M., (2006). *Growing Space: The Potential for Urban Agriculture in the City of Vancouver*. University of British Columbia, Canada.

Kareemulla, K., Rizvi, R. H., Kumar, K., Dwivedi, R. P., & Singh, R., (2005). Popular agroforestry systems in Western Uttar Pradesh: A socio-economic analysis. *Trees and Livelihood, 15*(4), 375–381.

Kedaitiene, A., & Martinaviciene, R., (2005). *Rural Areas in Lithuania: Significant Development, 99th Seminar of the EAAE: The Future of Rural in Global Agri-Food System*. Copenhagen, Denmark.

Kenney, W. A., Van Wassenaer, P. J. E., & Satel, A. L., (2011). Criteria and indicators for strategic urban forest planning and management. *Arboriculture and Urban Forestry, 37*(3), 108–117.

Kortright, R., & Wakefield, S., (2011). Edible backyards: A qualitative study of household food growing and its contributions to food security. *Agriculture and human Values, 28*, 39–53.

Kumar, B. M., & Nair, P. K. R., (2004). The enigma of tropical home gardens. *Agroforestry System, 61*, 135–152.

Kumar, B. M., (2006). Agroforestry: The new old paradigm for Asian food security. *Journal of Tropical Agriculture, 44*(1–2), 1–14.

Kupcak, V. (2011). Regional importance of forests and forestry for rural development, ACTA University of agriculture etsilvic. *Mendel. Brun., 4*, 137–142.

La Greca, P., La Rosa, D., Martinico, F., & Privitera, R., (2011). Agricultural and green infrastructures: The role of non-urbanized areas for eco-sustainable planning in a Metropolitan Region. *Environmental Pollution, 159*, 2193–2202.

Lal, R., (2015). Sequestering carbon and increasing productivity by conservation agriculture. *Journal Soil Water Conservation, 70*, 55A–62A.

Landuse Consultant, (2004). *Making the Links: Green Space and Quality of Life, Land Use Consultants*.

Lang, T., (2010). Crisis. What crisis? The normality of the current food crisis. *Journal of Agrarian Change, 10*(1), 87–97.

Lawrence, R., Todd, R., Paul, J. P., Hendrickson, O., Leonard, R., Aswussen, L., & Cornell, D. L., (1984). Riparian forests as nutrients filter in agricultural watersheds. *Bioscience, 34*, 374–377.

Leakey, R. R. B., (1996). Definition of agroforestry revisited. *Agroforestry Today, 8*(1), 5–7.

Lerner, A. M., & Eakin, H., (2011). *An Obsolete Dichotomy?* Rethinking the rural-urban interface in terms of food security and production in the global south.

Lindquist, E. J., D'Annunzio, R., & Gerrand, A., (2012). Global forest land-use change 1990–2000, FAO forestry paper no. 169. In: *Food and Agriculture Organization of the United Nations and European Commission Joint Research Centre*. FAO, Rome.

Lowenfels, J., & Lewis, W., (2010). *Teaming with Microbes: The Organic Gardener's Guide to the Soil Food Web* (p. 220). timber Press, Portland.

Lutz, A. E., Swisher, M. E., & Brennan, M. A., (2010). *Defining Community Food Security, University of Florida, Institute of Food and Agricultural Sciences*. Retrieved from: http://edis.ifas.ufl.edu/pdffiles/WC/WC06400.pdf (Accessed on 9 October 2019).

Macias, T., (2008). Working toward a just, equitable, and local food system: The social impact of community-based agriculture. *Social Science Quarterly, 89*(5), 1086–1101.
Mark, H. (2017) *Why Conserve Small Forest Fragment and Individual Trees in Urban Areas*. The nature of cities, Gainesville.
Matoch, J., Schroth, G., Hills, T., & Hole, D., (2012). Integrating climate change adaptation and mitigation through agroforestry and ecosystem conservation. In: Nair, P. K. R., & Garrity, D., (eds.), *Agroforestry–the Future of Global Land Use* (pp. 105–126, 541). Springer, Dordrecht.
Maxwell, S., & Slater, R., (2003). Food policy old and new. *Development Policy Review, 21*(5/6), 531–533.
Mayer, H., & Knox, P., (2010). Small-town sustainability: Prospects in the second modernity. *European Planning Studies, 18*(1), 1546–1568.
McBeath, J., & McBeath, J. H., (2009). Environmental stressors and food security in China. *Journal of Chinese Political Science, 14*(1), 49–80.
McDonald, A. G., Bealey, W. J., Fowler, D., Dragosits, U., Skiba, U., Smith, R. I., Donovan, R. G., Brett, H. E., Hewitt, C. N., & Nemitz, E., (2007). Quantifying the effect of urban tree planting on concentrations and depositions of PM_{10} in two UK conurbations. *Atmospheric Environment, 41*(38), 8455–8467.
Menteus, J., Raes, D., & Herney, M., (2006). Green roofs as a tool for solving the rain-water runoff problems in the urbanized 21st Century? *Landscape and Urban Planning, 77*, 217–226.
Miller, (1988). *Urban Forest Planning and Managing Green Spaces*. Englewood Cliffs, New Jersey, Prentice-Hall.
Mohd-Yusof, M. J., (2012). The true colors of urban green spaces: Identifying and assessing the qualities of green spaces in Kuala Lumpur, Malaysia, *PhD Thesis*, Institute of Geography, The University of Edinburgh, UK.
Morgan, K., (2009). Feeding the city: The challenge of urban food planning. *International Planning Studies, 14*(4), 341–348.
Morgan, K., (2014). Nourishing the city: The rise of the urban food question in the global north. *Urban Studies, 1*(16), 1–17.
Morimoto, T., & Katoh, K., (2005). The effect of greenways connecting urban parks on Avifauna in the winter period. *Journal of the Japanese Institute of Landscape Architecture, 68*(5), 589–592.
Mougeot, L. J. A., (2000). Urban agriculture: Definition, presence, potentials and risks. In: Bakker, N., et al., (eds.), *Growing Cities, Growing Food: Urban Agriculture on the Policy Agenda* (pp. 1–42). German foundation for international development, Feldafing.
Muhammed, T. L., (2012). Review of NESREA act 2007 and regulations 2009–2011: A new dawn in environmental compliance and enforcement in Nigeria. *Law, Environment and Development Journal, 8*(1), 116–140.
Muschler, R. G., (2001). Árboles en cafetales.mo´dulo de ensen˜anzaagroforestal. *CATIE, Costa Rica. Proyecto Agroforestal CATIE/GTZ*, p. 137.
Muschler, R. G., (2004). Shade management and its effect on coffee growth and quality, In: Wintgens, J. N., (ed.), *Coffee: Growing, Processing, Sustainable Production* (pp. 391–418). A guide book for growers, processors, traders, and researchers. Wiley-VCH, Weinheim.
Nair, P. K. R., (2012). Climate change mitigation and adaptation: A low hanging fruit of agroforestry. In: Nair, P. K. R., & Garrity, D., (eds.), *Agroforestry–the Future of Global Land Use* (pp. 31–67). Springer, Dordrecht, 541 p.

National Research Council (NRC), (1991). *Forest Trees–Managing Global Genetic Resources, Executive Summary* (p. 248), USA. http://nap.edu/catalog/1582.html (Accessed on 9 October 2019).
Nisbet, T. R., Orr, H., & Broadmeadow, S., (2004). *A Guide to Using Woodland for Sediment Control*. Forest Research, Farnham.
Nowak, D. J., (1994). Air pollution removal by Chicago's urban forest. Results of the Chicago urban forest climate project, United States Department of Agriculture.
Nowak, D. J., Dwyer, J. F., & Childs, G., (1996). *The Benefits and Costs of Urban Greening, Proceedings of Urban Greening Seminar Held at Mexico-City*.
Nowak, O. J., Crane, D. E., & Stevens, J. C., (2006). Air pollution removal by urban trees and shrubs in the United States. *Urban Forestry and Urban Greening, 4*, 115–123.
O'Neill, G. A., Dawson, C., Sotelo-Montes, L., Guarino, I., & Weber, J. C., (2001). Strategies for genetic conservation of trees in the Peruvian Amazon. *Biodiversity and Conservation, 10*, 837–850.
Officha, M. C., Onwuemesi, F. E., & Akanwa, A. O., (2012). Problems and prospect of open spaces management in Nigeria: The way forward. *World Journal of Environmental Biosciences, 2*(1), 7-12.
Officha, M. C., Onwuemesi, F. E., & Akanwa, A. O., (2017a). Drawbacks facing sustainable development and management of open recreational spaces in Owerri. Nigeria. *Inter-National Journal of Multidisciplinary Research, 3*(11), 59–67.
Officha, M. C., Onwuemesi, F. E., & Akanwa, A. O., (2017b). Problems of open/recreational space delivery and management in Onitsha, Nigeria: Implications for sustainable best practice. *International Journal of Research and Development, 2*(11), 32–41.
Ogbuene, E. B., (2010). Environmental consequences of rainfall variation and deforestation in South-eastern Nigeria. *International Journal of Water and Soil Resources Research, 1*, 1–3.
Ogbuene, E. B., (2012). Impact of temperature and rainfall disparity on human comfort index in Enugu urban environment, Enugu State, Nigeria. *Journal of Environmental Issues and Agriculture in Developing Countries, 4*(1), 23–40.
Okojia, G. O., (1993). *Agro Forestry Education for Sustainable Rural Livelihoods: A Case Study of the Nyabyeya Forestry College*, Masindi, Uganda.
Olarinde, T., & Orecho, S. M., (2015). Evolution of environmental policies in Uganda and Nigeria: A developing country perspective. *TECHNICO LISB, 9*, 1–14.
Ong, C. K., Black, C. R., Marshall, F. M., & Corlett, J. E., (1996). Principles of resource capture and utilization of light and water. In: Ong, C. K., & Huxley, P., (eds.), *Tree-Crop Interactions – A Physiological Approach* (pp. 73–158). CAB Int'l, Wallingford.
ONS, (2010). Labor market unemployment: People. *Science, 327*(5967), 812–818. http://www.statistics.ov.UK/cci/nugget.asp?id=12 (Accessed on 9 October 2019).
Peters, C. J., Bills, N. L., Wilkins, J. L., & Fick, G. W., (2008). Foodshed analysis and its relevance to sustainability. *Renewable Agriculture and Food Systems, 24*(1), 1–17.
Peters, K. A., (2011). Creating a sustainable urban agriculture revolution, *Journal of Environ-mental) Law and Litigation, 25*(203), 203–348.
Pezo, D., & Ibrahim, M., (2001). Sistemassilvopastoriles.módulo de enseñanzaagroforestal No. 2. CATIE. Seriemateriales de enseñanza No. 44. Proyectoagroforestal CATIE-GTZ. Turrialba, Costa Rica, p. 275.
Plyler, W., (2012). Nearby nature: A logical framework for building integrated agriculture. *Dissertation*. West Virginia University, Morgantown, WV.

Potchter, O., Cohen, P., & Britain, A., (2006). Climatic behavior of various urban parks during hot and human summer in the Mediterranean City of Tel Aviv, Israel. *International Journal of Climate, 26*(12), 1695–1711.

Reid, R. S., Thornton, P. K., Mc Crabb, G. J., Kruska, R. L., Atieno, F., & Jones, P. G., (2004). Is it possible to mitigate greenhouse gas emissions in pastoral ecosystems of the tropics? *Environ. Dev. Sustain., 6*, 91–109.

Reif, D., (1992). Human issues in horticulture. *Hort-Technology, 2*(2), 159–171.

Reynolds, B., (2009). Feeding a world city: The London food strategy, *International Planning Studies, 14*(4), 417–424.

Rudd, H., Vala, J., & Schaefer, V., (2002). Importance of backyard habitat in a comprehensive biodiversity conservation strategy: A connectivity analysis of urban green spaces. *Restoration Ecology, 10*(2), 368–375.

Samra, J. S., Kareemulla, K., Marwaha, P. S., & Gena, H. C., (2005). *Agroforestry and Livelihood Promotion by Cooperatives* (p. 104). National research center for agroforestry, Jhansi, India.

Saxena, N. C., (2000). *Farm and Agroforestry in India–Policy and Legal Issues* (p. 50). Planning Commission, Government of India.

Scottish Enterprise, (2008). *Additionality and Economic Impact Assessment Guidance*. Scottish Enterprise, Glasgow. Available at: http://www.scottish-enterprises.com/evaluations-impact.htm (Accessed on 9 October 2019).

Shan, X., (2009). Urban green space in Guangzhou, China: Attitude preference, use pattern and assessment. *PhD Thesis*. The University of Hong Kong.

Shashua-Bar, L., & Hoffman, M. E., (2000). Vegetation as a climatic component in the design of an urban street: An empirical model for predicting the cooling effects of urban green areas. *Trees, Energy and Building, 31*(3), 221–235.

Slee, B., & Snowdon, P., (1999). Rural development forestry in the United Kingdom. *Forestry, 3*(72), 273–284.

Smit, S., Anna, R., & Janis, B., (1996). *Urban Agriculture: Opportunity for Sustainable Cities in Sub-Saharan Africa*. Report to the World Bank.

Sonnino, R., (2009). Feeding the city: Towards a new research and planning Agenda. *International Planning Studies, 14*(4), 425–435.

Sorensen, M., Smit, J., Barzetti, V., Williams, J., & Kerpi, K., (1997). *Good Practices for Urban Greening*. A report prepared for the environmental division of the social programs and sustainable development, Department of the inter-American development bank. ENV-109, Washington, D. C.

Stamets, P., (2005). *Mycelium Running: How Mushrooms Can Help Save the World Ten Speed Press* (p. 344). Berkeley.

Steel, C., (2008). *Hungry City*. Chatto and Windus, London.

Stewart, G. H., Ignatieva, M. E., Meurk, C. D., & Earl, R. D., (2004). The re-emergence of indigenous forest in an urban. *Urban Forestry and Urban Greening, 2*(3), 149–158.

Stovin, V. R., Jorgensen, A., & Clayden, A., (2008). Street trees and stormwater management. *The Arboricultural Journal, 30*, 1–4.

Sullivan, W. C., Kuo, F. E., & Depooter, S. F., (2004). The fruit of urban nature: Vital neighborhood space. *Environment and Behavior, 36*(5), 678–706.

Swanwick, C., (2009). Society's attitudes to and preferences for land and landscape. *Land Use Policy, 26*, 562–575.

Taylor, L. S., (2010). Multifunctional urban agriculture for sustainable land use planning in the United States. *Sustainability, 2*(8), 2499–2522.

The Tree Protection Legislation, (2006). *The Planning Department Directors*. Bulletin No 2006–01.
Tibbatts, D., (2002). *The Benefits of Parks and Green Space*. The Urban Parks Forum.
Tiwary, A., Sinnett, D., Peachey, C. J., Chalabi, Z., Vardoulakis, S., Fletcher, T., Leonards, G., Grundy, C., Azapagie, A., & Hutchings, T. R., (2009). An integrated tool to assess the role of new planting in PMIO capture and the human health benefits: A case study in London. *Environmental Pollution, 157*, 2645–2653.
Ulrich, R. S., (1990). The role of trees in wellbeing and health. *Proceedings of the Fourth Urban Forestry: Make Our Cities Safe for Trees Held in St Louis, M.O.* Washington DC. American Forestry Association.
UNDP, (1996). *Urban Agriculture: Food Jobs and Sustainable Cities*. New York: UNDP.
UNDP, (2005). *Local Action for Global Goals*. Water and sanitation in the world's cities. London. Earthscan.
United Nations Environmental Programme, (2009). *Climate Change Science Compendium 2009*. [Place unknown].
Van Rooyen, C. J., De Waal, D., Gouws, A., Van Zeyl, P., Rust, L., Kriek, N., McCrystal, I., & Grobler, N., (1995). *Towards an Urban and Rural Development Strategy for the Gauteng Province, with Emphases on Agriculture and Conservation: An Inventory*.
Venn, S. J., & Niemela, J. K., (2004). Ecology in a multidisciplinary study of urban space. The URGE project. *Boreal Environment, 9*, 479–489.
Weldon, S., Bacley, C., & Brien, O. L., (2007). *New Pathways to Health and Well-Being: Summary of Research to Understand and Overcome Barriers to Accessing Woodland*. Forestry Commission Scotland.
WHO/UNICEF, (2004). *Meeting the MDG Water and Sanitation Target*. Geneva: WHO/UNICEF Joint monitoring programme for water supply and sanitation.
Wolf, K. L., (2004). Trees and business district preferences: A case study of Athens, Georgia, U. S. *Journal of Arboriculture, 30*(6), 336–346.
Wong, H. H., Tan, P. Y., & Chen, Y., (2007). Study of thermal performance of extensive rooftop greening systems in the tropical climate. *Building and Environment, 42*, 25–54.
World Commission on Forests and Sustainable Development (WCFSD), (1999). *Summary Report, International Institute for Sustainable Development (IISD)*. Winnipeg, Manitoba, Canada.
Yang, J., McBride, J., Zhou, J., & Sun, Z., (2005). The urban forest in Beijing and its role in air pollution. *Urban Forestry and Urban Greening, 3*, 65–78.
Yoveva, A., Gocheva, B., Voykova, G., Borrisov, B., & Spassov, A., (2000). Urban agriculture in an economy in transition. Ternative food systems. *Humboldt Journal of Social Relations, 33*(1/2), 4–30.
Yu, C., & Hien, W. N., (2006). Thermal benefits of city parks. *Energy and Building, 38*, 105–120.
Zakariah, S., Lamptey, G. M., & Maxwell, D., (1998). Urban agriculture in Accra: A descriptive analysis. In: Armar-Klemesu, M., & Maxwell, D., (eds.), *Urban Agriculture in the GREATER ACCRA Metropolitan Area: Report to IDRC*. Legon: NMIMR.

CHAPTER 4

Mitigation of Climate Change Through Carbon Sequestration in Agricultural Soils

ZIA UR RAHMAN FAROOQI,[1] MUHAMMAD SABIR,[2]
MUHAMMAD ZIA-UR-REHMAN,[2] and MUHAMMAD MAHROZ HUSSAIN[1]

[1]*Doctoral Student, Institute of Soil and Environmental Science, University of Agriculture, Faisalabad–38040, Pakistan, Mobile: +923156040622, E-mail: ziaa2600@gmail.com (Z. U. R. Farooqi); Mobile: +923217251329, E-mail: hmahroz@gmail.com (M. M. Hussain)*

[2]*Assistant Professor, Institute of Soil and Environmental Science, University of Agriculture, Faisalabad–38040, Pakistan, Mobile: +923336545518, E-mail: cmsuaf@gmail.com (M. Sabir); Mobile: +923216637127, E-mail: ziasindhu1399@gmail.com (M. Zia-ur-Rehman)*

ABSTRACT

Climate change (CC) is an inevitable phenomenon in the world owning to the use of fossil fuels for energy production. It has numerous consequences related to it ranging from effects on crops by affecting yield by creating water shortage and disturbing soil health, decreasing livestock productivity, rising sea level, biodiversity loss and glaciers melting. Different methods are employed to mitigate its effects including carbon sequestration (CS) in its different sinks which are most effective and reliable method. It can be applied by managing agriculture fields with crop rotations, green manuring and nutrient management, organic farming, cover cropping, fertilizer, and irrigation management, desert soilization, cover cropping, erosion control, maintaining soil biota, agroforestry, reforestation, afforestation, and REDD+ to increase soil organic matter (SOM) in soils which helps to carbon (C) in soils. Non-agricultural approaches are also used to enhance CS like rainwater harvesting, water desalinization, geological storage of C, C trading, urban planning, C emissions management and using alternative fuel technologies.

4.1 INTRODUCTION

Climate change (CC) is one of the most serious phenomena now a day due to its hard impacts on agriculture, soil, climate, and environment. It is attributed to different anthropogenic activities responsible for the release of greenhouse gasses (GHGs), which triggers the greenhouse effect and leading to CC. It is a variation in the statistical distribution of weather conditions that lasts for a period preferably decades or millions of years (http://www.ipcc.ch/ipccreports/tar/wg2/index.php?idp=689). Because CC is affecting the environment, it is even hard to fight with the challenges posed by it. It is spoiling the natural eco-systems which provide us oxygen, water for drinking and other uses, food, and raw materials for industry (McNutt, 2013). Recent hazardous trends of CC are causing the disturbance to more than 1,700 animal species and causing the ecological zone shifting of average 6.1 km decade^{-1} and spring advancement of 2.3 days earlier decade^{-1} (Parmesan and Yohe, 2003). CC impacts did not stop here, but there will be an increased risk of extreme weather events like drought, storms, floods, and deforestation due to forest fires and drought posed by extreme weather events (Lindner et al., 2010). Scavia et al., (2002) assessed the CC impacts on marine and coastal eco-system and discussed its impacts on estuaries, coastal wetlands, coral reefs, and nearby ecosystems. They said that sea-level rise, changes in precipitations, increase in ocean temperatures, changes in circulation patterns, frequency, and intensity of storms and altered concentrations of C are affecting the marine ecosystem by dissolving coral reefs, causing glacier melting, biodiversity loss, and migration.

It is also estimated that agricultural crops will face a reduction in yield, which will aggravate the problem of food security. Based on the general models, results showed that Asia will suffer food shortage by 2030 (Lobell et al., 2008). CC will affect food availability and stability of food systems, short-term variability in water supply, and weather conditions (Wheeler and Braun, 2013). CC impacts to food crops will include temperature-induced yield losses, which will be 30–46% at the end of this century and 63–82% by the end of the next century (Schlenker and Roberts, 2009). Another study suggested 37% yield loss in the next 20–80 years, and if C concentration increases by 450–550 ppm, it will cause deleterious effects to grain quality (Erda et al., 2005). As CC is a threat to agriculture and crops, it will reduce crops yields up to 8% by 2050 inclusive of 17% reduction in wheat yield, 5% in maize, 15% in sorghum and 10% yield reduction will be observed in millet (Knox et al., 2012).

CC impacts on other natural resources are also studied; for example, water resources are being washed away with the increasing temperature, and glaciers are melting at an alarming speed, which will end the freshwater reservoirs (Piao et al., 2010; Christensen and Lettenmaier, 2006). It is estimated that by 2025, around 5 billion out of 8 billion people will face water shortage due to rise in temperature associated with CC, which will change the amount of precipitation, reduction in snowfall time, shift of snow falling and area of snow melting. By 2050, the world population will decrease due to deaths related to water shortage and its bad quality (Arnell, 1999). Coristine (2016) assessed the CC impacts on biodiversity. It was stated that biodiversity extinction, urbanization induced habitat loss are the potential risks for animal species. Habitat loss due to its compartmentalization through infrastructure development, biodiversity migration and shift to more suitable areas are the other consequences of CC.

Due to the increase in atmospheric concentration of C, humans are facing different consequences which are discussed above. Thus, there is a need to adopt or mitigate these consequences by restoring the sinks for C and exploring the new, reliable, and economical ways to sequester C in them to reduce C in atmosphere. Agricultural land use, recommended management practices, restoration of marginally degraded lands to normal lands, using conservation tillage, cover crops, nutrient management, crop rotations, agroforestry, green manuring, organic farming, desert soilization (conversion of desert sand into fertile soil) and soil microbe management are the strategies to control CC impacts. By using conservation tillage, 50–1000 kg C is sequestered in 1 ha per year. Carbon sequestration (CS) is totally a win-win strategy as it restores marginal soils improve soil health and CS potential, biomass production, and yield of crop grown on it (Lal, 2004; Bonan, 2008).

Agroforestry, afforestation, reforestation, and REDD+ (reducing emissions from deforestation and degradation) are the strategies by which we can maintain C levels to a bearable concentration, different engineering, and trade-related techniques are also used for this purpose like building equipment for rainwater harvesting, water conservation strategies like drip irrigation, water desalinization and storing C in deep soil horizons through geological storage. Trade-related strategies include C trading in which C emitter pays to the company or organizations who reduces its concentrations in atmosphere, urban planning, developing equipment which capture GHG emissions and using alternative fuels which emit less or no C in atmosphere (Lal et al., 2007).

The purpose of this chapter to highlight the CC, its causes, and sources, which contribute to it. Its current and projected effects on soil, crops, and other life forms were also presented in this chapter. It also highlights the prospects and opportunities and ways of CS in degraded soils by rehabilitating them, forests, and other technologies like REDD+ and using them as C sinks. Additionally, these soils and plants grown on them can be used for crop production hence reducing danger of food security and rendering C in the atmosphere.

4.2 CAUSES OF CLIMATE CHANGE (CC)

It is the most important phenomenon governing changes in longer-term weather conditions over an area for longer period. Different factors/agents contribute to the phenomenon of CC. These factors need better understanding to deal with this phenomenon effectively. Below are the different factors which affect the phenomenon of CC.

4.2.1 GREEN HOUSE GASES

Earth exchanges heat energy with atmosphere. When sun energy or heat energy approaches the earth, it is trapped in the atmosphere due to presence of envelope of GHGs present around the atmosphere. This heat is lost at night, but some of it remains in the earth surface and cause warming, called global warming, which leads to CC (USEPA, 2016). GHGs traps heat coming from the sun and called GHGs. These are present in atmosphere and different GHG has different potential to trap heat as even small change in their concentration for some time shows the effects which indicate the change in their behavior. There are many GHGs like water vapors, carbon dioxide (CO_2), methane (CH_4), nitrous oxide (N_2O), chlorofluorocarbons (CFCs), etc. Among these, water vapors are the most abundant one and responsible for most of the havoc (De Klein et al., 2008).

Its sources are the rains and moisture present in the atmosphere. CO_2 is the second most abundant and it is also a very important part of the atmosphere. Its sources in atmosphere include animal respiration, volcano eruptions, different human activities such as deforestation and burning fossil fuels. Methane is the third abundant but more heat-absorbing GHG than C;

however, its concentration is low as compared to C (St. Louis et al., 2000). Its sources include decomposition of municipal wastes in landfill sites, paddy rice fields, ruminant digestion and different manures in agriculture. N_2O is produced by soil tillage and cultivation and use of fertilizers, fossil fuel combustion, nitric acid production and biomass burning. CFCs are industrial chemicals widely used in refrigeration products. They are also lethal for ozone layer (Plummer and Busenberg, 2000). Water vapors are the most abundant GHG. It is produced due to elevated temperatures in atmosphere and moisture present in it. CO_2 is found in environment because of the C cycle, plant, and animal respiration, natural volcanic eruptions and ocean-atmosphere exchange, different anthropogenic activities like fossil fuels burning and changes in land use and deforestation. Its concentration increased from 280 ppm concentration in pre-industrial area to 408 ppm now a day (NASA, 2018).

Methane is emitted by natural wetlands, agricultural activities, i.e., paddy rice and from livestock and fossil fuel extraction. It is also emitted through lakes and water reservoirs up to 103 Tg CH_4 $year^{-1}$ (Bastviken et al., 2011). It is an important GHG originate from anaerobic processes (Keppler et al., 2006). Wetlands are also a big source of CH_4 as subtropical and temperate wetlands emit nearly 5 $Tgyr^{-1}$, northern wetlands 38 $Tgyr^{-1}$, and 4 $Tgyr^{-1}$ from dry tundra (Bartlett and Harriss, 1993).

Nitrous oxide is produced due to fertilizer application in agriculture, tillage activities, and natural biological processes, fossil fuel burning. Its concentrations also rose up to 20% as compared to the pre-industrial era (Reay, 2015). Nitrous oxide added to the atmosphere by 3 main sources, i.e., by direct emissions from agricultural soils, emissions from livestock, and by enhanced agricultural activities. Nitrous oxide input by anthropogenic activities includes synthetic fertilizer, animal wastes, and increased biological N-fixation. Its emissions from agricultural soils are 2.1 Tg and same amount from livestock (Mosier et al., 1998). Fertilizer based emissions of it ranges from 0.1–1.0 TgN. It is estimated that by using 100 Tg N fertilizers, there will be 3 TgN emissions in the year 2000. Currently, 0.3–18.4 kg N ha^{-1} is emitted due to fertilizer application (Dobbie et al., 1999).

CFCs came from air conditioner and refrigerator coolants, fire extinguishers, pesticides, and aerosol propellants. They have longer resident time in atmosphere and their effects may remain for decades.

There is gradual increasing habit in GHGs emissions and all of them are continuously emitted by major developed and developing countries regardless of Kyoto Protocol (St. Louis et al., 2000). GHGs emissions, their top contributors are presented in Table 4.1.

TABLE 4.1 Sources and Effects of CO_2

Sr. No.	Source	Percentage (%)	Effects on Environment	References
1.	Power Generation	28	Ocean acidification (Hoegh-Guldberg et al., 2007), changes climate scenario (McNutt, 2013), plant diseases (Garrett et al., 2015), on vegetation (Anderegg, 2015).	USEPA (2016)
2.	Transport	28		
3.	Industry	22		
4.	Commercial and residential	11		
5.	Agriculture	9		

Currently, annual deforestation rates are increasing due to land-use change and need of land for industries, agriculture, and residential purposes. As trees absorb C from the atmosphere and use it for preparation, when trees are less, more C will go to atmosphere and cause global warming and CC (Santilli et al., 2005).

As coal is dangerous in terms of burning and emitting C in the atmosphere and contributing in causing CC, growing concern about eradicating its will be a good option. As many alternative fuels are synthesized and available in the market, which can comprehensively taking the place of fossil fuels, including coal and other fossil fuels (Marland and Rotty, 1984). Global warming from GHG is increased by 25% since 1850 due to the overuse of fossil fuels and land-use change, and it is estimated that it will increase by 2–6°C in the next century. Sea level rises will be 0.5–1.5 m (Schneider, 1989). Figure 4.1 shows countries with the highest deforestation rates.

Industries like cement, petroleum, and chemical emit C in excess as C is used in terms of fossil fuels in them for energy generation and emitted to atmosphere causing CC (Hulme et al., 1999). Agriculture sector has about 24% share of the total anthropogenic GHG emission by manufacture and use of agricultural inputs and farm machinery and the extensive tillage which decompose organic matter and release C in atmosphere, paddy rice release CH_4 (Lenka et al., 2015), livestock also contributes by releasing NO_x and SO_x gasses up to 18% anthropogenic GHGs emissions (Pitesky et al., 2009).

4.3 CLIMATE CHANGE (CC) IMPACTS

4.3.1 AGRICULTURE

Agriculture crops, livestock, and fisheries depend upon climate patterns. They require a specific range for their individual survival as climate affects

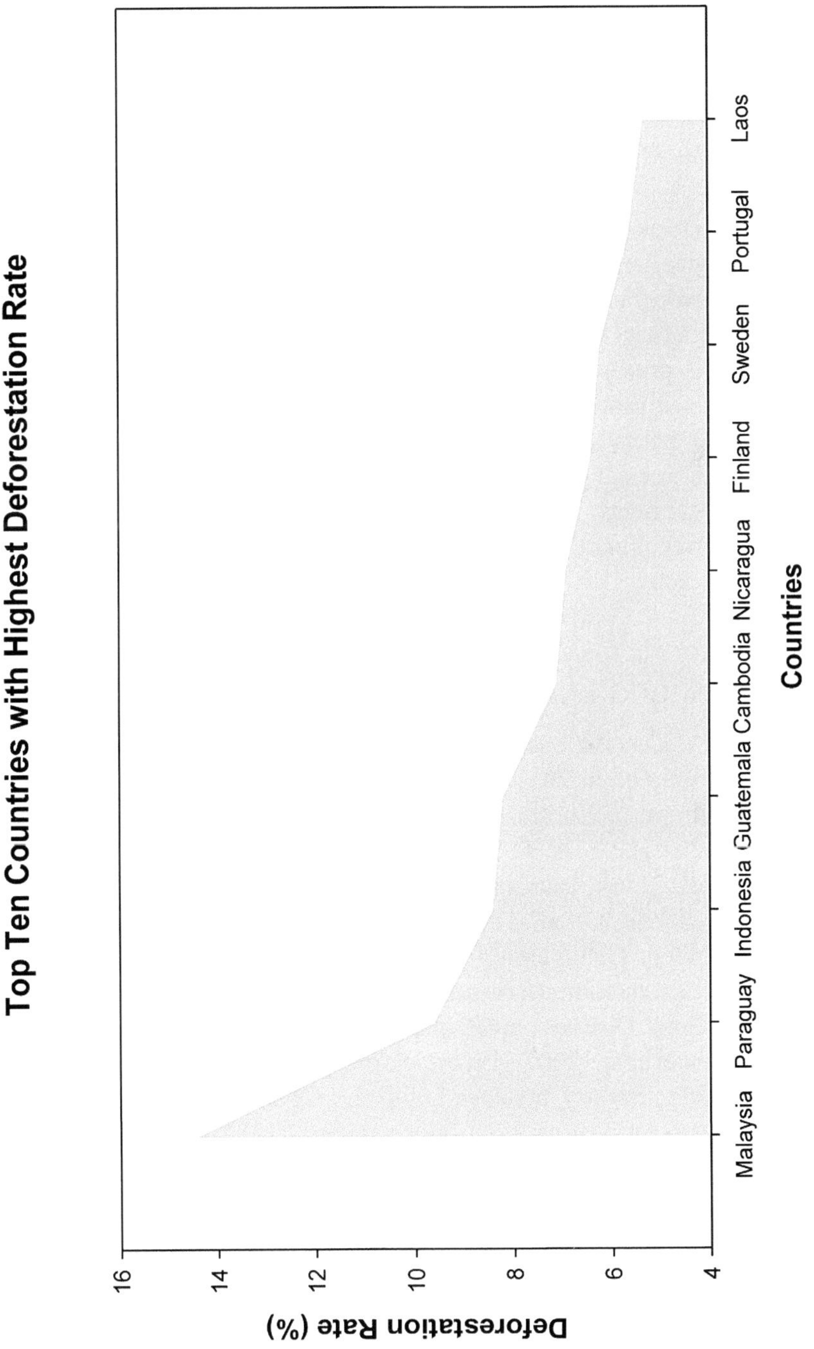

FIGURE 4.1 Countries with highest rate of deforestation.

nutrient levels, soil moisture, water availability, and other soil conditions. Climatic change can also cause droughts and floods, which can pose extreme challenges for survival of plants and livestock and food security.

4.3.1.1 IMPACTS ON CROPS

Agricultural crops are sensitive to high temperatures, high CO_2 concentration, droughts, and other extreme weather events. In some areas of the world, high temperatures may be beneficial for crops. HighCO_2 levels can also prove beneficial for crop yields because this is used in food making in photosynthesis process, but it can be detrimental as it reduces nutritional value as reduced protein and nitrogen contents, reduced grain and forage quality is also observed (Conroy et al., 1994). Extremely high temperature and precipitation can reduce crop seed germination. In addition to weather events, weeds, insects, pests, and pathogenic fungi become active in high temperature, wet climate and high C, and it also reduces the nutritional value of most food crops.

4.3.1.2 IMPACTS ON LIVESTOCK

CC affects the pastures and grasslands and thus can affect the livestock directly. As observed in 2011, when high temperatures cost USD 1 billion loss to agricultural producers. Increase in CO_2 compromises the quality of forage and nutrients in it (Lenart et al., 2002). High temperature affects animals by heat stress, causing them diseases, reducing fertility and milk production (Klinedinst et al., 1993). Worldwide, livestock production emits the fifth part of total GHGs emissions, but the effect on livestock is more than this. Livestock production decreases when the temperature increases beyond their bearing limit. Diseases, meat, and milk quality are also decreased likewise (McMichael et al., 2007). Livestock production will decrease 20–30%, which will lead to poverty, meat, and milk shortage in the country. Total loss to agriculture sector will be 2–15 billion dollars at the 21st century (http://www.lrrd.cipav.org.co/lrrd16/1/kris161.htm).

4.3.1.3 IMPACTS ON FISHERIES

But due to the higher concentrations of CO_2, the acidity of water bodies changes and the pH which is the fit for aquatic life is disturbed and fish dies, e.g., Shellfish (Knutzen, 1981). In addition to this, higher temperatures also contribute to add in danger to aquatic life. Some diseases and parasites of fish are also associated with CC like oyster parasites and salmon diseases. The temperature change can affect fish reproduction rate and lifecycle (Knutzen, 1981; Bamber, 1990).

4.3.2 IMPACTS ON ENVIRONMENT

CC alters food security scenario, rise in temperature and alters precipitation, causes extreme weather events and water scarcity and reduces agricultural productivity, drought, and weeds, insects, and diseases are also spread. Weeds cause up to 34% loss to agriculture, insects cause 18% and 16% from diseases. Livestock production also affected due to disease and pest prevalence. Soil contamination and its erosion, water resources are also affected by CC. Precipitation and temperature affect water availability to crops and determine their yield. In all the above, CC also affects oceans, forests, wildlife, and their habitat, soil ecosystem, and polar regions by reducing area under ice and glaciers.

4.3.2.1 OCEANS

Even oceans' capacity to absorb a lot of heat and C is also affected by CC. CO_2 absorbed by oceans increases their acidity, which affects marine life, and temperature of seawater is also increased (Stachowicz et al., 2002). Plankton, corals, fish, polar bears, walruses, seals, sea lions, penguins, and seabirds are victims of CC. The major effects of CC on oceans include coral bleaching in which high temperature water causes coral death, stormy weathers, and severe effects to coastal ecosystems. Loss in biodiversity, migration to other areas, sea level rises due to melting of glaciers present in oceans, changed profiles of ocean temperature due to which cause low oxygen levels at ocean depth and acidic oceans which cause death of aquatic life and dissolution of calcium carbonate ($CaCO_3$) shells of some of the aquatic life are also the adverse effects of CC (Fabry et al., 2009).

4.3.2.2 FORESTS

Different types of forests are present around the world. These include tropical rain forests, boreal forests, Amazon forests, Sundarbans forests, Tongass, and taiga forests. These are on the verge to be cut down for space to agriculture and living purposes. Due to CC, change in temperature and precipitation are now controlling the number of forests in the world as there was about 4 billion ha of area under forests covering 30% of the world's area, but now, this area is decreasing at the rate of 6 million ha per year since 1990. Brazil and Indonesia the two major countries having their large area under forests losing 4.9 million ha forests annually (Pan et al., 2011).

4.3.2.3 WILDLIFE

CC over the extended period of past 30 years resulted in numerous species shifting to other ecological zones and extinction too. If the present scenario exists, then it is predicted that by 2050, 15–37% of world's biodiversity will go to extinction (Thomas et al., 2004). Global warming and CC cause ecosystem shift, causing habitat loss and shift to other regions, the greatest cause of biodiversity loss. According to Intergovernmental Panel on Climate Change (IPCC), increase of 1.5°C temperature will cause to loss of 20–30% of animal and plant species. Currently, some of the species are gone extinct or struggling for their survival including rhino, gorilla, turtle, tiger, elephant, whale, dolphin, panda, lion, hippopotamus, bear, leopard, etc. (https://www.worldwildlife.org/initiatives/wildlife-and-climate-change).

4.3.2.4 SOIL

It takes thousands of years for a process in which fertile soil is formed, but it is much easier that a fertile soil is degraded by factors like climate which deteriorates soil structure, stability, water holding capacity, nutrient availability and accelerates erosion (Karmakar et al., 2016). Some other major soil health affects soil organic matter (SOM) contents, structural stability, decreases in vegetation cover, soil structure degradation, decreased porosity, increased water runoff and erosion (http://www.fao.org/docrep/w5183e/w5183e05.htm). CC will modify the vigor of the hydrological cycle by implicating in less or more frequent rainfall, its intensity, precipitation, plant biomass production, plant residue decomposition rates, soil microbial activity, evapo-transpiration rates and land use change which will increase soil erosion and nutrient loss and soil health will be deteriorated. Temperature, solar radiation, and CO_2 will also contribute to it (Nearing et al., 2004). It is estimated that erosion will be increased up to 10–310% and soil loss due to this will be increased from 33–274% in 2040 as compared to the previous years (O'Neal et al., 2005).

4.3.2.5 POLAR REGIONS

Temperature affects arctic climate badly by melting glaciers and changing hydrological cycle (Hinzman et al., 2005). This scenario resulting in abrupt changes in weather and climate. For example, less sea ice results in altered temperature and salinity of seawater, which affect ocean circulation patterns (thermohaline circulations), which ultimately determine the global climate. These circulations determine the weather of small and climate of large scale. As in Europe, there will be winter and at the same time in Asia, there will summer. But this temperature difference has been decreased and leading to extreme weather events, i.e., extreme, and high intensity snowfalls, droughts, and heatwaves. Some of these extreme events were observed in 2009–10 and 2010–11 when severe winter and extreme snowfall events was observed in China, United States and Europe (Clark et al., 2002).

4.4 CLIMATE CHANGE (CC) MITIGATION

Due to the adverse impacts of increasing C concentration on environment in terms of CC, it should be controlled and mitigated. CC can be mitigated using various techniques however economical and long-lasting options which ensure food security and food safety should be adopted. One of the most economical and easily adoptable techniques is CS. It incorporates C as well as produces food for people and helps to maintain environment in equilibrium. Like other methods, there are no barriers, limits, and costs problems in this case. However, we need to study comprehensively to estimate the global costs and benefits of adaptation. It is a term, frequently used today as it is an economical and long-lasting solution of CC. Every day, people gather and talk about the prospects and challenges related to it. It is the process of removing excessive C from atmosphere and increasing its concentration in different sinks (Smit et al., 2014).

4.4.1 MAJOR SINKS

C sinks play an important role in mitigating the prevailing CC and helps to reduce C concentrations in atmosphere. They act as C holding tanks and don not allow it to go to atmosphere. Major C sinks include soil, forests, and oceans. Soil is considered the biggest sink as the soil, vegetation grown on it and deep soil horizons act as sinks separately.

4.4.1.1 AGRICULTURE

Agriculture is the producer and sinks for C from atmosphere, but its contribution in C removal or sequestration is more than the production. Different agriculture techniques are used for this purpose like agroforestry, using different soil amendments to enhance SOM and enhancing sequestration potential, crop rotation, green manuring, no or zero tillage farming, cover cropping, by maintaining soil microbes and even crop plats itself are store of C.

4.4.1.2 SOILS

Soil is considered the biggest land-based sink of C as it stores C in the form of plants and trees grown in soil. SOM is the main component that is responsible for CS in soil. Its higher contents in the soil are good and help in CS. Deficiency of SOM in loss of C from soil and deteriorated soil properties, thereby affecting its potential to support plant growth and CS. So, it is the first step in soil CS to maintain SOM in soil to hold C in its growth and support the plant. In the presence of SOM, soil properties are improved like water holding capacity, nutrients availability to plant, soil pH, soil microbes, and all other physical, chemical, and biological properties. In the presence of good soil health and microbes, C in actively stored and re-circulated between soil-plant-microbes (McNeill and Winiwarter, 2004; Ontl and Schulte, 2012).

4.4.1.3 CROPS

It is a known fact that agricultural lands are extensively managed for CS, but crops are also used for CS and they have the potential to take part in overcoming CC. The total CS potential of crops is 0.75–1Pg yr^{-1}. Crop plants take up C for photosynthesis and make food which they use and some of them exudates into the soil and sequester there or used by microbes for their food (http://dbkgroup.org/carbonsequestration/rootsystem.html). Now, the crops are also bred for having the extended potential of taking up C (Kell, 2012), maximize growth, biomass, fuel, etc. No-till agriculture also helps in improving C storage in plants and soil. CS potential of different plants/trees is given in Table 4.2.

TABLE 4.2 Major Carbon Sinks and Their Potential to Sequester Carbon

Sr. No.	Sink	Total Potential	References
1.	Soil	2500 (P g year^{-1})	Bellassen and Luyssaert, 2014; Schlesinger, 1992; Goodale et al., 2002
2.	Forests	756 (Tg year^{-1})	
3.	Oceans	39112 (Gt Year^{-1})	
4.	Atmosphere	~800 (P g year^{-1})	

The process of CS in soil involves the uptake of C by plants and its use in their food-making process of photosynthesis. During photosynthesis, C from atmospheric is transformed into plant food, which is necessary for their growth (Woodrow and Berry, 1988). Through this process, C present in the atmosphere becomes part of the plant. Perennial trees which are Long-lived plants can keep the C sequestered in them for longer period (Swift et al., 1979). As part of the C cycle, the C present in the decomposing plant material and present within the soil is retained in the soil, or is consumed by soil organisms.

4.4.1.4 AGROFORESTRY

Agro-forestry is being considered one of good options to stabilize the C concentrations in atmosphere to increase C sink and reducing it to combat CC. Agro-forestry systems are very effective in C concentrations reduction because C is extensively used in photosynthesis by crops and trees for biomass production. Total of 2.2 Pg C can be stored in these plants for the period of 50 years. That's why area under agroforestry systems (AFs) is increasing and C storage could be as much as 300 Mg C ha^{-1} in top 1 m depth of the agricultural land (Lorenz and Lal, 2014). According to Torres et al., (2017), agroforestry systems can sequester C as much as 54.6 t C ha^{-1} when there are 17–44 trees are planted ha^{-1}, which was lower than the total planted in the systems.

4.4.1.5 FORESTS

Forests are the best sinks for C as they take up C directly from atmosphere, use it in photosynthesis and accumulated it as their integral part of the body. Forest trees contain C in their every form, i.e., vegetation which contains C 35–65% of dry weight, deadwood, and litter is plant debris, leaves, fruits, seeds, roots, and their dead biomass. Overall, 19% of the total world C is stored in the forests and 81% to soil. Out of the 19% C stored in forest plants, 31% remains in plant biomass, and 69% goes to soil. But this state is different in tropical forests, where approximately 50% is stored in plant biomass, and rest half goes to soil (FAO, 2003). According to other estimation, total world's forests have absorbed about 2 pg C yr^{-1} which is almost 30% of the C produced annually due to anthropogenic activities (Bellassen and Luyssaert, 2014).

4.4.1.6 OCEAN

Oceans and seas cover over 70% area of earth up to the depth of 3,800 meters. That's why these are known as the largest sink of C by area and accumulates about 7 Pg C yr^{-1}. But there is a problem using oceans as sink of C as oceans become acidified and life in it gets threatened in acidic water. Coral reefs are also dissolved which are a great ecosystem for the aquatic life there. Ocean CS aims to store C in ocean depth to minimize negative impacts on the surface of the ocean and promotion of photosynthesis (Chow, 2014).

4.4.2 MANAGEMENT PRACTICES FOR CARBON SEQUESTRATION (CS)

Soils are ultimate and economical option for CS to mitigate CC. Agricultural soils are responsible for storing C in two ways, i.e., in soils and in crops grown on them. Potential of CS can also be enhanced by adding different organic amendments like farm manure, biochar, etc., different practices like crop rotations, green manuring, nutrient management, organic farming, tillage management, cover cropping, desert soilization, erosion control and maintenance of soil biota. Data from a study showed that crop grown on agricultural soils store C up to 0.32 ± 0.08 Mg ha^{-1} yr^{-1} in top 22 cm soil. It is estimated that potential for CS in agricultural soils are between 0.12 ± 0.03 Pg C yr^{-1} which accounts about 8% of the total C emissions in the world (Fujisaki et al., 2018).

4.4.2.1 CROP ROTATIONS

Soils are considered as major C pool but these are tilled extensively which results in the loss of soil structure and texture and stores less C in it. Different management practices can restore soil structure and texture and increase CS. Crop rotation is one of the management practices which can improve soil structure and can restore the potential of the soil to sequester C (Holeplass et al., 2004). Soil CS rates are increased when different management practices are applied to crop. In an experiment it was noted that crop rotation can enhance CS up to 14 ± 11 g C m^{-2} yr^{-1} and estimated to be increased to 27–430 kg C ha^{-1} $year^{-1}$ (McConkey, 2003). SOM is also increased while rotating crop under no tillage. Other different soil properties are also enhanced as increased water infiltration, conservation of nutrients and erosion control (Hernanz et al., 2009).

4.4.2.2 GREEN MANURING AND NUTRIENT MANAGEMENT

Green manures improve the soil fertility which enhances potential of the soil to sequester more C and thus plays role in mitigation of CC. Organic matter causes the C fraction to be increased in soil due to green manure as it provides organic matter and food for the soil, plants, and microbes residing in it. Leguminous crops are good option for green manuring as they contribute in N fixation and increase nutrients in soil. Cultivation of leguminous crops and farm manure addition and nutrient management improve CS in soil up to 1.5 and 2.0 times higher than normal (Choudhury et al., 2018). As animal manures are 40–60% composed of C, therefore its application increase CS rates, recycles, and provide nutrients to crops (Franzluebbers and Doraiswamy, 2007; Liu et al., 2013).

4.4.2.3 ORGANIC FARMING

Traditionally, organic farming was performed for food and fodder production; however, gradually it was replaced with modern agriculture where use of chemical fertilizers and pesticides is common. Now, the world is aware about the consequences of chemical or synthetic fertilizers use and use of non-conventional practices in agriculture. World is now diverting to conventional practices of agriculture and organic farming which can give more, safe, and quality yield up to many folds and sequesters 2–3 times C in soils (Liu et al., 2013). It is believed to enhance soil fertility by increasing SOM contents in addition to CS whose rate increases by 2.2% annually (Leifeld and Fuhrer, 2010). Soils supplied with organic amendments have 26% more potential for CS and have 13% more SOM and sequester more Cin soil and plants.

4.4.2.4 FERTILIZER, IRRIGATION, AND TILLAGE MANAGEMENT

It is estimated that total potential of soils for CS is up to 0.6–1.2 Pg C yr^{-1} which is equal to 15% of the total fossil fuel emissions (Lal, 2006). An adequate amount of fertilizer, irrigation, and zero or minimum tillage are responsible for soil health. When soil health is good, it will sequester more C in it due to the more SOM and microbial community in it. Right dose of N fertilizer increases SOM and sequesters up to 1250 g of g of Cm^{-2} in soils, 1740 g C m^{-2} in soils under zero-tillage and 22–83 g C m^{-2} annually in properly irrigated soils and increase plant biomass production up to 200 g C m^{-2} $year^{-1}$ (Schlesinger, 1999). In another study, results indicated that CS is significantly improved by tillage and irrigation management and optimal N fertilizer application. Up to 0.46%, CS was achieved when soil was least tilled (Wang et al., 2014).

4.4.2.5 COVER CROPPING

Agriculture is the sector that can sequester a major portion of the C emission, but when it is well managed and organized (Conant, 2011). For example, cover cropping can enhance the C sequestration in soils with that feature with an average of 0.08–0.32 Mg ha^{-1} yr^{-1} in soil in depth of 22 cm, 1.5–16.7 Mg ha^{-1} according to another study and 0.32 Mg C ha^{-1} yr^{-1} in the third study. It is also estimated that by cover cropping, there is a potential of total 0.12±0.03 Pg C yr^{-1} in soils, which equals 8% of the GHG emissions from agriculture (Poeplau and Don, 2015).

4.4.2.6 DESERT SOILIZATION

A large area of the earth is under desert eco-system, which is not fertile and productive. A narrow range of plant species can grow there and contribute least in CS. Scientists were investigating to convert this area into a fertile land and now they made it possible to made desert a fertile land and any crop can be grown. This is a big break through by which CS, food security and efficient use of the available resources is being done. In addition to this, economic benefits are also immense (Zhang and Huisingh, 2018). When water-based soil paste is added in sand changes into a rheological state called "wet soil." After moisture evaporates from this paste, this paste changes to solid state called "dry soil" or soilized sand which possess all the physical and chemical properties of natural soil as there is no significant difference between the "soilized" sand and natural soil in terms of their mechanical properties and ecological attributes and suitable for the growth of plants, thus making contribution in CC mitigation and food production (Yi and Zhao, 2016).

4.4.2.7 EROSION CONTROL

Soil erosion carries away top fertile soil which supports plants for their growth. Without nutrients, soils are considered as unsupportive for plants survival as they don't contain nutrients. As eroded soils can't support plant growth so, they do not support C accumulation in plants and soils and made no contribution in CS. There are many ways to deal with the losses faced by soil erosion like re-vegetation, no-till farming, use of mulches, cover cropping, etc. (Lenka et al., 2012). It is estimated that by adopting zero tillage and avoiding soil erosion, agricultural soils in EU-15 can sequester 16–19 Mt C year^{-1}. It is also being enhanced by increasing organic matter inputs, perennial tree plantation, promotion of organic farming and adopting zero tillage (Freibauer et al., 2004).

4.4.2.8 MAINTAINING SOIL BIOTA

Soil biota carry out the most important functions in the soil including formation and decomposition of organic matter, release CO_2 (Kaiser et al., 2015), nutrient cycling, pollutant degradation and soil structure improvement (Gupta, 2011). A single microbial product that is organic matter has a lot of benefits. Some benefits include improvement in retention of water and nutrients and reduce erosion by soil aggregation and this leads to good soil structure and texture. As organic matter is the key component which involve in the storage of C in the soil and plants, their presence should be necessary in the soil to carry-out the CS. Sequestration of C in low structured and textured soils will require the rehabilitation of these soils and increasing the organic matter to considerable amount and then C can be sequestered through trees/ crops grown in rehabilitated soils (DeDeyn et al., 2008).

4.4.3 THROUGH FORESTRY

Soils are the major C sinks in the world and sequester about 1550 PgC in it (Zdruli et al., 2017). But due to deforestation, land degradation through soil burning, plowing, intensive tillage, grazing, and erosion, this pool is losing its importance. It can be managed and rehabilitated through soil water conservation and management, maintaining soil fertility, conservation tillage, reducing runoff, controlling soil erosion, strategies to increase micro-aggregation formation in soil, soil fertility enhancement, improved crops, cover crops, fallowing, and cultivation of deep-rooted crops (Lal and Kimble et al., 1997). Trees plantations in urban areas of US is a common practice and C sequester through them is estimated to be 25.6 million tons with benefit of USD 2 billion value (Nowak et al., 2013). The CS potential of trees is different for different tree species. There are some strategies which can be adopted when addressing CC using forests; (1) Management of already present forests, (2) Avoiding/prohibiting deforestation, (3) Preservation of forests, and (4) Afforestation on abandoned lands or where forests are not present. Actively managed forests can sequester C and provide fuel and timber, help us in cleaning air and water, provide habitat for wildlife, give recreational opportunities, preserve the natural beauty and load soils in nutrients (Lal, 2005; Lorenz and Lal, 2009).

4.4.3.1 AFFORESTATION

Afforestation is the conversion of barren, non-agricultural, and non-forested area into forest. Reforestation of an area can sequester about113–138.2 g C m^{-2} yr^{-1} in top 20cm of the soil in total of 60 years after their establishment and Afforestation can sequester 0.4–1.2 tC ha^{-1} yr^{-1} in boreal, 1.5–4.5 tC ha^{-1} yr^{-1} in temperate and 4–8 tC ha^{-1} yr^{-1} in tropical regions (Obersteiner et al., 2006). It is estimated that 345 million ha of area can be used for plantation with maximum potential of 1.48 $Gtyr^{-1}$ in the next 60 years after their establishment. In which 1.14 Gt will be stored above-ground and 0.34 Gt below-ground. In the period of 1995–2095, a total of 104 Gt of C would be sequestered (Nilsson and Schopfhauser, 1995). Conclusively, afforestation has the great capability of CS and improves our understanding regarding CS potentials associated with afforestation projects in the world (Shi et al., 2015).

4.4.3.2 REFORESTATION

Human has disturbed almost half of the tropical biome already. Some of these are in its recovery stage but mostly gone. It is required that one-fourth part of a country should be a forest, but due to extensive deforestation for agriculture and residential purposes, no country can match this criterion. Results are being observed as CC as there is no other efficient absorber of C as trees are. To mitigate the CC impacts, reforestation of abandoned areas has been gaining importance help in creating more C sinks. Results of a study revealed that reforestation has the potential to store C as longer as 40–80 years, but more research is needed to determine the potential of reforestation for longer-term CS (Silver et al., 2000). Reforestation is being carried out in almost every part of the world to tackle CC involving plantation of trees on deforested area. Natural and managed reforestation has been thought to be the easiest way to offset increasing C concentration in the atmosphere. It is estimated that C is accumulated up to 0.41 Mg $ha^{-1}yr^{-1}$ can be achieved using this technique (Silver et al., 2004).

4.4.3.3 AGROFORESTRY

Agroforestry is growing trees and crops in each-other's combination and it is estimated that 1 billion ha of agricultural land is under it and about 0.29–15.21 Mg ha^{-1} yr^{-1} of C is stored by this technique (Nair et al., 2010). AFs can mitigate CC better than ocean and other sinks because they sequester C as well as provide food and increase farm income, restore soil health and structure, maintain soil biota, control flood water speed, protect soil from being washed away during floods and provide fire and timber wood and reduce pressure on natural forests (Albrecht and Kandji, 2003). Agroforestry allows crop, livestock, and trees on one site which can contribute up to 2.2 Pg C in total 50 years of time span. Soil C storage through agroforestry may be 300 Mg C ha^{-1} in the upper 1 m depth of the soil (Lorenz and Lal, 2014). Agroforestry is the integrated method of sustainable land use due to its economic and environmental benefits. As it mitigates C through CS and owing to the fact, the area under it is increasing which is currently 1,023 mha. Degraded and unproductive lands are also used for agroforestry due to the profitability and economic returns (Nair et al., 2009).

4.4.3.4 REDD+

Reducing emissions from deforestation and degradation, plus related pro-forest activities (REDD+) aims to reduce those amounts of C which comes due to deforestation and forest degradation. These account about 10% in global warming. As forests are a big sink of C but these turned into sources when they are burnt or under forest fires or land under forests is cleared (deforestation). There is a need to reduce C emissions 50% by 2020 and there is a need to reduce them up to 0% by 2030 to help to mitigate the CC. There are almost 850 million ha forests which are degraded or removed and 2.3 million hectares of forests are being degraded every year. There is a need to protect these forests by their proper management, different conservation techniques that promote natural regeneration, reducing fire and grazing threats, planting islands of forests and even planting mixed-species plantations. National level accountability, capacity building, monitoring of forests and incentives will also enhance importance of trees among people (Figure 4.2). A community-based REDD+ can also be established as done in Ludikhola and Kayarkhola, Nepal. After calculating the costs and benefits, it was declared that the benefits of REDD+ are about $7994 and $152 per area as compared to $4815, $29 of cost (Pandit et al., 2017). It is gaining importance with increasing risks of CC in all over the world. To estimate the total potential and methods to sequester C by using this method, Gibbs et al., (2007) has developed a map and updated forest biomass C databases and created the first complete set of forest C stocks at national-level which will enhance the range of globally consistent estimates.

4.4.4 CARBON SEQUESTRATION (CS) THROUGH ENGINEERING TECHNIQUES

4.4.4.1 RAINWATER HARVESTING

Rainwater can be collected, stored, and harvested in small dams. It has very good potential to provide agricultural irrigational water and which is a need of time. Water use efficiency of crops is increased by adopting this strategy up to 14.3 $kg\,ha^{-1}m^{-1}$ (Akhtar et al., 2015). Rainwater harvesting can reduce C emissions in Mexico City and benefit-cost analysis report shows that 1.23 $kWh\ m^{-3}$ electricity is saved by rainwater harvesting which is used to pump water to Mexico City (Valdez et al., 2016). The energy used in pumping, transporting, and conveying water can be reduced by adopting rainwater harvesting techniques which will help in CC mitigation, water scarcity due to it and irrigational water need for crops will be saved. In addition to this, water shortage in upcoming time will be addressed properly.

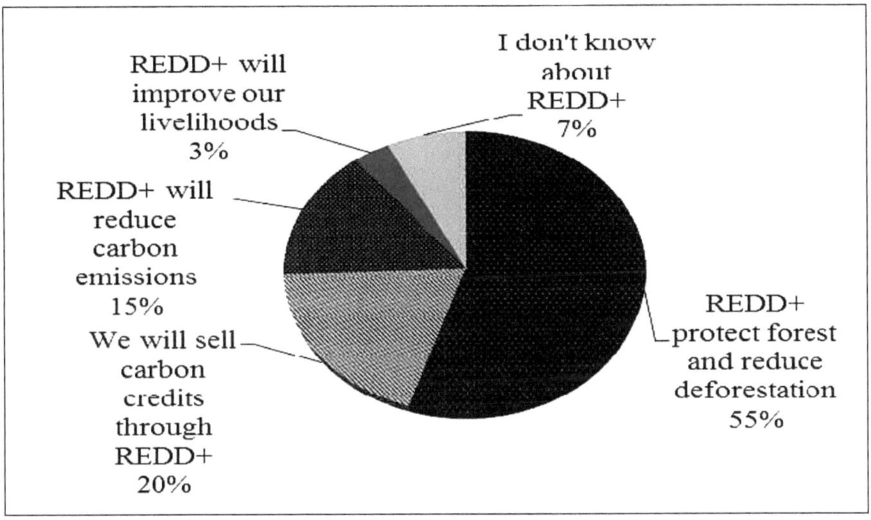

FIGURE 4.2 People comments and awareness level of people about REDD+ and its contribution towards C sequestration (Used with permission from Ojija, 2015).

4.4.4.2 WATER CONSERVATION TECHNIQUES

Runoff, evaporation, and high temperature causes water losses and runoff water, erodes, and carries away the fertile land and all the nutrients required for the plant growth and soil stability. This loss decreases the ability of soil to sequester C and decreases already sequestered C. The best way to mitigate this loss is water conservation. Water conservation reduce soil erosion which improve soil properties and increase crop yield up to 2–3% and increase CS 4 to 5.8 g C kg^{-1} of soil with the increase in this rate of 0.18 g C kg y^{-1} (Stroosnijder and Hoogmoed, 2004). We can save water in simple daily tasks to large irrigation in agriculture. Adopting water saving devices, soil management, adopting organic farming, using weather apps for determining the best time for irrigation, drought tolerant crops cultivation, dry farming, using compost and mulching, conservation tillage, laser leveling of agricultural soils, drip irrigation systems, using xeriscaping technique in your lawns which is using plants that are native to the area and use less water and water recycling in which instead of putting used water in drain, this water is collected and filtered or treated and then put back to the system.

4.4.4.3 WATER DESALINIZATION

CC is accelerated when more water is pumped using electricity and this electricity is generated through fossil fuels (Newmark et al., 2010). In addition to this, water is used in energy production and this energy emits less C. Approximately 30 billion m^3/year is desalinized to produce energy combat the increased energy consumption and C emissions. In addition to this, another technique in which C is added into ocean (ocean sequestration) and then this water is desalinized using magnesium chloride ($MgCl_2$) which neutralize the ocean water due to its alkaline nature and removes C by conversion of magnesium oxide (MgO) to bicarbonate. This process in known as Davies Process (Aines et al., 2011).

4.4.4.4 GEOLOGICAL STORAGE

In this mechanism, C is extracted as a by-product from ongoing industrial processes and transported to the place of sequestration mainly deep into soil. It can be stored in abandoned oil and gas reservoirs, un-mineable coal seams and deep saline reservoirs. A Report by IPCC indicates that about 99% of the sequestered C can be hold- up in these reservoirs for more than 100 years (Solomon, S., Bindoff, N., Brasseur, G., Technical Summary by IPCC). As C storage capacity of saline aquifers is much greater than other sinks, C can be stored there for longer periods of time. Deep saline aquifers store C at 800 m or more depth (Shukla and Vishal, 2016). Geological carbon sequestration (GCS) also uses off-the-shelf technology from the hydrocarbon industry and stores C in underground reservoirs at a very large scale and can be used commercially. But GCS requires the better understanding of the processes that help in trapping C and can store it to underground by improved means and monitoring it while injection to the reservoir (Friedmann, 2007).

Geological C storage involves the C capture, transport, injection of C in geological sites. Considering the current trend in the world's economic and population growth, this practice to sequester C deep in soil will continue to rise which lead towards CC mitigation.

4.4.5 OTHER STRATEGIES

4.4.5.1 CARBON TRADING

C trading is projected to be one of the largest markets in the world due to the part of the Kyoto Protocol. C Trading is the business in which high C emitting industries pay to those who remove C from the atmosphere to help in CC mitigation. Permits are issued to industries for C emissions, and the price is paid to them who help in off-setting the C emitted by them. There are two basic ways which are C taxes and direct regulations. Taxes on emissions and regulation or limits are set for C emissions for industries. The world's biggest C trading system is in the European Union (EU) called the European Union Emissions Trading System (EU-ETS). Market price of \$15–20 ton^{-1} of C is charged in developed countries (http://www.synapse-energy.com/sites/default/files/2016-Synapse-CO2-Price-Forecast-66–008.pdf). Developing nations are also being encouraged to engage in it and clean development mechanisms (CDM) are being developed.

4.4.5.2 URBAN PLANNING

Cities have 89% of GHGs in buildings, 72% in the form of motor vehicles and 55% is associated with waste. That's why urban planning to adapt or cope with CC is a good option. As GHGs can affects the city by urban heat islands and results in temperatures rise, low rainfall patterns. As 60% of the population in the world live in cities, and cities are vulnerable to these impacts, so urban planning and territorial governance play a crucial role in this context.

It is a voluntarily based approach in which private sector ties with local development authorities to address CC adaptation. These shifts of society to CC adaptation due to the observed impacts of ongoing extreme weather and climate events ensures that the urban planning and building management can contribute in it. This planning specifically focuses on the spatial planning and building the eco-cities like project which is being developed at the University of Manchester's School of Environment and Development. These cities will be climatically well suitable for living. Planning and design related to CC hazards, vulnerabilities, and responses in these cities and ways of structures to cope them are the main points to consider.

4.4.5.3 FOSSIL FUEL EMISSIONS MANAGEMENT

Emission control and management includes reduction technologies for nitrogen oxides (NO_x), sulfur oxides (SO_x) and mercury (Hg). C free or low-C sources for energy are available and use of solar power, wind power, geothermal energy, hydropower, and nuclear power are good options. Further, switching from high-C fuels to low-C fuels such as natural gas also reduces the emission (Powlson et al., 2005). Another option for reducing C is the Regional GHGs Initiative (RGGI), USA in which there is a target of reducing C emissions up to 10% by 2018. It is a system in which participants are issued with permits to emit C up to a specific limit. They also can purchase C-offset credits to help meet a C reduction or off-set target. Volunteer based approaches are also common in which industries agree to achieve a given C off-set target (Fuss et al., 2014).

4.4.5.4 ALTERNATIVE FUEL TECHNOLOGIES

Gasoline and diesel are fossil fuels which are used in automobiles extracted from crude oil. They are composed of hydrogen (H_2) and oxygen (O_2) and produce CO_2 and other GHGs and leads to global warming and CC. But today, it is possible to produce fuels which have same chemical composition but emit less or no pollutants like biodiesel which is a mixture of H_2 and CO and commonly known as alternative fuels and considered the best as they are eco-friendly and has no impact on environment. These are synthesized domestically from non-petroleum sources. Ethanol is most common fuel known as biofuel which is produced from corn, sugarcane, etc. Biodiesel is used as alternative to diesel produced from animal fats and oils. Natural gas is also a good alternative. Some of the alternatives are:

1. Ethanol is alcohol-based fuel prepared by fermentation and distillation of oils and extracts of corn, barley, and wheat crops. It can be used separately or mixed with petrol to enhance the efficiency. *Advantage of ethanol* is that the materials required for its production are easily available and cheap. It is also safe and readily biodegradable.
2. Natural gas is also considered as one of the best alternative fuels that produce least amount of GHGs and widely available for use in house businesses. Cars and other transport vehicles can use this as compressed natural gas (CNG).

3. Electricity is used to charge the battery-powered electric and fuel-cell vehicles. Battery powered electric vehicles can travel on the energy stored in batteries by plugging the vehicle into an electrical source. Fuel-cell is also used for electricity generation and fuel-celled cars are also developed. Positive thing is that the electrical energy is highly efficient and we already have an extensive electricity network.
4. Hydrogen is mixed with natural gas to make alternative fuel for vehicles using internal combustion engines (ICE). It is also used in fuel-celled vehicles that run on electricity produced by the sun rays. It has no bad emissions.
5. Propane also called liquefied petroleum gas (LPG) is formed by combining natural gas and crude oil. It is widely used in domestic sector for cooking and heating and an alternative fuel for vehicles. It produces least GHGs emissions than fossil fuels.
6. Fuel methanol (M85) is specially prepared for fuel purposes which is a blend of 85% methanol and 15% gasoline. It is used as a vehicle fuel as it has high-octane number and has lower C emissions than conventional gasoline. Fuel methanol-gasoline blends have widespread use in China and have been introduced in several countries outside of China. Methanol is also emerging as a clean-burning fuel for shipping industry. It is cost-effective and complies with shipping industry's stringent emissions regulations.

4.5 CONCLUSIONS

Greenhouse effect is the natural phenomena but humans are the responsible for the escalation of it leading to the global warming and CC by their un-wise activities of industrialization, exploited, and non-judicial use of fossil fuels. Due to the CC, natural environment is facing different types of extreme weather events like un-expected and high intensity weather events of rains and snowfall, agro-ecological zone shifting, shortening/lengthening of crops growing seasons, reduction in crops yield. Loss of organic matter in soils due to high temperature results in reduction in soil quality and health, structure, and texture degradation. It also results in shifting, migration, and habitat loss of biodiversity of animal species, marine life disturbance and coral reef loss via ocean acidification, deforestation due to forest fires and land use change and industrialization, sea level rise, glacier melting, water quality deterioration and food security issues. All are due to a single phenomenon, CC.

CC mitigation or the solution of all above problems lies in the reduction of the C concentrations in the atmosphere. There are many techniques to accomplish this including C storage in oceans, forests or geological sequestration which require engineering and heavy mechanization. Forests require a large area to grow and lot of time to mature and sequester C in them. There is a better, economical, and reliable way to sequester C in soils as well as crops grown on them. Because crops have shorter life span and give food, these are the suitable way of CS. CS can also be done in degraded or marginally degraded soils which is again a benefit that soils are rehabilitated and used as crop production as well as sink for C.

There are some other techniques, which helps in CS like agroforestry in which trees are grown in combination with agricultural crops, crop rotation, organic farming, nutrient management, zero or low tillage, cover cropping, afforestation, reforestation, REDD+, rainwater harvesting and its saving technologies, water desalinization, desert soilization to enhance C pool, C trading to support its reduction, urban planning to support less C emissions, using C capturing technologies and alternative fuels which emit less or no C.

KEYWORDS

- **agriculture**
- **agroforestry**
- **carbon sequestration**
- **climate change**
- **mitigation**

REFERENCES

Aines, R. D., Wolery, T. J., Bourcier, W. L., Wolfe, T., & Hausmann, C., (2011). Fresh water generation from aquifer-pressured carbon storage: Feasibility of treating saline formation waters. *Energy Procedia*, *4*, 2269–2276.

Akhtar, M., Hassan, F., Ahmed, M., Hayat, R., & Stöckle, C. O., (2015). Is rainwater harvesting an option for designing sustainable cropping patterns for rainfed agriculture? *Land Degradation and Development*, *27*(3), 630–640.

Albrecht, A., & Kandji, S. T., (2003). Carbon sequestration in tropical agroforestry systems. *Agriculture, Ecosystems and Environment*, *99*(1–3), 15–27.

Anderegg, W. R., (2015). Spatial and temporal variation in plant hydraulic traits and their relevance for climate change impacts on vegetation. *New Phytologist, 205*(3), 1008–1014.
Arnell, N., (1999). Climate change and global water resources. *Global Environmental Change, 9*, 31–49.
Bamber, R., (1990). The effects of acidic seawater on three species of lamellibranch mollusk. *Journal of Experimental Marine Biology and Ecology, 143*, 181–191.
Bartlett, K. B. F., & Harriss, R. C., (1993). Review and assessment of methane emissions from wetlands. *Chemosphere, 26*(1–4), 261–320.
Bastviken, D., Tranvik, L. J., Downing, J. A., Crill, P. M., & Enrich-Prast, A., (2011). Freshwater methane emissions offset the continental carbon sink. *Science, 331*(6013), 50.
Bellassen, V., & Luyssaert, S., (2014). Carbon sequestration: Managing forests in uncertain times. *Nature, 506*(7487), 153–155.
Bonan, G. B., (2008). Forests and climate change: Forcings, feedbacks, and the climate benefits of forests. *Science, 320*(5882), 1444–1449.
Choudhury, G., Yaduvanshi, N. P. S., Chaudhari, S. K., Sharma, D. R., Sharma, D. K., Nayak, D. C., & Singh, S. K. (2018). Effect of nutrient management on soil organic carbon sequestration, fertility, and productivity under rice-wheat cropping system in semi-reclaimed sodic soils of North India. *Environmental Monitoring and Assessment, 190*(3).
Chow, A., (2014). *Ocean Carbon Sequestration by Direct Injection.* CO_2 sequestration and valorization, in-tech.
Christensen, N., & Lettenmaier, D. P., (2006). A multi-model ensemble approach to assessment of climate change impacts on the hydrology and water resources of the Colorado River basin. *Hydrology and Earth System Sciences Discussions, 3*, 3727–3770.
Clark, P. U., Pisias, N. G., Stocker, T. F., & Weaver, A. J., (2002). The role of the thermohaline circulation in abrupt climate change. *Nature, 415*, 863.
Conant, R. T., (2011). Sequestration through forestry and agriculture. *Wiley Interdisciplinary Reviews: Climate Change, 2*(2), 238–254.
Conroy, J. P., Seneweera, S., Basra, A. S., Rogers, G., & Nissen-Wooller, B., (1994). Influence of rising atmospheric CO_2 concentrations and temperature on growth, yield and grain quality of cereal crops. *Functional Plant Biology, 21*(6), 741–758.
Coristine, L. F., (2016). *Climate Change Impacts on Biodiversity (Doctoral Dissertation).* Universitéd'Ottawa / University of Ottawa.
De Deyn, G. B., Cornelissen, J. H., & Bardgett, R. D., (2008). Plant functional traits and soil carbon sequestration in contrasting biomes. *Ecology Letters, 11*(5), 516–531.
De Klein, C., Pinares-Patino, C., & Waghorn, G., (2008) *Greenhouse Gas Emissions Environmental Impacts of Pasture-Based Farming*, pp. 1–32.
Dobbie, K. E., McTaggart, I. P., & Smith, K. A., (1999). Nitrous oxide emissions from intensive agricultural systems: Variations between crops and seasons, key driving variables, and mean emission factors. *Journal of Geophysical Research: Atmospheres, 104*, 26891–26899.
Erda, L., Wei, X., Hui, J., Yinlong, X., Yue, L., Liping, B., & Liyong, X., (2005). Climate change impacts on crop yield and quality with CO_2 fertilization in China. *Philosophical Transactions of the Royal Society B: Biological Sciences, 360*(1463), 2149–2154.
Fabry, V. J., McClintock, J. B., Mathis, J. T., & Grebmeier, J. M., (2009). Ocean acidification at high latitudes: The bellwether. *Oceanography, 22*(4), 160–171.
Food and Agricultural Organization (FAO), (2003). Data available at: http://www.fao.org/docrep/005/ac836e/AC836E03.htm (Accessed on 9 October 2019).

Franzluebbers, A. J., & Doraiswamy, P. C., (2007). Carbon sequestration and land degradation. In: Sivakumar, M. V. K., & Ndiang'ui, N., (eds.), *Climate and Land Degradation* (pp. 343–358). Berlin, Heidelberg: Springer Berlin Heidelberg.

Freibauer, A., Rounsevell, M. D., Smith, P., & Verhagen, J., (2004). Carbon sequestration in the agricultural soils of Europe. *Geoderma, 122*(1), 1–23.

Friedmann, S. J., (2007). Geological carbon dioxide sequestration. *Elements, 3*(3), 179–184.

Fujisaki, K., Chevallier, T., Chapuis-Lardy, L., Albrecht, A., Razafimbelo, T., Masse, D., & Chotte, J. L., (2018). Soil carbon stock changes in tropical croplands are mainly driven by carbon inputs: A synthesis. *Agriculture, Ecosystems and Environment, 259*, 147–158.

Fuss, S., Canadell, J. G., Peters, G. P., Tavoni, M., Andrew, R. M., Ciais, P., & Yamagata, Y., (2014). Betting on negative emissions. *Nature Climate Change, 4*(10), 850–853.

Garrett, K. A., Nita, M., De Wolf, E. D., Esker, P. D., Gomez-Montano, L., & Sparks, A. H., (2015). Plant pathogens as indicators of climate change. In: *Climate Change* (2nd edn., pp. 325–338).

Gibbs, H. K., Brown, S., Niles, J. O., & Foley, J. A., (2007). Monitoring and estimating tropical forest carbon stocks: Making REDD a reality. *Environmental Research Letters, 2*(4), 045023.

Goodale, C. L., Apps, M. J., Birdsey, R. A., Field, C. B., Heath, L. S., Houghton, R. A., & Nabuurs, G. J., (2002). Forest carbon sinks in the Northern Hemisphere. *Ecological Applications, 12*(3), 891–899.

Gupta, V., (2011). Microbes and soil structure. In: *Encyclopedia of Agrophysics* (pp. 470–472). Springer Netherlands.

Hernanz, J. L., Sánchez-Girón, V., & Navarrete, L., (2009). Soil carbon sequestration and stratification in a cereal/leguminous crop rotation with three tillage systems in semiarid conditions. *Agriculture, Ecosystems and Environment, 133*(1/2), 114–122.

Hinzman, L. D., Bettez, N. D., Bolton, W. R., Chapin, F. S., Dyurgerov, M. B., Fastie, C. L., & Jensen, A. M., (2005). Evidence and implications of recent climate change in northern Alaska and other arctic regions. *Climatic Change, 72*(3), 251–298.

Hoegh-Guldberg, O., Mumby, P. J., Hooten, A. J., Steneck, R. S., Greenfield, P., Gomez, E., & Knowlton, N. (2007). Coral reefs under rapid climate change and ocean acidification. *Science, 318*(5857), 1737–1742.

Holeplass, H., Singh, B. R., & Lal, R., (2004). Carbon sequestration in soil aggregates under different crop rotations and nitrogen fertilization in an inceptisol in southeastern Norway. *Nutrient Cycling in Agroecosystems, 70*(2), 167–177.

Hulme, M., Barrow, E. M., Arnell, N. W., Harrison, P. A., Johns, T. C., & Downing, T. E., (1999). Relative impacts of human-induced climate change and natural climate variability. *Nature, 397*(6721), 688–691.

Kaiser, C., Franklin, O., Richter, A., & Dieckmann, U., (2015). Social dynamics within decomposer communities lead to nitrogen retention and organic matter build-up in soils. *Nature Communications, 6*(1).

Karmakar, R., Das, I., Dutta, D., & Rakshit, A., (2016). Potential effects of climate change on soil properties: A review. *Science International, 4*(2), 51–73.

Kell, D. B., (2012). Large-scale sequestration of atmospheric carbon via plant roots in natural and agricultural ecosystems: Why and how. *Philosophical Transactions of the Royal Society B: Biological Sciences, 367*(1595), 1589–1597.

Keppler, F., Hamilton, J. T. G., Braß, M., & Röckmann, T., (2006). Methane emissions from terrestrial plants under aerobic conditions. *Nature, 439*(7073), 187–191.

Klinedinst, P. L., Wilhite, D. A., Hahn, G. L., & Hubbard, K. G., (1993). The potential effects of climate change on summer season dairy cattle milk production and reproduction. *Climatic Change, 23*, 21–36.

Knox, J., Hess, T., Daccache, A., & Wheeler, T., (2012). Climate change impacts on crop productivity in Africa and South Asia. *Environmental Research Letters, 7*(3), 034032.

Knutzen, J., (1981). Effects of decreased pH on marine organisms. *Marine Pollution Bulletin, 12*, 25–29.

Lal, R., & Kimble, J. M., (1997). *Nutrient Cycling in Agroecosystems, 49*(1/3), 243–253.

Lal, R., (2004). Soil carbon sequestration impacts on global climate change and food security. *Science, 304*(5677), 1623–1627.

Lal, R., (2005). Forest soils and carbon sequestration. *Forest Ecology and Management, 220*(1–3), 242–258.

Lal, R., (2006). Carbon management in agricultural soils. *Mitigation and Adaptation Strategies for Global Change, 12*(2), 303–322.

Lal, R., Follett, R. F., Stewart, B. A., & Kimble, J. M., (2007). Soil carbon sequestration to mitigate climate change and advance food security. *Soil Science, 172*(12), 943–956.

Leifeld, J., & Fuhrer, J., (2010). Organic farming and soil carbon sequestration: What do we really know about the benefits? *AMBIO, 39*(8), 585–599.

Lenart, E. A., Bowyer, R. T., Hoef, J. V., & Ruess, R. W., (2002) Climate change and caribou: Effects of summer weather on forage. *Canadian Journal of Zoology, 80*, 664–678.

Lenka, N. K., Dass, A., Sudhishri, S., & Patnaik, U. S., (2012). Soil carbon sequestration and erosion control potential of hedgerows and grass filter strips in sloping agricultural lands of eastern India. *Agriculture, Ecosystems and Environment, 158*, 31–40.

Lenka, S., Lenka, N. K., Sejian, V., & Mohanty, M., (2015). Contribution of agriculture sector to climate change. *Climate Change Impact on Livestock: Adaptation and Mitigation*, pp. 37–48.

Lindner, M., Maroschek, M., Netherer, S., Kremer, A., Barbati, A., Garcia-Gonzalo, J., & Marchetti, M. (2010). Climate change impacts, adaptive capacity, and vulnerability of European forest ecosystems. *Forest Ecology and Management, 259*(4), 698–709.

Liu, E., Yan, C., Mei, X., Zhang, Y., & Fan, T., (2013). Long-term effect of manure and fertilizer on soil organic carbon pools in dryland farming in northwest China. *PLoS One, 8*(2), e56536.

Lobell, D. B., Burke, M. B., Tebaldi, C., Mastrandrea, M. D., Falcon, W. P., & Naylor, R. L., (2008). Prioritizing climate change adaptation needs for food security in 2030. *Science, 319*, 607–610.

Lorenz, K., & Lal, R., (2009). The importance of carbon sequestration in forest ecosystems. *Carbon Sequestration in Forest Ecosystems*, 241–270.

Lorenz, K., & Lal, R., (2014). Soil organic carbon sequestration in agroforestry systems: A review. *Agronomy for Sustainable Development, 34*(2), 443–454.

Marland, G., & Rotty, R. M., (1984). Carbon dioxide emissions from fossil fuels: A procedure for estimation and results for 1950–1982. *Tellus B, 36*, 232–261.

McConkey, B., (2003). Crop rotation and tillage impact on carbon sequestration in Canadian prairie soils. *Soil and Tillage Research, 74*(1), 81–90.

McMichael, A. J., Powles, J. W., Butler, C. D., & Uauy, R., (2007). Food, livestock production, energy, climate change, and health. *The Lancet, 370*(9594), 1253–1263.

McNeill, J. R., & Winiwarter, V., (2004). Breaking the sod: Humankind, history, and soil. *Science, 304*(5677), 1627–1629.

McNutt, M., (2013). Climate change impacts. *Science, 341*(6145), 435–435.

Mosier, A., Kroeze, C., Nevison, C., Oenema, O., Seitzinger, S., & Van Cleemput, O., (1998). *Nutrient Cycling in Agroecosystems, 52*(2/3), 225–248.

Nair, P. K. R., Kumar, B. M., & Nair, V. D., (2009) Agroforestry as a strategy for carbon sequestration. *Journal of Plant Nutrition and Soil Science, 172*, 10–23.

Nair, P. K. R., Nair, V. D., Kumar, B. M., & Showalter, J. M., (2010). Carbon sequestration in agroforestry systems. *Advances in Agronomy,* 237–307.

National Aeronautical and Space Administration (NASA), (2018). Data available at: https://climate.nasa.gov/vital-signs/carbon-dioxide/ (Accessed on 9 October 2019).

Nearing, M. A., Pruski, F. F., & O' Neal, M. R., (2004). Expected climate change impacts on soil erosion rates: a review. *Journal of Soil and Water Conservation, 59*(1), 43–50.

Newmark, R. L., Friedmann, S. J., & Carroll, S. A., (2010). Water challenges for geologic carbon capture and sequestration. *Environmental Management, 45*(4), 651–661.

Nilsson, S., & Schopfhauser, W., (1995). The carbon-sequestration potential of a global afforestation program. *Climatic Change, 30*(3), 267–293.

Nowak, D. J., Greenfield, E. J., Hoehn, R. E., & Lapoint, E., (2013). Carbon storage and sequestration by trees in urban and community areas of the United States. *Environmental Pollution, 178*, 229–236.

O'Neal, M. R., Nearing, M., Vining, R. C., Southworth, J., & Pfeifer, R. A., (2005). Climate change impacts on soil erosion in Midwest United States with changes in crop management. *Catena, 61*, 165–184.

Obersteiner, M., Alexandrov, G., Benítez, P. C., McCallum, I., Kraxner, F., Riahi, K., & Yamagata, Y. (2006). Global supply of biomass for energy and carbon sequestration from afforestation/reforestation activities. *Mitigation and Adaptation Strategies for Global Change, 11*(5/6), 1003–1021.

Ojija, F., (2015). Assessment of current state and impact of REDD+ on livelihood of local people in Rungwe district, Tanzania. *International Journal of Scientific and Technology Research, 04*(04).

Ontl, T. A., & Schulte, L. A., (2012). Soil carbon storage. *Nature Education Knowledge, 3*(10), 35.

Pan, Y., Birdsey, R. A., Fang, J., Houghton, R., Kauppi, P. E., Kurz, W. A., & Ciais, P., (2011). A large and persistent carbon sink in the world's forests. *Science,* 1201609.

Pandit, R., Neupane, P. R., & Wagle, B. H., (2017). Economics of carbon sequestration in community forests: Evidence from REDD+ piloting in Nepal. *Journal of Forest Economics, 26*, 9–29.

Parmesan, C., & Yohe, G., (2003). A globally coherent fingerprint of climate change impacts across natural systems. *Nature, 421*(6918), 37–42.

Piao, S., Ciais, P., Huang, Y., Shen, Z., Peng, S., Li, J., & Fang, J., (2010). The impacts of climate change on water resources and agriculture in China. *Nature, 467*(7311), 43–51.

Pitesky, M. E., Stackhouse, K. R., & Mitloehner, F. M., (2009). Clearing the air. *Advances in Agronomy,* 1–40.

Plummer, L. N., & Busenberg, E., (2000). Chlorofluorocarbons. In: *Environmental Tracers in Subsurface Hydrology* (pp. 441–478). Springer.

Poeplau, C., & Don, A., (2015). Carbon sequestration in agricultural soils via cultivation of cover crops – A meta-analysis. *Agriculture, Ecosystems and Environment, 200*, 33–41.

Powlson, D. S., Riche, A. B., & Shield, I., (2005). Biofuels and other approaches for decreasing fossil fuel emissions from agriculture. *Annals of Applied Biology, 146*(2), 193–201.

Raich, J. W., & Schlesinger, W. H., (1992). The global carbon dioxide flux in soil respiration and its relationship to vegetation and climate. *Tellus B., 44*, 81–99.

Reay, D., (2015). Nitrous oxide sources. In: *Nitrogen and Climate Change* (pp. 49–68). Springer.

Santilli, M., Moutinho, P., Schwartzman, S., Nepstad, D., Curran, L., & Nobre, C., (2005). Tropical deforestation and the Kyoto protocol. *Climatic Change, 71*(3), 267–276.

Scavia, D., Field, J. C., Boesch, D. F., Buddemeier, R. W., Burkett, V., Cayan, D. R., & Titus, J. G., (2002). Climate change impacts on U. S. coastal and marine ecosystems. *Estuaries, 25*(2), 149–164.

Schlenker, W., & Roberts, M. J., (2009). Nonlinear temperature effects indicate severe damages to U. S. crop yields under climate change. *Proceedings of the National Academy of Sciences, 106*(37), 15594–15598.

Schlesinger, W. H., (1999). Carbon and agriculture carbon sequestration in soils. *Science, 284*(5423), 2095–2095.

Schneider, S. H., (1989). The greenhouse effect: Science and policy. *Science, 243*(4892), 771–781.

Shi, S., Han, P., Zhang, P., Ding, F., & Ma, C., (2015). The impact of afforestation on soil organic carbon sequestration on the Qinghai plateau, China. *Plos One, 10*(2), e0116591.

ShuklaPotdar, R., & Vishal, V., (2016). Trapping mechanism of CO2 storage in deep saline aquifers: Brief review. *Geologic Carbon Sequestration,* 47–58.

Silver, W. L., Kueppers, L. M., Lugo, A. E., Ostertag, R., & Virginia, M., (2004). Carbon sequestration and plant community dynamics following reforestation of tropical pasture. *Ecological Applications, 14*(4), 1115–1127.

Silver, W. L., Ostertag, R., & Lugo, A. E., (2000). The potential for carbon sequestration through reforestation of abandoned tropical agricultural and pasture lands. *Restoration Ecology, 8*(4), 394–407.

Smit, B., Reimer, J. A., Oldenburg, C. M., & Bourg, I. C., (2014). *Introduction to Carbon Capture and Sequestration.* Imperial College Press, London.

St. Louis, V. L., Kelly, C. A., Duchemin, É., Rudd, J. W., & Rosenberg, D. M., (2000). Reservoir surfaces as sources of greenhouse gases to the atmosphere: A global estimate. *AIBS Bulletin, 50,* 766–775.

Stachowicz, J. J., Terwin, J. R., Whitlatch, R. B., & Osman, R. W., (2002). Linking climate change and biological invasions: Ocean warming facilitates nonindigenous species invasions. *Proceedings of the National Academy of Sciences, 99*(24), 15497–15500.

Stroosnijder, L., & Hoogmoed, W. B., (2004) Contribution of soil and water conservation to carbon sequestration in semi-arid. *Africa Bulletin Réseau Erosion, 23,* 523–539.

Swift, M. J., Heal, O. W., & Anderson, J. M., (1979). *Decomposition in Terrestrial Ecosystems* (Vol. 5). University of California Press.

Thomas, C. D., Cameron, A., Green, R. E., Bakkenes, M., Beaumont, L. J., Collingham, Y. C., & Williams, S. E., (2004). Extinction risk from climate change. *Nature, 427*(6970), 145–148.

Torres, C. M. M. E., Jacovine, L. A. G., Nolasco De Olivera, N. S., Fraisse, C. W., Soares, C. P. B., De Castro, N. F., & Lemes, P. G., (2017). Greenhouse gas emissions and carbon sequestration by agroforestry systems in southeastern Brazil. *Scientific Reports, 7*(1), 16738.

United States Environmental Protection Agency (US-EPA), (2016). Data available at: https://www3.epa.gov/climatechange//kids/basics/today/greenhouse-effect.html (Accessed on 9 October 2019).

United States Environmental Protection Agency, (2016). Data available at: https://www.epa.gov/ghgemissions/sources-greenhouse-gas-emissions (Accessed on 9 October 2019).

Valdez, M. C., Adler, I., Barrett, M., Ochoa, R., & Pérez, A., (2016). The water-energy-carbon nexus: Optimizing rainwater harvesting in Mexico City. *Environmental Processes, 3*(2), 307–323.

Wang, G. C., Wang, E. L., Huang, Y., & Xu, J. J., (2014). Soil carbon sequestration potential as affected by management practices in Northern China: A simulation study. *Pedosphere, 24*(4), 529–543.

Wheeler, T., & Von Braun, J., (2013). Climate change impacts on global food security. *Science, 341*(6145), 508–513.

Woodrow, I. E., & Berry, J., (1988). Enzymatic regulation of photosynthetic CO_2, fixation in C_3 plants. *Annual Review of Plant Physiology and Plant Molecular Biology, 39*, 533–594.

Yi, Z., & Zhao, C., (2016). Desert soilization: An eco-mechanical solution to desertification. *Engineering, 2*(3), 270–273.

Zdruli, P., Lal, R., Cherlet, M., & Kapur, S., (2017). *New World Atlas of Desertification and Issues of Carbon Sequestration, Organic Carbon Stocks, Nutrient Depletion and Implications for Food Security, 15*, 13–25.

Zhang, Z., & Huisingh, D., (2018). Combating desertification in China: Monitoring, control, management and revegetation. *Journal of Cleaner Production, 182*, 765–775.

CHAPTER 5

Agroforestry: Soil Organic Carbon and Its Carbon Sequestration Potential

NONGMAITHEM RAJU SINGH,[1] DHIRAJ KUMAR,[2] K. K. RAO,[1] and B. P. BHATT[1]

[1]ICAR-Research Complex for Eastern Region, Patna, Bihar–800014, India, E-mail: rajuforestry@gmail.com (N. R. Singh)

[2]ICAR-Central Agroforestry Research Institute, Jhansi (U.P)–284003, India

ABSTRACT

Challenges to the lives of human being and other living communities by changing climate due to increase in greenhouse gases (GHGs) concentration, majority of carbon dioxide (CO_2) are leading in a miserable way. There calls a diversified land-use system to face the alarming issue of crop productivity as well as to challenge the effect of climate changes (CCs). Moreover, the releasing of CO_2 from the soil due to intensive cultivation on limited land resources without considering its future land degradation is also adding another challenge to farming community. In this context, the practice of agroforestry has been realized and shown promising land use system. Agroforestry has started gaining attention across the globe and the aspect of carbon sequestration (CS) potential recognized a big asset in terms of CC mitigation approach. The practices of agroforestry help to improve the soil physico-chemical and biological properties by continuous addition of litter in the soil surface. Since, soil organic carbon (SOC) is having the largest contribution in carbon pool among the terrestrial ecosystem, which is estimated to be over 1550 Pg C at 1m soil depth. Considering the potential of soil ecosystem to store carbon, it is attracting considerable attention to curb the issues of CC in near future. The practices of agroforestry involving the minimal disturbance of soil and continuous cover of litter helps in stabilizing the soil organic and making the room for vast CS opportunities in the soil.

It is believed that other ecological, biological, and edaphic factors, several social factors such as adoption of different management practices like application of fertilizers, irrigation supply, application of pesticides, herbicides, etc. could also affect the SOC sequestration potential under agroforestry system (AFs) by influencing the soil aggregates stability. In this context, several studies conducted in different places of world, however, their reports have shown large variation in estimating the CS potential in AFs across the world due to non-homogeneous estimation.

5.1 INTRODUCTION

Today, the world is facing a multitude of challenges, more prominently of climate change (CC) issues led to hunger, food security, scarcity of water, soil-related issues like degradation, depletion of soil fertility, and loss of biodiversity. When it comes to solution measure many of the scientists realized the introduction of agroforestry practices, where a traditional agricultural system can be adapted by inclusion of woody perennials like trees and shrubs with annual crops. The concept agroforestry system (AFs) with the introduction of multipurpose species has become immensely important and need to be devised in different regions of the country for higher productivity and soil nutrient enrichment. Thus, the introduction of agroforestry, which is a form of multiple land/use system, is considered more resilient than monocropping system in challenging the CC phenomena with a view to improve the food and livelihood security.

Indeed, the practice of agroforestry is old but the science remained the new one. Agroforestry attempts to manage the land use management system in which woody perennials or animals components are integrated with agricultural crops on the same unit of area. From this, it is clearly understand that AFs are more diverse than the conventional agricultural systems, thereby aiming to utilize the resources efficiently and maximize the productivity. However, sometimes, the complexity and management intervention add a new challenge to the farmers. The on-farm technologies, transfer of improved technologies to the farmers, motivational training programmes, will help the farmers to adopt and practices agroforestry in sound manner. The successfully adoption of AFs by farmers is mainly determined by economic advantages of the system (Nair, 2007). Biodiversity conservation and protection can be enhanced by incorporation of trees in agricultural landscapes, degraded lands, watersheds areas, sand dunes, etc. Sustainable development aims the use of resources without hampering its production and soil productivity, and AFs has the ability to achieve through diversification of the system, maximizing the productivity, improving the soil overall properties. Agroforestry's potential and opportunities to challenge the menace of CC phenomena through the process of carbon sequestration (CS) will be highly optimistic and appreciated works.

India's National Forest Policy (1952) ignited the importance of social forestry, which also includes agroforestry practices. It is worth mentioning that NFP (1952) also set a national goal of maintaining 33% of land area under forest cover. In this aspect, agroforestry could serve the national importance by increasing the country forest area and this system is more encouraging as well as expanding constantly to different places of the country. According to DARE (2014) areas of wasteland and unutilized land area of about 53 m ha could be converted to agroforestry areas in the near foreseeable future.

5.2 HISTORICAL STATUS OF AGROFORESTRY

The tradition of incorporating agricultural crops with trees has been evolved since ancient times. The chapter of agroforestry is reckoned to be associated with the evolution of cultivation of agriculture. It was believed that during old time until middle ages in European regions, forests were clear felled, burn, and cultivated the crops (King, 1968). In the Asian region, particularly in the Philippines traditional form of AFs such as shifting cultivation was evident to be practiced by the people of Hanunoo (Conklin, 1953). With the passage of time, several forms of farming system alike with AFs have been extended. Growing of agricultural crops under scattered trees in southern Nigeria, was reported by Forde (1937). It is also believed that home garden/homestead system a typical multistoried and diversified form of AFs had been practiced from a long time ago in many parts of the world. It was during the 1970s and 1980s the agroforestry and its related activities began to gain institutional attention. Since then, many of the international organization started to initiate the research in agroforestry, notably, the International Development Research Centre (IDRC), Canada, Food and Agriculture Organization (FAO), Rome, Swedish International Development Authority (SIDA), Sweden. Later on, the International Centre for Research in Agroforestry (ICRAF), Kenya has been established in 1977 for agroforestry research and its related works. Subsequently, the International Institute of Tropical Agriculture (IITA) has also started the work on agroforestry and its related activities (Nair, 1979).

India has a long tradition in adopting agroforestry practices. It is evident from the past history and ancient Vedas, people of the country have the socio-religious intact of raising trees, worshipping, caring for and respecting trees.

In India, the first agroforestry seminar was held in 1979 at Imphal, Manipur. The different participants in this seminar had ignited the importance of agroforestry in relation to forestry and made promised to start the scientific agroforestry in India. With due time, All India Coordinated Research Project on Agroforestry (AICRAF) was initiated in 1983 by ICAR at 20 centers and now it is expanded to 37 coordinating centers across the country including ICAR institutes and State Agricultural Universities. In 1988, National Research Centre for Agroforestry came into existence at Jhansi (Uttar Pradesh). Now the centre has renamed as Central Agroforestry Research Institute (CAFRI) in December, 2014.

The National Agriculture Policy, (2000) emphasized the importance of agroforestry and clearly stated that, "Agriculture has become a relatively unrewarding profession due to generally unfavorable price regime and low value addition, causing abandoning of farming and increasing migration from rural areas." Hence the Policy stresses, "Farmers will be encouraged to take up farm/agroforestry for higher income generation by evolving technology, extension, and credit support packages and removing constraints to the development of agro forestry." The adoption of agroforestry practices can be accelerated by an implementation of improved technology; however, the level of technology transfer and dissemination of information is still lagging behind. Agroforestry will play a significant role in achieving the nation target of one-third cover of forest area. More emphasis on agroforestry research and development activities have been made in recent years and to address the problems encountered while employing the agroforestry and its technology, the Government of India, set up agroforestry policy in 2014.

5.3 EXTEND OF AGROFORESTRY SYSTEMS (AFS)

Agroforestry practices have been in vogue since time immemorial, indicating the practices of agroforestry is old but transform into new acceptable way of land management practices with due course of time. The practices of agroforestry have been distributed across the world in various forms of practice such as scattered trees in grassland to highly diversified home garden system. The adoption and extend of different agroforestry practices depends on the climatic, ecological, and socio-economic condition of the particular place. The adherence and positive attitude of people towards the agroforestry technology and their disseminations will have significant impact on extend of agroforestry. So, the distribution or extent of agroforestry can be influenced by various factors broadly of demographic, climatic, and socioeconomic status of the region (Nair, 1993). For instance, home garden systems are predominantly adopted in the areas of tropical humid; shelter belts and windbreaks, pasture-based AFs in arid and semiarid regions; improved fallow, trees for soil and water conservation, etc. It is clear that the use of different AFs had abound with particular region. However, agroforestry in temperate region also employed the systems followed in the tropical or subtropical region, principality, and norm for establishment of agroforestry remains same, but their structure and function may vary. It is due to difference in environment, social, and economic characteristics, the scale and adoption of AFs vary in different region across the world (Jose et al., 2004).

The magnitude of integrating trees in farms or agricultural lands tends to vary across the different regions. In the last few decades, the area and distribution of agroforestry practices have been extended with fair results, markedly with the aim to achieve sustainable development. These results showed agroforestry as a very promising landuse system with expected land area coverage of 1.6×10^9 ha in the foreseeable future across the world (Nair and Garrity, 2012). Similarly, World Bank (2004) also highlighted the importance of agroforestry in which they estimated 1.2 billion rural people are currently engaged in agroforestry practices on their farms, communities, and thus become a source for livelihood. The area under AFs have increased during recent few decades, however, the quantification of area under agroforestry becomes biased and difficult to delineate. So, many authors have produced different estimates of area under agroforestry across the world. For example, Nair et al., (2009) projected total area of agroforestry in world is 1023 M ha and out of this silvopastoral systems covers an area of about 516 M ha. In India also, the estimation of area under agroforestry has been incorporated for first time in State Reports of Forestry, 2013, which is recorded as 11.54 m ha (FSI, 2013). However, after one year, Dhyani et al., (2014) claimed the projected area of agroforestry in India is of about 25.32 M ha. Zomer et al., (2007) estimated the total area under agroforestry in India of about 7.4 M ha.

5.4 SCOPE AND POTENTIAL OF AGROFORESTRY

The concepts of agroforestry, entails deliberate mixing of trees/animals with agricultural crops have come up to meet the diversified demand of farmers in terms of productivity as well as maintaining the fertility of soil at same time agroforestry a sustainable landuse system has been gaining popularity and considered a viable option for livelihood security and CC mitigation and adaptation (Singh et al., 2017).

The adoption of agroforestry in farmland and agricultural landscapes have proven the benefits of ecological and socioeconomic development of the farmers, contributing to the overall development of the national economy (Gao et al., 2014; Gold and Garrett, 2009). The potentiality of agroforestry while transforming the lives of rural farmers especially the marginal one through diversification and sustainability need to be addressed. Broadly, agroforestry practices encompass several components related to different basic and applied sciences. Thus, in simple sense, it is a multi-disciplinary approach to achieve a common goal.

The scope and potential of AFs have been elaborated by many of the researchers and approach of AFs to achieve sustainable development is a prominent one. There is a tremendous scope of agroforestry for maximizing the food production system while conserving the farmers land area, thereby increased the diversification of farmland (Singh and Jhariya, 2016; Jhariya et al., 2015, 2018). With the realization of importance of agroforestry in meeting the demands of food and livelihood security, not the government organization but also several NGOs and many organizations across the world have engaged in agroforestry activities with the aim to achieve sustainable development.

Agroforestry platform has created potentiality to rehabilitate the ecological disturbance areas, deforested, and already eroded watersheds, etc. reclaiming wastelands, degraded areas. Shifting cultivation or sequential system a predominantly traditional way of agricultural practiced in North-east India, where woody perennials are clear felled, burned, thereby making the land becomes unproductive and unsustainability way of farming. In this condition, during the fallow period, agroforestry trees species which is having the capacity to fix atmospheric nitrogen can be successfully grown and helped in regaining the soil fertility. So, agroforestry practices are considered as a viable alternative to this kind of unproductive and destructive way of farming. Agroforestry has been attracting as a viable option in terms of CC mitigation and adaption by taking the advantages of agroforestry can sequesters huge amount of carbon in biomass (above and below) as well in soil ecosystem. Agroforestry practices like wind breaks, shelter belts, silvi-pasture, would likely to decrease the effect of drought, wind erosion, fodder scarcity in arid and semi-arid regions (Swaminathan, 1987). Records of experiences from agroforestry initiatives worldwide, the potentialities of agroforestry have been summarized as follows:

- AFs being a diversified form of nature, it will provide the multifarious need of food, fodder, fuel, fiber, etc. to the farmers.

- This system will enable to enhance the productivity of the farmland as well as safeguarding the livelihood of the farmers.
- Employment generation by engaging the youth in agroforestry activities.
- Conservation of natural resources.
- AFs will improve the ecosystem services.
- Improvement of sand dunes areas by planting of shelterbelts, windbreaks.
- AFs have a huge CS potential to in the above and below ground biomass.
- It will improve the stabilization of watershed areas by introducing deep-rooted grasses species along with the plantation of tree species, thus ultimately conserve the soil resources by controlling soil erosion.
- Improvement of soil quality and productivity through litter addition of litter and decomposition process.
- Due to nutrient pumping from a deeper layer and the continuous cover of soil will help to minimize the loss of nutrient from the soil.
- Improvement of farm biodiversity and beautification.
- Microclimate amelioration this will enhance the growth of agricultural crops.

5.5 QUANTIFYING THE CARBON SEQUESTRATION (CS) UNDER AGROFORESTRY SYSTEM (AFS)

Quantification in terms of measurement of CS of AFs is generally focused on aboveground and belowground biomass. These measurements can be done by using on-site direct measurement of biomass, indirect remote sensing techniques and modeling. Direct on-site measurement of aboveground biomass is generally performed by harvesting the tree or partially harvest and measurement. In simply, the carbon content of biomass is generally considered as 50% of the total biomass (Nair, 2012). The total destructive method of biomass estimation is a time consuming and costly process but show accuracy and mostly employed to validate the results of another method. In this method, the harvested trees are divided into different parts such as above-ground parts like barks, leaves, branches, bole, and fruits are measured separately for each of the parts and then combined to form the total aboveground biomass (Beer et al., 1990). Allometric equations can be developed by using the data where regression models are used by adopting independent and dependent variables. Tree growth parameters like diameter, height are generally used as independent variables. However, this method is lacking of accuracy since they are very general in nature and they are, at best, approximations.

Belowground biomass carbon estimation involves root biomass as well as soil carbon. Belowground biomass measurement especially for root is very arduous and fatigue works which require standardized method for estimation (Ingram and Fernandes, 2001). Root biomass measurement can be done by using spatially distributed soil cores or pits or complete excavation of roots. So, in the easiest way the root biomass is calculated by measuring the root to shoot ratio. However, these ratios vary significantly among the species, different ecological zones as well as edaphic factors of the site. Likewise the aboveground biomass, the carbon content in the root biomass can be estimated by using the factor or direct estimation in the laboratory analysis. Soil carbon content can be estimated in laboratory by using standard procedure. The amount of carbon content in soil is usually expressed either in% or g/kg which can be later extrapolated to express usually as Mg ha^{-1}. Soil carbon stock can also be derived by using the values of soil bulk density, soil carbon content and soil depth and generally expressed as Mg ha^{-1}.

An indirect remote sensing technique used in biomass estimation is one of the emerging fields in forestry/agroforestry aspects. Various techniques of remote sensing such as photograph taken from aircraft, air borne space, sensors from satellite, etc., are commonly used. The imagery taken from different sources are processed and interpreted the information. Among the different sensor used for biomass estimation purpose, light detection and ranging is becoming a very essential and most accurate technology (Lu et al., 2014). Other than LIDAR, more technological advancement with the development of new techniques like use of high resolution cameras, data recorders, installation of GPS, etc. on remote sensing platform brings another level of agroforestry inventory works (Brown, 2002). By all this method, the tree height, crown, and number of stem per hectare measurement under AFs become easier and more accurate one.

Over the last few years, several carbon stocks modeling have been developed by using the simulation technique. In this regard, some of the commonly used models developed are PROCOMAP, CO2FIX, CENTURY, ROTH, etc. All these models have their own assumption and requirements for data input and output system. Among these models, CO2FIX have been extensively used in agroforestry projects because it is easy to use, can predict the carbon stock for single or multiple species. In case of soil carbon stocks simulation, ROTH, and CENTURY are commonly used, and ROTH is very simple to use due to simple data input requirement. Although, these models are commonly used for various projects at national or worldwide level, there still problem persists mainly for their input data requirement, and attention should be give while selecting a proper model for a particular use (Ravindranath and Ostwald, 2008).

5.6 OVERVIEW OF CARBON SEQUESTRATION (CS) POTENTIAL OF AGROFORESTRY SYSTEM (AFS)

The concentration of carbon dioxide (CO_2) level in the atmosphere is rising at alarming rate and now the level of CO_2 has increased to 40% (400 ppm) in comparison with the preindustrial times (IPCC, 2013). The substantial increase in CO_2 concentration in the atmosphere is mainly due to human anthropogenic activities like the combustion of fossil fuel derived materials, which emits several pollutants to the atmosphere, majority of CO_2 have been causing a major problem in this present day of civilization. Besides, this conversion of natural ecosystem for agricultural purposes and other developmental activities also led to the increased in atmospheric greenhouse gases (GHGs) (Tilman et al., 2002). In general, global warming refers to the substantial increase of GHGs in the atmosphere. Global warming harmful action have threaten the lives of many people around the world such as rise in sea level; increased the chances of natural calamities like flood, drought, etc.; increase of wild fires; intensity and frequency of tropical storms, cyclones, thunderstorms; distributional pattern of rainfall and snowfall; decreased the productivity of marine ecosystem; shifting of human population; likelihood to increase insect and pest incidence; human health related issues, etc. melting of glaciers particularly in the areas of Antarctica.

Since from early 1970's it was realized that long term storage of carbon in the terrestrial ecosystem could pave a way to reduce the impacts of climate aberration through the adoption of trees in farmlands, forest conservation and other degraded lands becomes the viable option in long run process (Montagini and Nair, 2004). After that United Nations Framework Convention on Climate Change (UNFCCC) has emphasized the importance of sinks' in stabilizing the CO_2 level and carbon sink potential of different land use system. Among the different land use systems, the carbon sink potentiality of tree-based land use system has proved the highest and promising one. The programmes of REDD under UNFCCC have focused on the reduction of emissions released mainly due to deforestation and forest degradation. In this context, agroforestry is started gaining a hot spot over worldwide for their ability to have more resiliency and flexibility towards CC effects as well as diversification of the farm and farmers income.

CS with respect to agroforestry denotes the removal and long term storing of atmospheric carbon in the tree biomass through the process of photosynthesis (Nair and Nair, 2003). In simple term, CS in any system implies that the carbon gained through photosynthesis minus the carbon lost through respiration and is usually denoted by net system productivity (Montagini and Nair, 2004). Large extend of carbon are stored in the soil ecosystem thought the improvement of soil organic carbon (SOC) under AFs. Agroforestry being a diversified and more complex than the traditional monocropping system will have the tendency to store and capture more atmospheric carbon in the plant and soil biomass. In general, under monocropping or intensive agriculture system, the crop duration is very short and the amount of carbon accumulation in the plant biomass is less as compared to multitier AFs. On the other hand, agroforestry where the inclusion of woody perennials is necessary and this incorporation of woody perennials in the farmlands and other intercropping areas will help in accumulating more carbon both in above and belowground biomass. Carbon accumulation or sequestered in the tree bole form the major part of the aboveground biomass while the carbon sequestered through root biomass and soil carbon build up represented the belowground biomass.

Agroforestry practices can be adopted to enhance and strengthen the CS potential with the aim to meet the CC adaptation and mitigation. The structure and functional prospective of any AFs will influence the potentiality of CS in that particular system. Besides this, climatic, and edaphic factors of the region where agroforestry has been adopted will also affect the quantity of carbon to be sequestered in the system (Jose, 2009). Tree-based farming system can be encouraged to achieve more C storage and adopting such system would yield more biomass accumulation in vegetation and soils. IPCC (2000) has mentioned that agroforestry practices will provide a vast platform in terms of CS potential. One such of the estimation revealed that agroforestry have the potential to mitigate the global terrestrial carbon content by removing 1.1–2.2 Pg C in the coming next five decades (Solomon et al., 2007). The inclusion of woody perennials as being the essential component in any agroforestry practices helps in maximizing the biomass production in particular and at the same time trees have the potential to share the nutrients requirement by companion crops while improving the system carbon storage potential. In fact, one of the estimate provides by IPCC (2000) has estimated that agroforestry will provide a vast opportunity for CS prospective of nearly 586,000 Mg C/yr by converting unproductive croplands and grasslands of area expanding of about 630 M ha around the world by 2040.

AFs distributed across the world have shown varied structure and functional attributes which made the system difference in system components and their arrangement to suit the prescribed objectives. Apart from this the adoption and distribution of AFs are mainly influenced by the climatic, social, economic factors of the region. Interestingly, these factors have significant impact on variation in the biomass production and carbon allocation even under same agroforestry practices but located in different places. For example, Dixon et al., (1993) had claimed that the carbon storage potential in dry lowlands agrisilvicultural systems have higher potential as compared to humid tropical low in south American condition. However, under Asian condition, agrisilvicultural systems practiced in humid tropical areas have the highest carbon storage potential (Krankina and Dixon, 1994).

Similarly, Schroeder (1993) estimated the carbon storage potential of silvipastoral system practiced in North America but under different ecological conditions. It was found that silvipastoral system of humid tropical low area (151 Mg C ha^{-1}) had the highest carbon storage potential as compared to humid tropical low (39.5 Mg C ha^{-1}), humid tropical high (143.5 Mg C ha^{-1}) and dry low lands (132.5 Mg C ha^{-1}) that under North American condition. The observation and estimated found in different literature had revealed that most of the agroforestry related studies have been focused in the tropical regions of the world. However, in the last few decades' attention towards temperate agroforestry studies most particularly CS has been made.

People around the world have started to realize the importance of agroforestry as a large potential carbon sink in its tree biomass as well as in soil ecosystem. On an average, tropical AFs have the potential to store 2.1×10^9 Mg/C/year in its aboveground biomass, while in case of temperate region the reported value was to be around 1.9×10^9 Mg/C/year. CS potential for different AFs around the world has been estimated by many of the researchers. Among those reports, some of the prominent values are represented in Table 5.1. There is variation in the calculated and estimated values for CS in different AFs is mainly arise due to easy method of estimation or method is not standardized as "the extent of C sequestered in any AFs will depend on a number of site-specific biological, climatic, soil, and management factors" (Nair et al., 2009).

5.7 SOIL ORGANIC CARBON (SOC) SEQUESTRATION UNDER AGROFORESTRY SYSTEMS (AFS)

Soil ecosystem is having the largest contribution in carbon pool among the terrestrial ecosystem (Amundson, 2001; Jobbágy and Jackson, 2000). It is estimated that around 1550 Pg C in the form of SOC have been stored in the soil within one-meter depth (Lal, 2008). Owing to its large reservoir of carbon in the soil, it is providing a huge scope for sequestering C in the soil thus emerging an option for challenging the menace of global CC. It is true and scientifically proven that the continuous addition CO_2 in the atmosphere through various sources have been the primary cause of global warming. On this context, it is expected that the amount of CO_2 concentration in the atmosphere will be doubled by the end of 21st century with a temperature rise between 1.5 °C and 4.5 °C (Smith et al., 2003).

TABLE 5.1 Reported Carbon Sequestration Value (Above and Below Ground) for Different Agroforestry Systems Across the World

Agroforestry Systems	C Stock (Mg $ha^{-1}yr^{-1}$)	References
Fodder bank, Mali, W African Sahel	0.29	Takimoto et al., 2008
Live fence, Mali, W African Sahel	0.59	Takimoto et al., 2008
Parklands, Mali, W African Sahel	1.09	Takimoto et al., 2008
Silvipastoralism, Kurukshetra, India	1.37	Kaur et al., 2002
Silvipastoralism, Kerala, India	6.55	Kumar et al., 1998
Silvipasture, India	6.72	NRCAF, 2007
Silvipasture, W Oregan, USA	1.11	Sharrow and Ismail, 2004
Tree-based intercropping, Canada	0.83	Peichl et al., 2006
Poplar-based AFS, Punjab, India	9.24	Chauhan et al., 2015
Agrisilviculture, Chhattisgarh, India	1.26	Swamy and Puri, 2005
Cocoa agroforests, Papua New Guinea	5.85	Duguma et al., 2001
Cocoa agroforests, Turrialba, Costa Rica	11.08	Beer et al., 1990
Shaded coffee, SW Togo	6.31	Dossa et al., 2008
Rotational woodlots, Tanzania	3.70	Kimaro, 2009
Home and outfield gardens, Panama	4.29	Kirby and Potvin, 2007
Indonesian homegardens, Sumatra	8.00	Roshetko et al., 2002
Bamboo homegarden, Assam, India	1.32	Nath and Das, 2011
Midhills Agroforestry, Nepal	0.97	Pandit et al., 2014
Poplar based system, India	15.81	Arora et al., 2014

The present scenario under CC and intensive cultivation on limited land resources without considering its future land degradation soil becomes fully deprived of soil organic matter (SOM). This constraint caused decline in soil fertility as well as in crop productivity and ultimately increased in atmospheric CO_2 concentration (Lal, 2004; Banerjee et al., 2006). Thus, the availability of SOM in the soil by supplying the plant nutrients will help in increasing the plant productivity as well as maintaining the soil quality. So, it is important to understand the status of SOC content in the soil and in most of the times, the amount and form of soil carbon tends to change with the adoption of different agricultural management practices like tillage operation, irrigation, incorporation of manures, etc. and thus have an impact on SOC storage in soil. Since, different physical, chemical, and biological processes of soil taken place over period of time associated under different land use have also effect on the magnitude of the change in C storage in soil (IPCC, 2000).

Therefore, potentiality of different land use system on long-term C storage has an immense role while tapping the importance of particular land use system on carbon storage potential in comparison with other land use system. For example, Nagaraja et al., (2016) in Karnataka assessed the soil carbon stocks (0–50 cm) for natural and man-made land use system and found that the carbon (C) tree-based land use system, i.e., mixed forest have the maximum carbon stocks (89.20 t/ha). However, on the other side, man-made system intensively managed horticultural systems namely, grapes plantation (85.52 t ha^{-1}) and pomegranate plantation (78.78 t ha^{-1}) can also attained higher levels of C stock but the value is comparatively low as compared to natural system. From this study, it was clearly seen that less and minimum disturbance in the soil system has helped in improving the SOC and adoption of various soil and crop management practices influence on the SOC balance in the soil. Similarly, Bessah et al., (2016) also found that the influence of different land use pattern on soil organic carbon stocks (SOCS) and found that savanna woodland (30.02 t/ha) recorded the highest mean SOCS while the cashew plantation (22.01 t/ha) have attained the lowest SOCS in 0–30 cm depth. This study has revealed that different plant species have their own pattern of litter fall accumulation and thus significantly influence on the SOC built up process in the soil. Besides, this, incorporation of several management practices and the level of disturbance in soil system have also effect on the SOC accumulation and availability in the soil.

The role of trees in stabilizing or beneficial effect upon the soil properties in any AFs has been discussed by many authors. The improvement of SOM in AFs is primarily due to the continuous addition of litter on the agroforestry floor along with the degeneration of roots (fine and coarse) and at the same time it also help in improving the ambient environment for soil microbes (Araújo et al., 2012). On the basis of their rate and degree of decomposition, different soil organic pools are classified and make them complex constituent of soil (Weil and Magdoff, 2004). Among the different form of soil organic pools, liable form of SOC has significant influence on the soil productivity and is frequently changed with the application of different land management practices (Duval et al., 2013). Agroforestry as compared to annual or mono-cropping systems have the tendency to sequester more carbon accumulation in the system, reflecting an efficient nutrient cycling in AFs.

Lal and Follett (2009) have suggested that the soil organic CS potential under AFs can be increased by the adoption of following measures:

i. Conservation effective measures that reduce losses of nutrients and water;
ii. Increase biomass production; and
iii. Protect SOC against losses through enhancing biological, chemical, and physical stabilization mechanisms.

Thereby, help in decreasing the cultivation intensity which results in an increase in SOC levels in AFs (Nair et al., 2010). Different management practices like application of fertilizers, irrigation supply, application of pesticides, herbicides, etc. could also affect the SOC sequestration potential under AFs by influencing the soil aggregates stability. However, the adoptions of these management practices are site-specific that would affect the overall performance of SOC in AFs (Lorentz and Lal, 2014).

The assessment of biomass production potential in any system will determine the system output and carbon allocation rate in different components of the system. In general, AFs as compared to monocropping produced higher biomass (above and belowground) due to inclusion of woody perennials in agroforestry practices and thus able to sequester large amount of carbon in the system. Generally, the above-ground parts such as bole, leaves, branches, petioles, fruits have contributed more than 75% of the total biomass production, thereby indicating the majority of carbon balance in the system is allocated in above-ground biomass (Ravindranath and Ostwald, 2008). On an average the bole biomass contributes the maximum portion of above-ground biomass production. The production of roots including fine and coarse roots in the AFs represent the below-ground biomass production. It is stated that higher below-ground root biomass activities in AFs helps in improving the SOC production, which ultimately increased in the CS potential of soil (Benbi et al., 2012). It is highly desirable and expected that species having the deep and wide root system would able to contribute more carbon accumulation in the soil (Kell, 2012; Lorenz and Lal, 2010). So, for producing higher below-ground biomass in AFs, the selection of desirable deep-rooted and extensive as well as locally adaptable tree species will be highly recommended. Besides this, application of several management practices that will encourage the production of more belowground biomass like root pruning, trenching, fertilizer doses will also be benefitted in producing more CS (Nair et al., 2009). The potentiality of soil carbon stock varied greatly across the ecological zone as well as different AFs (Tables 5.2 and 5.3).

TABLE 5.2 Some of the Reported Soil Carbon Stock Under Different Agroforestry Systems

Country	Agroforestry Systems	Soil Carbon Stock (Mg ha^{-1})	Soil Depth (cm)	References
India	Agrisilviculture (Eucalyptus + paddy-wheat)	57.48	45	Rahangdale and Pathak, 2016
India	Agrisilviculture (Eucalyptus + pigeopea)	59.92	45	Rahangdale and Pathak, 2016
India	Agrihortisilviculture	56.70	30	Singh et al., 2018
Uganda	Arabica coffee agroforestry system	54.54	30	Tumwebaze and Byakagaba, 2016
Uganda	Robusta coffee agroforestry system	57.56	30	Tumwebaze and Byakagaba, 2016
Ethoipia	Enset-Coffee indigenous agroforestry system	177.8±44.5	60	Negash and Starr, 2015
Ethoipia	Fruit-Coffee indigenous agroforestry system	178.8±50.5	60	Negash and Starr, 2015
France	Silvorarable (without grazing)	46.7±1.0	30	Cardinael et al., 2017
France	Silvopastoral (grazing)	110.2±6.1	30	Cardinael et al., 2017
Canada	Hybrid poplar shelterbelt	143.4	50	Dhillion and Rees, 2016
India	Homegardens	119.3	100	Saha et al., 2009
Mexico	Indigenous agroforestry system (high tropical zone)	120.7	30	Soto-Pinto et al., 2010
Ethoipia	*Acacia etabica* woodland	43.00	60	Lemenih and Fisseha, 2003

5.8 IMPROVEMENT OF SOCS IN AGROFORESTRY

Soil organic carbon stock (SOCS) form a crucial part of global carbon budget. Unfortunately, due to mismanagement of agricultural practices leading to the loss of SOC in cultivated land, ultimately resulting in the increased of atmospheric carbon. It is more pronounced that SOC tends to changes with change in land-use systems. In other words, the quantity and quality of SOC present in a particular land management system tend to change when the system is changed to another form of management system. Therefore, the different management practices associated with different land-use systems has direct influence on the SOC status of the soil. The inclusion of trees in the agricultural system would like to increase in SOC stocks. Mulching, use of cover crops, no-tillage operations, etc. are some of the approaches made to increase SOC in any agroecosystem. It is generally believed that agroforestry land use system would likely to increase the SOC stocks and managing such systems that have the potential to increase the SOC can be simply call as a win-win strategy (Lal, 2004).

TABLE 5.3 Reported Soil Organic Carbon Sequestration Rates (Mg C ha^{-1} y^{-1}) in Different Agroforestry Systems

Country	Agroforestry Systems	Carbon Sequestration Rate (Mg ha^{-1} yr^{-1})	Age	Soil Depth (cm)	References
Canada	Alley cropping	0.69	13	60	Oelbermann et al., 2006
Costa Rica	Alley cropping	4.13	19	60	Oelbermann et al., 2006
India	Agrisilviculture (Poplar based)	2.63	3	30	Gupta et al., 2009
India	Agrisilviculture (Poplar based)	1.95	6	30	Gupta et al., 2009
India	Agrisilviculture (Poplar based)	1.62	7	30	Chauhan et al., 2010
Canada	Shelterbelts (different six species)	0.7	Varying	50	Dhillion and Rees, 2016

The process of leaf litter fall and accumulation in the agroforestry floor along with their decomposition process has accounted significant impact on soil improvement in AFs. However, the amount of nutrient released from the litter depends on the various factors like climate, abundance of soil microbes, species characteristics, land use pattern and their management activities, etc. (Yadav and Bisht, 2014). The understanding of soil nutrients dynamics in agroforestry is very crucial as it influences the amount of biomass production and availability of soil nutrients in soil and overall controlling the nutrient cycling in situ (Rawat and Singh, 1988).

Agroforestry trees species are capable of taking up the nutrients (nutrient pumping) and water from the layers that are usually not utilized by the herbaceous crops. Soil nutrient enrichment in agroforestry is mainly due to the consequence of continuous addition of litter, efficient nutrient cycling, improving the soil aggregates, atmospheric nitrogen fixation by nitrogen-fixing trees (Nair, 1993; Palm, 1995). There would be changes that may occur in soil nutrient profile and microbial activity as a result of leaf litter decomposition by tree species and interaction with inter crop in particular AFs and soil edapho-climatic conditions. Nutrient pumping capability of trees in agroforestry helps in improving the process of nutrient cycling, minimizing the leaching of nutrients, increasing the efficiency of food synthesis process (Veldkamp et al., 1999). AFs as compared to other terrestrial ecosystem have more CS potential by linking up the above-below ground components (Pandey, 2002). In general, the impact of soil disturbances and erosion in AFs has been considered as low as compared to intensive agricultural system, thereby improving the carbon stability in the soil. Conversion of forest area to AFs helps to restore the SOCS by 35% in the tropics, indicating AFs has improving the soil carbon stabilization in clear felled forest area (Palm et al., 1999). In the context of tropical ecosystem, conversion of degraded or waste-land area into AFs has resulted in the improvement of SOC which ultimately helps in increasing the SOC sequestration rate, soil aggregate improvement, increase in clay content and building up the soil litter layer (Lal, 2005).

According to Lorentz and Lal (2014) AF's carbon inputs can be maximized under agroforestry with the adoption of following management:

a. Retention of pruning materials for mulching and green manuring purposes.
b. Addition of cattle manures by allowing livestock to graze without hampering the system productivity.
c. Maintenance of agroforestry litter floor during the fallow periods.
d. Integration and management of trees litter along with animal production systems
e. Agricultural crops litter inputs and management.

Tumwebaze and Byakagaba (2016) at Uganda compared the SOC under coffee-based AFs and coffee monoculture. It was found that more SOC in coffee AFs (49.64–71.17 t C ha^{-1}) than coffee mono-crops (50.987–51.780 t C ha^{-1}). Cardinael et al., (2017) at France also claimed that AFs have more SOC stocks than control (tree less plot). In this study, they found that silvoarable systems have the capacity to store 0.24 Mg C ha^{-1} yr^{-1} in soil at

30 cm depth. It is proved that AFs have the potential to sequester more soil carbon than control (tree less plot).

5.9 CONCLUSION

AFs in the context of current CCs issues and pushing towards adaptation and mitigate options can be considered a viable one. The diversifying nature of agroforestry helps to meet the multifarious demands of the society. Additionally, the capacity to restore or rebuild the soil fertility and conservation in AFs has been elaborated and acceptable to land users. Trees in AFs help to improve the SOC stocks, resulting in maintaining the soil fertility and productivity. Since, soil contains the largest carbon pool among the terrestrial ecosystem, offering a huge potential for CS across the world. The continuous addition of litter, decomposition, and root activities in agroforestry making a room to improve SOC stocks in this system. The studies in relation to SOC sequestration potential of AFs show variable results thus indicating more studies in this aspect is very much needed. It is also more pronounced that land-use changes caused significant impact on SOC stocks and it is believed the SOC stocks increases when the treeless land-use system has shifted to tree-based system most preferably AFs. Considering the rapid expansion of agroforestry land-use system and more recent studies indicating AFs could provide a huge scope for improving the SOC sequestration while challenging the CC phenomenon.

KEYWORDS

- **agroforestry**
- **carbon sequestration**
- **climate change**
- **land-use system**
- **soil organic carbon**

REFERENCES

Amundson, R., (2001). The carbon budget in soils. *Annual Review of Earth and Planetary Sciences*, *29*, 535–562.

Araújo, A. S. F., Leite, L. F. C., Iwata, B. F., Lyra, Jr. M. A., Xavier, G. R., & Figueiredo, M. V. B. (2012). Microbiological process in agro-forestry system: A review. *Agronomy for Sustainable Development, 32*, 215–226. doi: 10.1007/s13593–011–0026–0.

Arora, G., Chaturvedi, S., Kaushal, R., Nain, A., Tewari, S., & Alam, N. M., (2014). Growth, biomass, carbon stocks, and sequestration in age series of *Populus deltoides* plantations in Tarai region of central Himalaya. *Turkish Journal of Agriculture and Forestry, 38*, 550–560.

Banerjee, B., Aggarwal, P. K., Pathak, H., Singh, A. K., & Chaudhary, A., (2006). Dynamics of organic carbon and microbial biomass in alluvial soil with tillage and amendments in rice-wheat systems. *Environmental Monitoring and Assessment, 119*, 173–189.

Beer, J., Bonnemann, A., Chavez, W., Fassbender, H. W., Imbach, A. C., & Martel, I., (1990). Modeling agroforestry systems of cacao (*Theobroma cacao*) with laurel (Cordia alliodora) or poro (*Erythrina poeppigiana*) in Costa Rica. *Agroforestry Systems, 12*, 229–249.

Benbi, D. K., Brar, K., Toor, A. S., Singh, P., & Singh, H., (2012). Soil carbon pools under poplar based agroforestry, rice-wheat, and maize-wheat cropping systems in semi-arid India. *Nutrient Cycling in Agroecosystems, 92*, 107–118.

Bessah, E., Bala, A., Agodzo, S. K., & Okhimamhe, A. A., (2016). Dynamics of soil organic carbon stocks in the Guinea savanna and transition agro-ecology under different land-use systems in Ghana. *Cogent Geoscience, 2*, 1140319.

Brown, S., (2002). Measuring carbon in forests: Current status and future challenges. *Environmental Pollution, 116*, 363–372.

Cardinael, R., Chevallier, T., Cambou, A., Beral, C., Barthès, B. G., Dupraz, C., Durand, C., Kouakoua, E., & Chenu, C., (2017). Increased soil organic carbon stocks under agroforestry: A survey of six different sites in France. *Agriculture, Ecosystems and Environment, 236*, 243-255.

Chauhan, S. K., Sharma, R., Singh, B., & Sharma, S. C., (2015). Biomass production, carbon sequestration and economics of on farm poplar plantations in Punjab, India. *Journal of Applied and Natural Sci*ence, *7*(1), 452–458.

Chauhan, S. K., Sharma, S. C., Chauhan, R., Gupta. R., & Ritu, (2010). Accounting poplar and wheat productivity for carbon sequestration agri-silvicultural system. *Indian Forester, 136*, 1174–1182.

Conklin, H. C., (1957). *Hanunoo Agriculture*. Rome: FAO.

Department of Agricultural Research and Education (DARE), (2014). January-March Report. http://dare.nic.in/node/166 (Accessed on 9 October 2019).

Dhillon, G. S., & Rees, K. C. J. V., (2016). Soil organic carbon sequestration by shelter-belt agroforestry systems in Saskatchewan. *Can. J. Soil Sci., 97*, 394–409. doi: 10.1139/cjss-2016–0094.

Dhyani, S. K., Handa, A. K., & Uma, (2014). Area under agroforestry in India: An assessment for present status and future perspective. *Indian Journal of Agroforestry, 15*(1), 1–11.

Dixon, R. K., Andrasko, K. J., Sussman, F. A., Lavinson, M. A., Trexler, M. C., & Vinson, T. S., (1993). Tropical forests: Their past, present and potential future role in the terrestrial carbon budget. *Water, Air, and Soil Pollution, 70*, 71–94.

Dossa, E. L., Fernandes, E. C. M., Reid, W. S., & Ezui, K., (2008). Above and below ground biomass, nutrient and carbon stocks contrasting an open-grown and a shaded coffee plantation. *Agroforestry Systems, 72*, 103–115.

Duguma, B., Gockowski, J., & Bakala, J., (2001). Smallholder cacao (*Theobroma cacao* Linn.) cultivation in agroforestry systems of West and Central Africa: Challenges and opportunities. *Agroforestry Systems, 51*, 177–188.

Duval, M. E., Galantini, J. A., Iglesias, J. O., Canelo, S., Martinez, J. M., & Wall, L., (2013). Analysis of organic fractions as indicators of soil quality under natural and cultivated systems. *Soil and Tillage Research, 131*, 1119.

Forde, D. C., (1937). Land and labor in a cross river village. *Geographical Journal* (Vol. XC, No.1).

FSI, (2013). India State of Forest Report 2013, Forest Survey of India, (Ministry of Environment and Forests), Dehradun, India.

Gao, J., Barbieri, C., & Valdivia, C., (2014). A socio-demographic examination of the perceived benefits of agroforestry. *Agroforestry Systems, 88*, 301–309.

Gold, M., & Garrett, H., (2009). Agroforestry nomenclature, concepts and practices. In: Garrett, H. E., (ed.), *North American Agroforestry: An Integrated Science and Practice* (pp. 45–56). America Society of Agronomy, Madison.

Gupta, N., Kukal, S. S., Bawa, S. S., & Dhaliwal, G. S., (2009). Soil organic carbon and aggregation under poplar based agroforestry system in relation to tree age and soil type. *Agroforestry Systems, 7*, 27–35.

IAASTD, (2008). *Agriculture at a Crossroads: Global Report*. International Assessment of Agricultural Knowledge, Science, and Technology for Development. Washington DC: Island Press.

Ingram, J. S. I., & Fernandes, E. C. M., (2001). Managing carbon sequestration in soils: Concepts and terminology. *Agriculture Ecosystem and Environment, 87*, 111–117.

IPCC (Intergovernmental Panel on Climate Change), (2000). *Land Use, Land-Use Change, and Forestry: A Special Report of the Intergovernmental Panel on Climate Change*. Cambridge University Press: Cambridge, UK.

IPCC (Intergovernmental Panel on Climate Change), (2013). *Climate Change: The Physical Science Basis. Contribution of Working Group I to the Fifth Assessment Report of the Intergovernmental Panel on Climate Change*. Cambridge University Press: Cambridge, UK.

Jhariya, M. K., Banerjee, A., Yadav, D. K., & Raj, A., (2018). Leguminous trees an innovative tool for soil sustainability. In: Meena, R. S., Das, A., Yadav, G. S., & Lal, R., (eds.), *Legumes for Soil Health and Sustainable Management* (pp. 315–345). Springer, ISBN 978–981–13–0253–4 (eBook), ISBN: 978–981–13–0252–7 (Hardcover). https://doi.org/10.1007/978–981–13–0253–4_10 (Accessed on 9 October 2019).

Jhariya, M. K., Bargali, S. S., & Raj, A., (2015). Possibilities and perspectives of agroforestry in Chhattisgarh. In: Miodrag, Z., (ed.), *Precious Forests-Precious Earth* (pp. 237–257). ISBN: 978–953–51–2175–6, In-Tech, doi: 10.5772/60841.

Jose, S., (2009). Agroforestry for ecosystem services and environmental benefits: An overview. *Agroforestry Systems, 76*, 1–10.

Jose, S., Gillespie, A. R., & Pallardy, S. G., (2004). Interspecific interactions in temperate agroforestry. *Agroforestry System, 61*, 237–255.

Kaur, B., Gupta, S. R., & Singh, G., (2002). Carbon storage and nitrogen cycling in silvopastoral systems on a sodic soil in northwestern India. *Agroforestry Systems, 54*, 21–29.

Kell, D. B., (2012). Large-scale sequestration of atmospheric carbon via plant roots in natural and agricultural ecosystems: Why and how. *Philosophical Transactions of the Royal Society of London B, 367*, 1589–1597. doi: 10.1098/rstb.2011.0244.

Kimaro, A. A., (2009). Sequential agroforestry systems for improving fuelwood supply and crop yield in semi-arid Tanzania. *PhD Thesis* (p. 124). University of Toronto, Toronto.

King, K. F. S., (1968). *Agri-Silviculture*. Bulletin No. 1, Department of Forestry, University of Ibadan, Nigeria.

Kirby, K. R., & Potvin, C., (2007). Variation in carbon storage among tree species: Implications for the management of a small-scale carbon sink project. *Forest Ecology and Management, 246*, 208–221.
Krankina, O. N., & Dixon, R. K., (1994). Forest management options to conserve and sequester terrestrial carbon in the Russian Federation. *World Resource Review, 6*, 88–101.
Kumar, B. M., Kumar, S. S., & Fisher, R. F., (1998). Intercropping teak with *Leucaena* increases tree growth and modifies soil characteristics. *Agroforestry Systems, 42*, 81–89.
Lal, R., & Follett, R. F., (2009). Soils and climate change. In: Lal, R., & Follett, R. F., (eds.), *Soil Carbon Sequestration and the Greenhouse Effect* (2nd edn., pp. xxi–xxviii). SSSA Special Publication 57. Madison, WI.
Lal, R., (2004). Soil carbon sequestration impacts on global climate change and food security. *Science, 304*, 1623–1627.
Lal, R., (2005). Soil carbon sequestration in natural and managed tropical forest ecosystems. *Journal of Sustainable Forestry, 21*, 1–30. doi: 10.1300/J091v21n01_01.
Lal, R., (2008). Carbon sequestration. *Philosophical Transactions of the Royal Society of London B, 363*, 815–830.
Lemenih, M., & Itanna, F., (2003). Soil carbon stocks and turnovers in various vegetation types and arable lands along an elevation gradient in southern Ethiopia. *Geoderma, 123*(1/2), 177–188.
Lorenz, K., & Lal, R., (2010). *Carbon Sequestration in Forest Ecosystems*. Springer, Dordrecht.
Lorenz, K., & Lal, R., (2014). Soil organic carbon sequestration in agroforestry systems: A review. *Agronomy for Sustainable Development, 34*, 443–454.
Lu, D., Chen, Q., Wang, G., Liu, L., Li, G., & Moran, E., (2014). A survey of remote sensing-based aboveground biomass estimation methods in forest ecosystems. *International Journal of Digital Earth, 9*(1), 63–105.
Montagini, F., & Nair, P. K. R., (2004). Carbon sequestration: An underexploited environmental benefit of agroforestry systems. *Agroforestry Systems, 61/62*, 281–295.
Nagaraja, M. S., Bhardwaj, A. K., Reddy, G. V. P., Parama, V. R. R., & Kaphaliya, B., (2016). Soil carbon stocks in natural and man-made agri-hortisilvipastural land use systems in dry zones of Southern India. *Journal of Soil and Water Conservation, 15*(3), 258–264.
Nair, P. K. R., & Garrity, D. P., (2012). Agroforestry-the future of global land use. *Advances in Agroforestry, 9*, 541. ISBN 978–94–007–4675–6.
Nair, P. K. R., & Nair, V. D., (2003). Carbon storage in North American agroforestry systems. In: Kimble, J., Heath, L. S., Birdsey, R. A., & Lal, R., (eds.), *The Potential of U. S. Forest Soils to Sequester Carbon and Mitigate the Greenhouse Effect* (pp. 333–346). CRC Press, Boca Raton, USA.
Nair, P. K. R., (1979). *Intensive Multiple Cropping with Coconuts in India*. Verlag Paul Parey, Berlin/Hamburg, Germany.
Nair, P. K. R., (1993). *An Introduction to Agroforestry* (p. 489). Kluwer, Dordrecht, The Netherlands.
Nair, P. K. R., (2007). The coming of age of agroforestry. *Journal of the Science of Food and Agriculture, 87*, 1613–1619.
Nair, P. K. R., (2012). Carbon sequestration studies in agroforestry systems: A reality-check. *Agroforestry Systems, 86*, 243–253.
Nair, P. K. R., Kumar, B. M., & Nair, V. D., (2009). Agroforestry as a strategy for carbon sequestration. *Journal of Plant Nutrition and Soil Science, 172*, 10–23.
Nair, P. K. R., Nair, V. D., Kumar, B. M., & Showalter, J. M., (2010). Carbon sequestration in agroforestry systems. *Advances in Agronomy, 108*, 237–307. doi: 10.1016/S0065–2113(10)08005–3.

Nath, A. J., & Das, A. K., (2011). Carbon storage and sequestration in bamboo-based smallholder homegardens of Barak Valley, Assam. *Current Science*, *100*(2), 229–233.

Negash, M., & Starr, M., (2015). Biomass and soil carbon stocks of indigenous agroforestry systems on the south-eastern Rift Valley escarpment, Ethiopia. *Plant Soil*, *393*, 95–107.

Newaj, R., & Dhyani, S. K., (2008). Agroforestry systems for carbon sequestration: Present status and scope. *Indian Journal of Agroforestry*, *10*(1), 1–9.

NRCAF, (2007). *Vision 2025*. National Research Centre for Agroforestry, Jhansi.

Oelbermann, M., Voroney, R. P., Gordon, A. M., Kass, D. C. L., Schlönvoigt, A. M., & Thevathasan, N. V., (2006). Soil carbon dynamics and residue stabilization in a Costa Rican and southern Canadian alley cropping system. *Agroforestry System*, *68*, 27–36. doi: 10.1007/s10457–005–5963–7.

Palm, C. A., (1995). Contribution of agroforestry trees to nutrient requirements of intercropped plants. *Agroforestry Systems*, *30*, 105–124.

Palm, C. A., Nziguheba, G., Gachengo, C., Gacheru, E., & Rao, M. R., (1999). Organic materials a source of phosphorus. *Agroforestry Forum*, *9*, 30–33.

Pandey, D. N., (2002). Carbon sequestration in agroforestry system. *Climate Policy*, *2*, 367–377.

Pandit, B. H., Shrestha, K. K., & Bhattarai, S., (2014). Sustainable local livelihoods through enhancing agroforestry systems in Nepal. *Journal of Forest and Livelihood*, *12*, 47–63.

Peichl, M., Thevathasan, N. V., Gordon, A. M., Huss, J., & Abohassan, R. A., (2006). Carbon sequestration potentials in temperate tree-based intercropping systems, southern Ontario, Canada. *Agroforestry System*, *66*, 243–257. doi: 10.1007/s10457–005–0361–8.

Pinto, L., Anzueto, M., Mendoza, J., Ferrer, G. J., & De Jong, B., (2010). Carbon sequestration through agroforestry in indigenous communities of Chiapas, Mexico. Soto-. *Agroforestry Systems*, *78*, 39–51.

Rahangdale, C. P., & Pathak, N. N., (2016). Dynamics of organic carbon stock in soil under agroforestry system. *Trends in Biosciences*, *9*(11), 662–667.

Ravindranath, N. H., & Ostwald, M., (2008). *Carbon Inventory Methods: Handbook for Greenhouse Gas Inventory, Carbon Mitigation and Round Wood Production Projects (Advance in Global Change Research-29)*. Springer science, Dordrecht, ISBN 978–1–4020–6546–0.

Rawat, Y. S., & Singh, J. S., (1988). Structure and function of oak forests in central Himalaya. I. Dry matter dynamics. *Annals of Botany*, *62*(4), 397–411.

Roshetko, M., Delaney, M., Hairiah, K., & Purnomosidhi, P., (2002). Carbon stocks in Indonesian homegarden systems: Can smallholder systems be targeted for increased carbon storage? *American Journal of Alternative Agriculture*, *17*, 125–137.

Saha, S. K., Nair, P. K. R., Nair, D., & Kumar, B. M., (2009). Soil carbon stock in relation to plant diversity of homegardens in Kerala, India. *Agroforestry Systems*, *76*, 53–65.

Schroeder, P., (1993). Agroforestry systems: Integrated land use to store and conserve carbon. *Climate Research*, *3*, 53–60.

Sharrow, S. H., & Ismail, S., (2004). Carbon and nitrogen storage in agroforests, tree plantations, and pastures in western Oregon, USA. *Agroforestry Systems*, *60*, 123–130.

Singh, M., Gupta, B., & Das, S. K., (2018). Soil organic carbon density under different agroforestry systems along an elevation gradient in north-western Himalaya. *Range Mgmt. and Agroforestry*, *39*(1), 8–13.

Singh, N. R., & Jhariya, M. K., (2016). Agroforestry and agrihorticulture for higher income and resource conservation. In: Narain, S., & Rawat, S. K., (ed.), *Innovative Technology for Sustainable Agriculture Development* (pp. 125–145). ISBN: 978–81–7622–375–1. Biotech Books, New Delhi, India.

Singh, N. R., Arunachalam, A., Bhusara, J. B., Dobriyal, M. J., & Gunaga, R. P., (2017). Diversification of agroforestry systems in Navsari district of South Gujarat. *Indian Journal of Hill Farming*, *30*(1), 70–72.

Smith, K. A., Ball, T., Conen, F., Dobbie, K. E., Massheder, J., & Rey, A., (2003). Exchange of greenhouse gases between soil and atmosphere: Interactions of soil physical and biological processes. *European Journal of Soil Science*, *54*, 779–791.

Solomon, S., Qin, D., Manning, M., Chen, Z., Marquis, M., Averyt, K. B., Tignor, M., & Miller, H. L., (2007). *Climate Change 2007: The Physical Science Basis*. Contribution of working group I to the fourth assessment report of the intergovernmental panel on climate change. Cambridge, UK and New York, NY, USA: Cambridge University Press (IPCC Fourth Assessment Report (AR4).

Swaminathan, M. S., (1987). The promise of agroforestry for ecological and nutritional security. In: Steppler, H. A., & Nair, P. K. R., (eds.), *Agroforestry: A Decade of Development* (pp. 25–42). International Council for Research in Agroforestry, Nairobi.

Swamy, S. L., & Puri, S., (2005). Biomass production and C-sequestration of *Gmelina arborea* in plantation and agroforestry system in India. *Agroforestry System*, *64*, 181–195.

Takimoto, A., Nair, P. K. R., & Alavalapati, J. R. R., (2008). Socioeconomic potential of carbon sequestration through agroforestry in the West African Sahel. *Mitigation and Adaptation Strategies for Global Change*, *13*, 745–761. doi: 10.1007/s11027–007–9140–3.

Tilman, D., Cassman, K. G., & Matson, P. A., (2002). Agricultural sustainability and intensive production practices. *Nature*, *418*, 671–678.

Tumwebaze, S. B., & Byakagaba, P., (2016). Soil organic carbon stocks under coffee agroforestry systems and coffee monoculture in Uganda. *Agriculture, Ecosystems and Environment*, *216*, 188–193.

Veldkamp, E., Davidson, E. A., Erickson, H. E., Keller, M., & Weitz, A. M., (1999). Soil nitrogen and nitrogen oxide emissions along a pasture chronosequence in the humid tropics of Costa Rica. *Soil Biology and Biochemistry*, *31*, 387–394.

Weil, R. R., & Magdoff, F., (2004). Significance of soil organic matter to soil quality and health. In: Magdoff, F., & Weil, R. R., (eds.), *Soil Organic Matter in Sustainable Agriculture* (p. 143). CRC Press, Boca Raton, Florida.

Yadav, R. P., & Bisht, J. K., (2014). Litter fall and potential nutrient returns from pecan nut (*Carya illinoinensis*) in agroforestry system in Indian Himalaya. *International Journal of Herbal Medicine, 2*, 51–52.

Zomer, R. J., Bossio, D. A., Trabucco, A., Yuanjie, L., Gupta, D. C., & Singh, V. P., (2007). *"Trees and Water: Smallholder Agroforestry on Irrigated Lands in Northern India."* International Water Management Institute, Colombo, Sri Lanka (Series: IWMI Research Reports, no. 122).

CHAPTER 6

Climate Change, Soil Health, and Food Security: A Critical Nexus

ABHISHEK RAJ,[1] MANOJ KUMAR JHARIYA,[2] DHIRAJ KUMAR YADAV,[2] ARNAB BANERJEE,[3] and PRABHAT RANJAN ORAON[4]

[1]*Department of Forestry, College of Agriculture, Indira Gandhi Krishi Vishwavidyalaya, Raipur–492012, Chhattisgarh, India, Mobile: +00-91-8269718066, E-mail: ranger0392@gmail.com*

[2]*Assistant Professor, University Teaching Department, Department of Farm Forestry, Sarguja Vishwavidyalaya, Ambikapur–497001, Chhattisgarh, India, Mobile: +00-91-9407004814, E-mail: manu9589@gmail.com (M. K. Jhariya); Mobile: +00-91-9926615061, E-mail: dheeraj_forestry@yahoo.com (D. K. Yadav)*

[3]*Assistant Professor, University Teaching Department, Department of Environmental Science, Sarguja Vishwavidyalaya, Ambikapur–497001, Chhattisgarh, India, Mobile: +00-91-9926470656, E-mail: arnabenvsc@yahoo.co.in*

[4]*Junior Scientist-cum-Assistant, Professor, Department of Silviculture and Agroforestry, Faculty of Forestry, Birsa Agricultural University, Ranchi–834006, Jharkhand, India, Mobile: +00-91-9431326222, E-mail: prabhat.ranjan.oraon@gmail.com*

ABSTRACT

Climate change (CC), which is a very challenging task today, affects the availability of food, water, and essential nutrients. Human and natural direct and indirect activities disturb our environment and ecosystem through changing climate due to an emission of several leading greenhouse gases (GHGs) in the atmosphere. Soil is the largest reservoir of natural organic carbon, essential nutrients and a harbor of various above ground and below-ground organism which directly and indirectly affects the food production system and human health under the era of changing the climate. Healthy soils have abundant organic carbon (organic matter) which is decomposed by the microorganism and release essential nutrients to plants and tree species on which all organism like animals and human depends for food. Therefore, a strong relationship exists among soil-food-health under the situation of changing the climate. Thus, healthy soil satisfies the human's food and nutritional needs along with CC mitigation. For healthy and quality soils, there is a need for better scientific management and farming practices which can solve the problem of extreme weather and food and nutrient insecurity under the changing climatic situations. Practices like agroforestry, conservation agriculture, non-till farming, mixed farming, integrated farming practices, overall we can say climate-smart agriculture and climate-smart soil management practices can minimize the problem of global warming and reduce the emission of GHGs to a considerable extent. Therefore, a great nexus among soil security, food security, and climate and health security, and these overall securities can be achieved through good management and scientific farming practices along with an implementation of existing several government policies, efficient non-government organizations works, and private agencies and organization.

6.1 INTRODUCTION

Increasing greenhouse gases (GHGs) in our environment is the global concern today, and the main cause of earth warming leads to climate change (CC). Carbon dioxide (CO_2) is the major potent gas that results in higher temperature and performs CO_2 fertilization effects on the plants and another biodiversity. Human plays a major role in disturbing the soil health through inducing the CC phenomenon through various anthropogenic activities and development. Therefore, the impact of human-induced CC on soil and other biodiversity are depicted in Figure 6.1. Industrial development, deforestation, faulty land use pattern/conversion, high application of inorganic/chemical fertilizers on cultivated land, rice cultivation etc. are the prime cause behind excessive emission of GHGs to the atmosphere resulting global warming which in turn affects all existing life on the earth and various agricultural based farming systems (Arora et al., 2011).

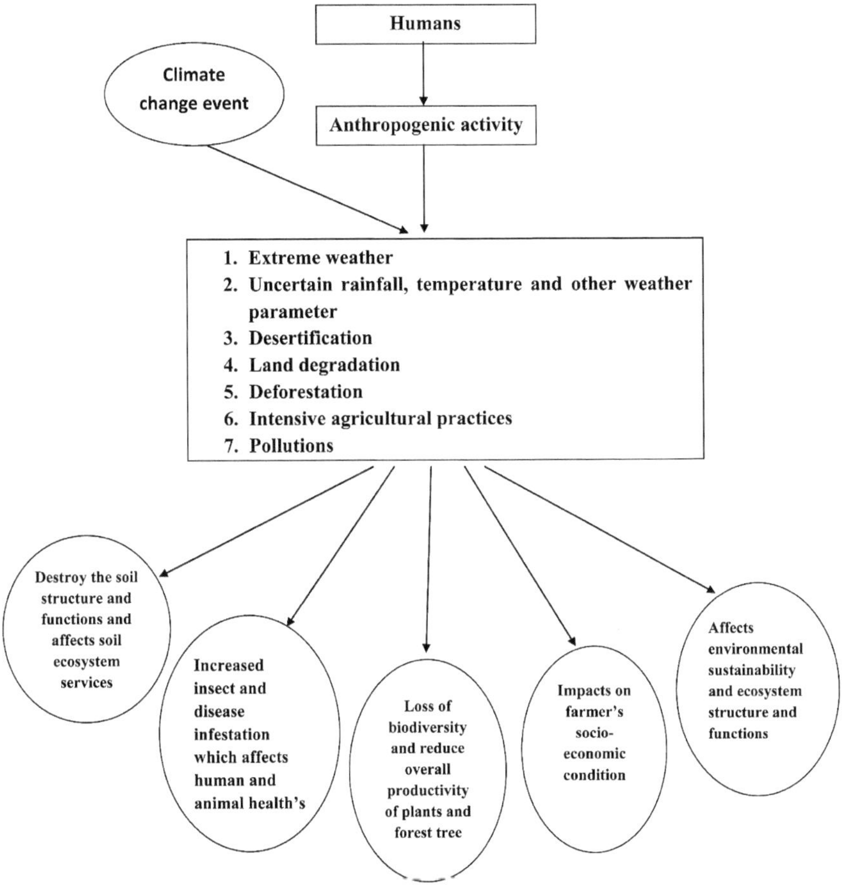

FIGURE 6.1 Impact of anthropogenic induced climate change on soil health and biodiversity.

Moreover, changing climate also affects the health system of livestock's and pasture palatability and quality (Thornton et al., 2009). Very few degree changes in temperature along with precipitation and humidity affects the water availability to plants, nutrient availability and its mobility, an occurrence of soil-inhabiting organism and their metabolic processes, emergence of infectious disease etc. which is major concern in the context of overall food productions and nutritional availability for cattle's and the burgeoning populations. Change in soil carbon stocks affects the presence of decomposing microbial populations and availability of other associated elements in the soil which in turn affects crop productivity. Soil is an invisible hero in the context of various ecosystem services and storage and sequestration of carbon is one of the greatest services provided by soil which not only minimize the excessive carbon in the atmosphere through absorption and fixation in both vegetation and soil but also maintain the soil carbon availability to plants and other organisms. In addition, adoption of various scientific and effective farming practices can enhance the storage capacity of carbon in both plant and tress component along with soil carbon pools. The adoption of the climate-smart farming system, i.e., climate-smart agriculture has the greatest potential in maintaining soil fertility and productivity of cropping systems with less emission of GHGs in the environment. As per one estimates, soil contributes higher carbon stocks percentage (81.1) in compared to vegetations (18.8%) out of total carbon stocks in all biomes. Similarly, different biomes have different carbon sink value which is shown in Figure 6.2, where boreal forest rank first in carbon stock value followed by tropical forest and least value is shown in cropland and tundra biomes (Watson et al., 2000).

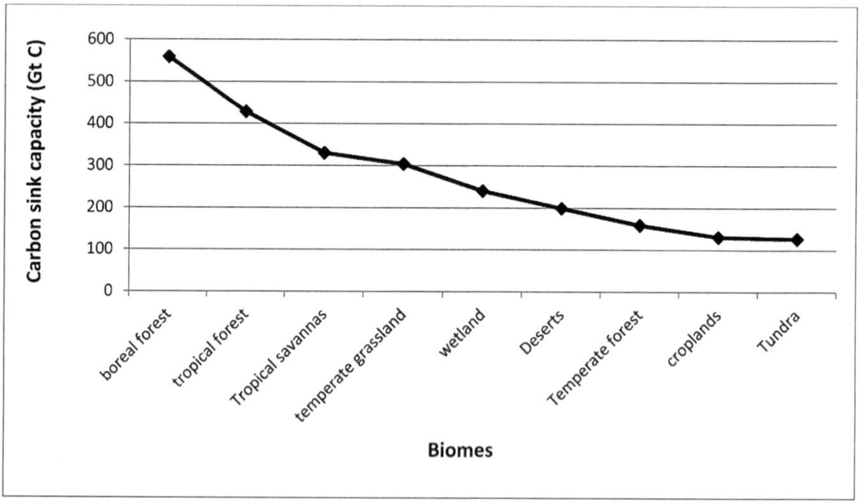

FIGURE 6.2 Carbon sinks value in different biomes.
(Adapted from Watson et al., 2000).

Likewise, agroforestry boosts the multiple productions along with the function of carbon storage and sequestration to mitigate CC issue (Jhariya et al., 2015; 2018; Singh and Jhariya, 2016; Parihaar, 2016). Thus, the undeniable advantage of agroforestry in term of soil, food, and ecological security is shown in Figures 6.3 and 6.4. Similarly, forest plays an incredible role in carbon sink value and as per one estimate forest stores higher carbon value (638 Giga tons) than the carbon which is present in our entire atmosphere (FAO, 2007; Rawat, 2010).

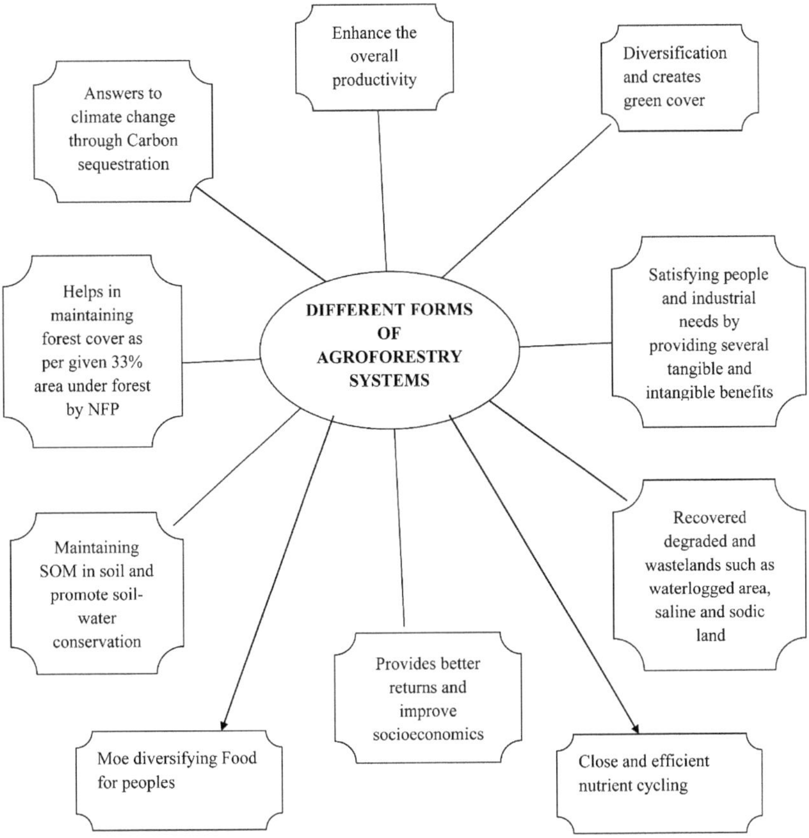

FIGURE 6.3 Undeniable advantages of agroforestry systems.

This chapter reviews the CC scenario, its impacts on food and nutrient availability by disturbing soil health and which in turn affects biodiversity and climate mitigation strategies through the climate-smart farming system. Similarly, climate-smart soil fertility management is another classical tool for achieving the food and nutritional demand, maintaining soil ecosystem services along with ecosystem and environmental health.

6.2 CLIMATE CHANGE (CC): AN UNDENIABLE TRUTH

CC is a major threat and undeniable issue today. Anthropogenic factors for economic development, industrialization, deforestation (for industrial timbers and non timber produce), high use of chemical fertilizers on cultivated field (for higher yield), etc. cause increasing concentration of CO_2 (which is the prime cause of CC) in the atmosphere and results changing climate that we are observed today. The extreme weather phenomenon leads to change in the pattern of rainfall and temperature, humidity, and other parameters which not only affects the whole plant, animal, and human biodiversity but also affects their morphological, phonological, and reproductive biology along with its distribution pattern, i.e., vegetation shifting phenomenon. A subtle change in weather data (temperature, humidity, etc.) results a big variation in global climate factors leads to global warming and CC phenomenon. As we aware of GHGs (viz., CO_2, N_2O, CH_4, and CFC, etc.) and its detrimental impacts on biodiversity, agricultural productivity and health issue of live-stocks and human along with economic losses are well known and reported from various parts of the world.

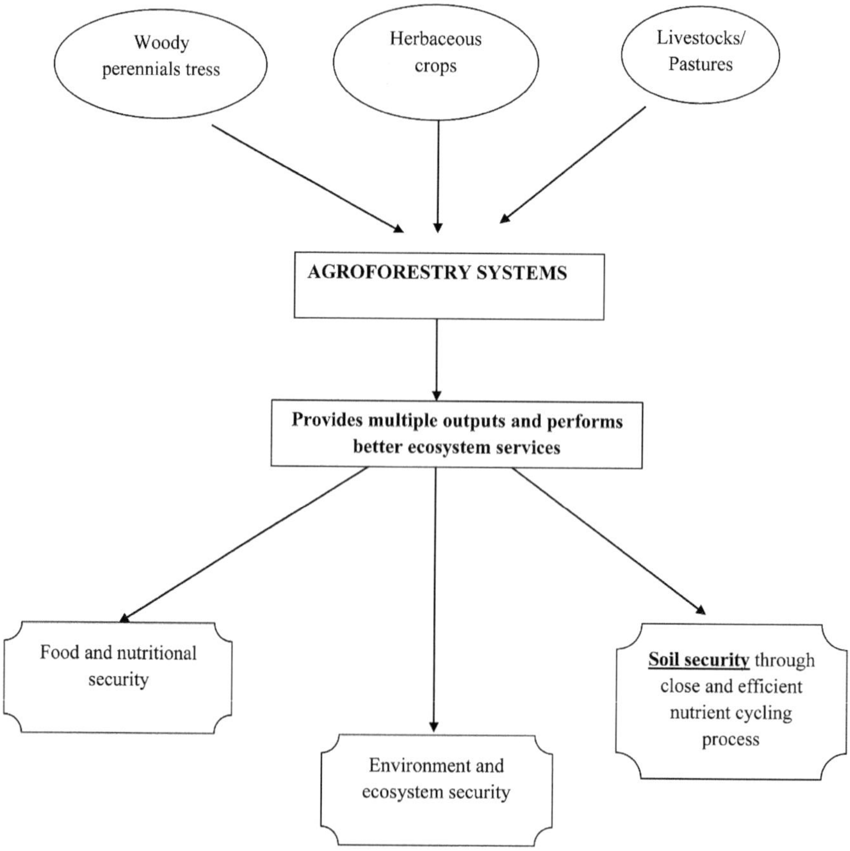

FIGURE 6.4 Agroforestry systems for ecological security.

6.3 IMPACTS ON AGROECOSYSTEMS

CC affects all ecosystems which pertaining and relating to agriculture and agricultural allied sectors/enterprises in term of loss of productivity and money too. Agroecosystem represents valuable natural resources which are very valuable and prerequisites treasure for the ecosystems but changing climates and extreme weather affects its viability, vitality, productivity, and socioeconomic aspects. A model has been represented in Figure 6.5, which shows the climatic impact on forest, agriculture, livestock's, humans, soils, etc. Change in diversity, richness, species susceptibility, soil productivity, its

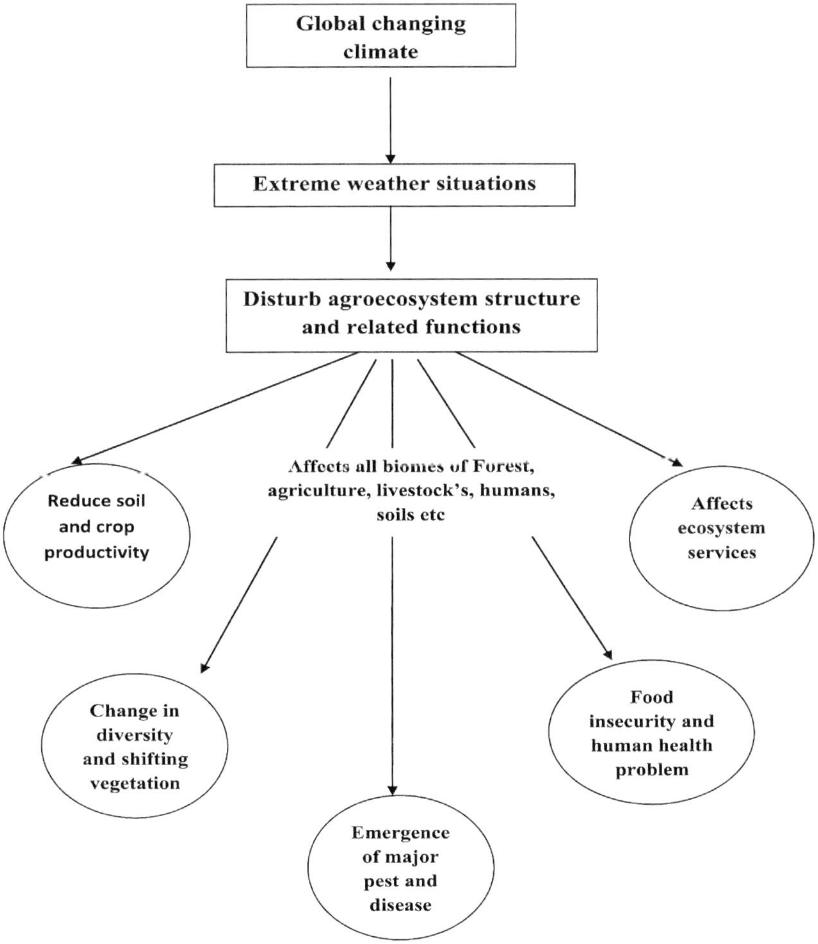

FIGURE 6.5 Climate change impact of agroecosystems.

ecosystem services, the emergence of infectious pest and disease, etc. are the major climatic impact due to extreme weather.

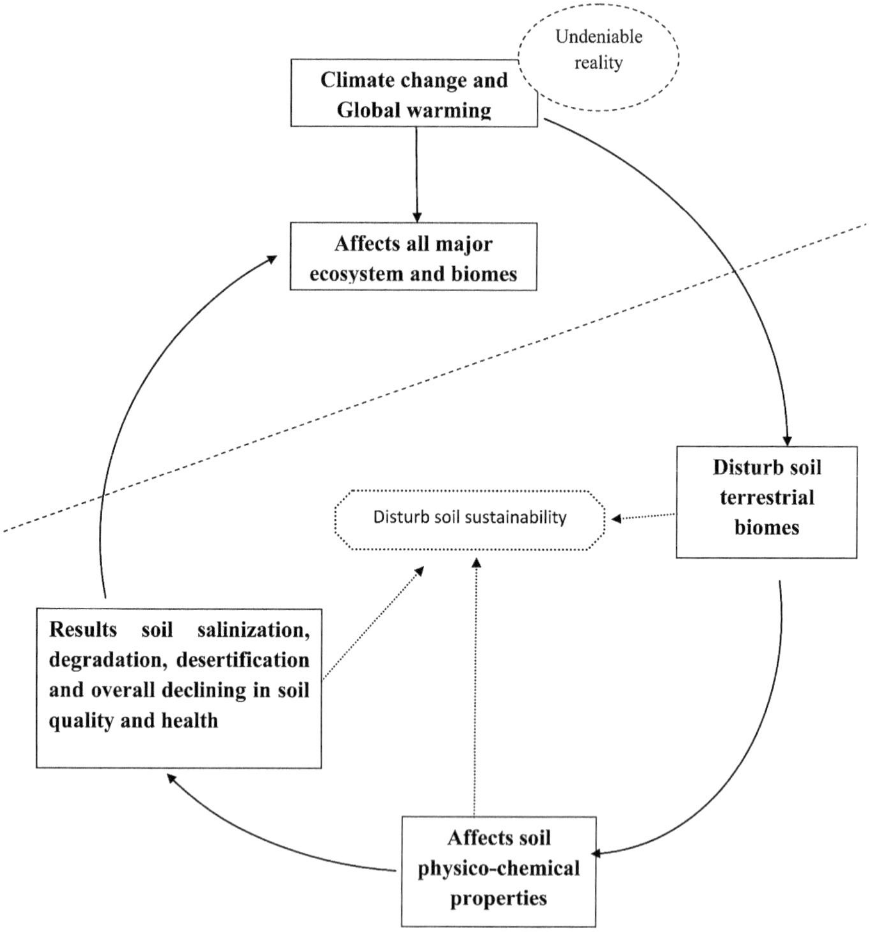

FIGURE 6.6 Soil sustainability under changing climate.

6.3.1 SOIL HEALTH AND SERVICES

Soil as we know the soul of infinite life and a great treasure beneath our feet represents an essential natural resource and works as the substratum for all living organism by holding on it. Soil provides an essential element to all herbaceous plant, perennial woody trees, livestock's (through pastures), humans, etc. and without it, the life can't exist. Today, soil sustainability and productivity get lost due to these extreme climatic events (Figure 6.6) by disturbing the soil physicochemical properties results decline in availability of basic essential nutrients to plants and in the absence of these elements plants does not able to sustain their life for more which critically induced soil salinization, degradation, desertification, and overall declining in soil quality and health and which affects all major biomes of the earth. There are great synergies among soil and different agroecosystem in term of providing ecosystem services and environmental sustainability (Jhariya et al., 2018). Soil performs four basic functions which support all structure and functions of agroecosystem viz., agriculture, forestry, livestock's, humans, etc. (Figure 6.7). For example, healthy and quality soil initiates proper functions of all agroecosystems components in term of the provision of tangible and intangible products and benefits, more than one output, economic, and livelihood security, food availability and environmental security, etc. Efficient nutrient cycling and ecosystem health are other intangible and indirect benefits through the better soil.

6.3.2 AGRICULTURE

World know, the CC through agriculture are happening due to faulty land management, mismanaged agricultural practices, monoculture practices, faulty land conversion, huge application of chemical fertilizers etc. which not only affects the soil and plants productivity (as declining in crops yields and tree productivity and land degradation as waste and un-utilizable land) but also affects human and environmental health through global warming and CC phenomenon (Rathore et al., 2014; Raj and Jhariya, 2017; Raj et al., 2018). Huge application of nitrogenous and phosphate fertilizers not only enters into the food chain but also release some GHGs to the atmosphere. Rice cultivation practices also release methane into the atmosphere which is also a cause of global warming (Hein et al., 1997; Kimura et al., 2004; Abhilash et al., 2015). Beside rice cultivation, burning of the biomass also contributes methane emissions; therefore, various studies have been conducted in this context which is depicted in Figure 6.8. As per one estimates, agriculture practices contributes an average of 5.6, 3.3 and 2.8 Gt equivalent/year of CO_2, N_2O and CH_4 respectively and considered as major destructive GHGs of the world (Vermeulen et al., 2013).

Therefore, some efficient agricultural practices are very essential and a first step towards the soil, human, and nature security along with the development of socioeconomic dimension. Certain practices like, climate-smart agriculture, agroforestry, conservation agriculture, no-till operation, mixed farming, integrated farming systems (IFS) etc. are the environmentally friendly and economically viable practices which meet out the problem of changing climate and helps in promoting security of environment through diversifying production which can satisfy the needs of people and nations. IFS represents a window of organic farming systems which is the good alternative and substitute of inorganic and intensive cultivation that not only improved the soil health and yield but also raise the quality of environment, ecosystems, and meets out the food and nutritional problems today (Figure 6.9).

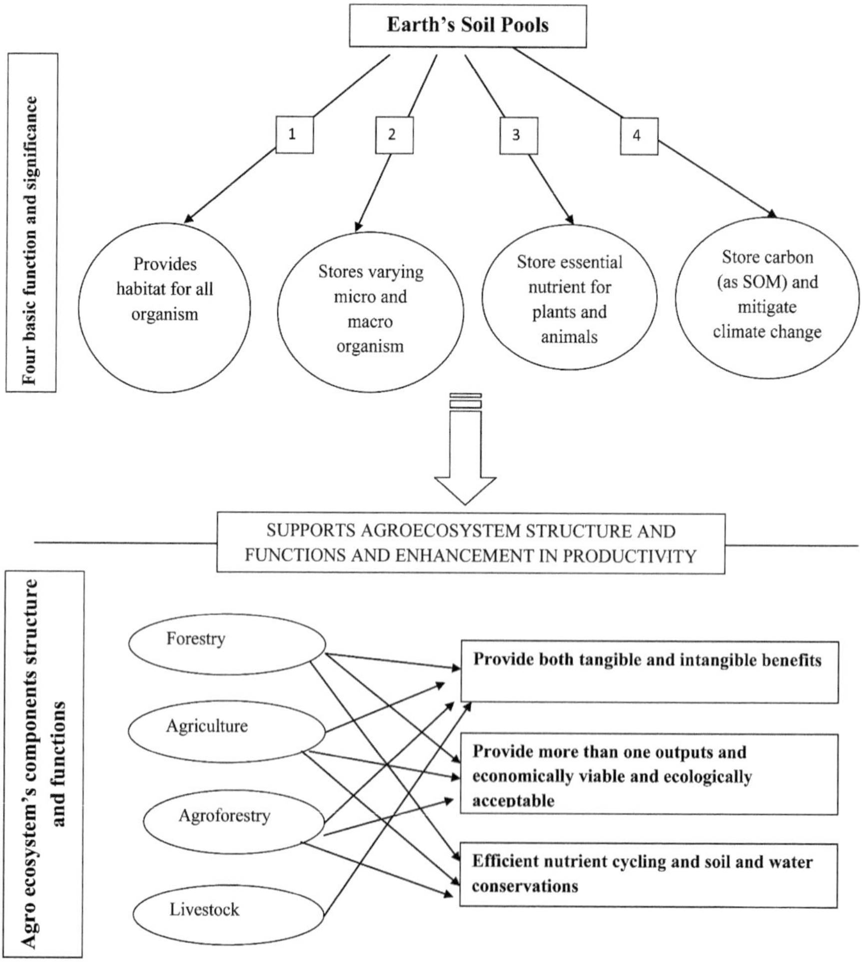

FIGURE 6.7 Synergies among soil and different agroecosystems.

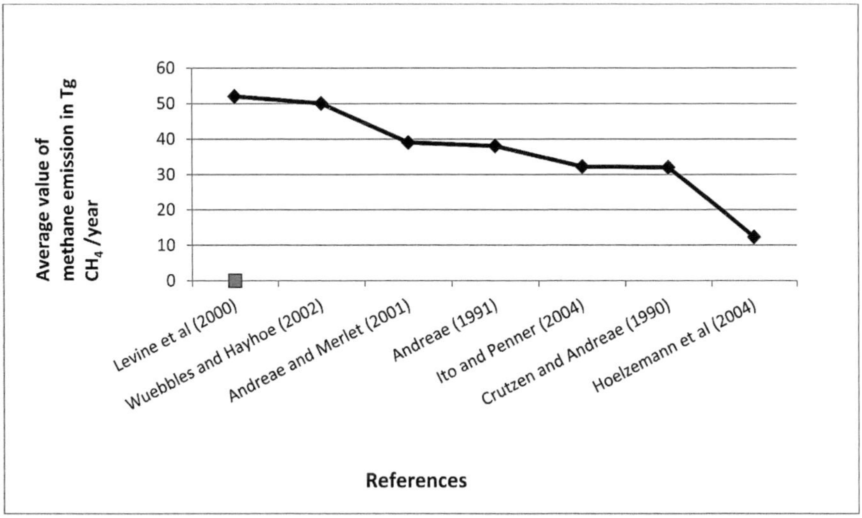

FIGURE 6.8 Methane emissions through the different biomass burning process.

6.3.3 FORESTRY

Forestry is the applied science and comprised timber and non-timber forest products (NTFPs) along with herbaceous flora, liana, climbers, and harbor of various wild animals which plays an important role in maintaining structure and functions of ecosystems. Forest provides both tangible (direct) and intangible (indirect) benefits to the whole biomes. Tangible benefits comprise timber and NTFPs (gum, Katha, tendu patta, resin, dye, etc.), which are valuable products and high economic value strengthen the socioeconomic conditions of tribal and rural people. Today, anthropogenic factors like deforestation, mining, industrialization, etc. affect the forest cover, which must be 33% as per given forest policy. Vegetation shifting, change in morphology, phenology, reproductive biology, disturbance in associated species, mortality of wild animals, etc. are the major impacts that we are seeing due to changing the climate and extreme weather patterns (Figure 6.10).

The major question is how changing climate affects forest ecosystem and its tangible and intangible benefits? Change in weather affects rainfall pattern, moisture availability, temperature, and humidity change etc. which affects the species phytosociological parameters (species distribution pattern, its density, richness, diversity), tree morphological, phonological, and reproductive parameters, carbon stock pattern and other physicochemical properties of soil, soil inhabiting organism and its life cycle along with nutrient cycling systems. However, the impact of CC is also seen in term of shifting of vegetation in the world. The climatic impact on population and diversity of trees species are depicted in Table 6.1. As per one estimate, 80% of plant and animal species on which they gathered data represents temperature-related shifts in physiological aspects (Root et al., 2003).

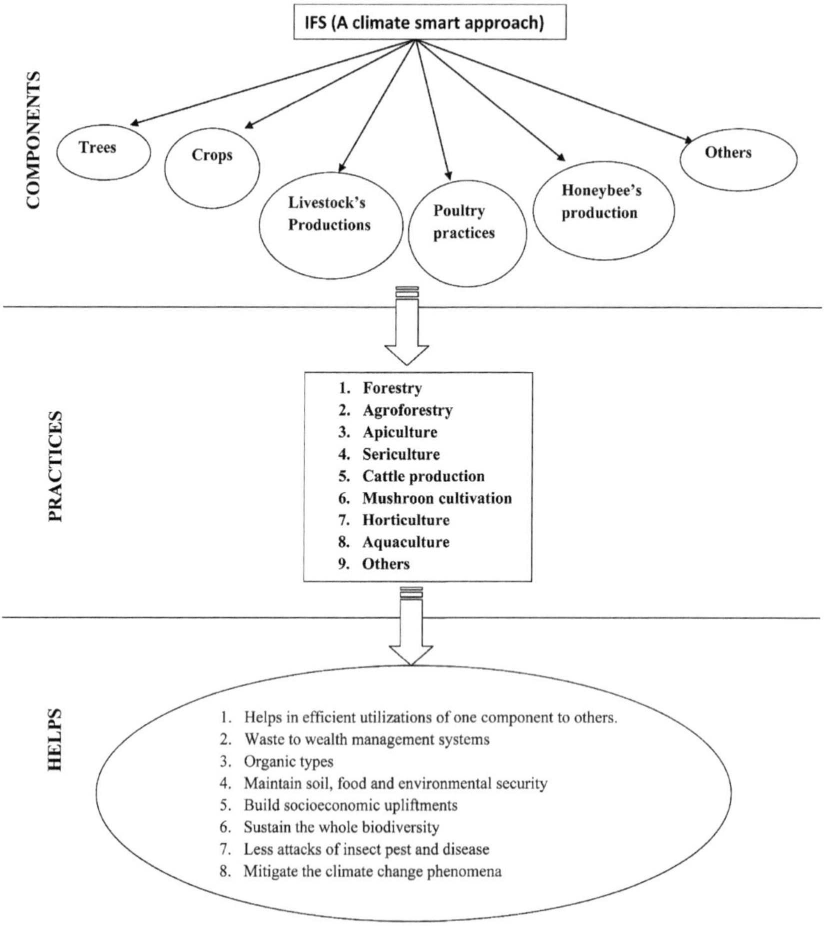

FIGURE 6.9 Integrated farming systems (IFS) for climate change mitigation.

6.3.4 LIVESTOCK

Globally, livestock is one of the important friendly component of the any farming systems as IFS and agroforestry systems (AFs) and make use of up to

60% contribution in biomass harvesting of the world (Krausmann et al., 2008; Weindl et al., 2015;), utilizing up to 30% of agricultural based water (Peden et al., 2007; Mekonnen and Hoekstra, 2010), play a lead role in the agricultural based efficient nitrogen cycling process (Bodirsky et al., 2012, 2014, Bouwman et al., 2013) along with the 12–18% share in the emission of anthropogenic GHGs in the atmosphere (Steinfeld et al., 2006; Westhoek et al., 2011).

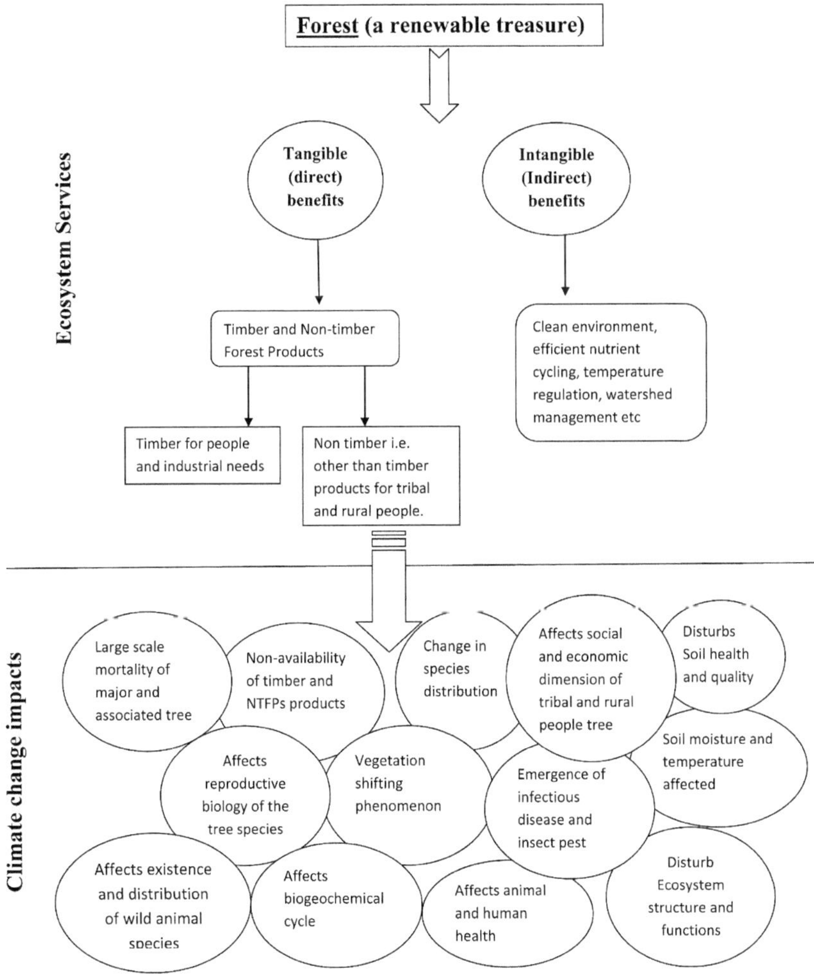

FIGURE 6.10 Forest services and climate change.

The important question is how changing climate has been affecting the livestock's and cattle's production and what is the real strategies behind

TABLE 6.1 Extreme Climatic Impacts on Tree Species Population

Events	Extreme Climatic Impacts	References
Declining the population of Birch (*Betula pendula*)	Due to `experiencing winter thaw cycles	Bourque et al., 2005
Birch species (*Betula pendula*) dieback	Extreme climate results thaw event	Allen, 2003
Widespread decline of *Acer saccharum* (Sugar maple) population	Extreme climate results thaw event and excessive fine root damage. Extreme weather also increases the population of insect pest, disease, and changes in soil nutrient.	Decker et al., 2003
Mortality of Oak species in England in the 19th century	Several factors drought, emergence of insect pest and deadly pathogens due to extreme climatic situations	Keyser and Brown, 2016
Ash (Fraxinus species) dieback in the Northeast since 1920	Several extreme climatic factors like drought and excessive freezing along with ash borer and Asian beetles which extensively damage this tree species	Poland and McCullough, 2006; Coble et al., 2017
Declined Red spruce (*Picearubens*) population in southern range margin in Massachusetts	Due to experiencing high temperature and other extreme climatic factors	Ribbons, 2014
Declining tree-ring value in the forest stand of *Picearubens* in the North America	Affects mean monthly and few degree of temperature	Cook et al., 1987; Ribbons, 2014

the viable transformation of this sector for optimal production without affecting livestock's and environmental health in a sustainable way? Many researchers have been studies the extreme climate impacts on livestock' health and related changes in the productivity of range and pasture land along with yields of edible pasture crops through the integrated approach (Thornton and Gerber, 2010; Ghahramani and Moore, 2013; Havlík et al., 2015). Rise in temperature results destructive heat storms which can impact on livestock's production, i.e., yield, and quality of milk, eggs, meat, etc. but disturbing the morphological, anatomical, and reproductive systems of animals (Lara and Rostagno, 2013). Livestock makes the soil more fertile through the addition of excreta as manure which decomposes by several soils inhabiting microorganism and forms soil organic matter (SOM) and humus which can maintain nutrient status in the soil and helps in maintaining a metabolic process of tree and plant species. In turn plant (forage and fodder plants) provides as a feed material to the livestock's and grazing animals to the considerable extent and enhance the livestock's productivity which provides milk and other resources to the people and people gets a considerable economic benefits results upliftments in socioeconomic and livelihood security. Thus, a model is developed in the linking concepts among livestock's health and productivity, soil fertility, and pasture/fodder production, which is shown in Figure 6.11.

6.4 FOOD SECURITY UNDER CHANGING CLIMATE

Today, we are aware of global climate impacts on the food system and its availability which is shortening day by day either due to extreme weather effects or due to burgeoning populations. Agriculture and allied sectors (forestry, fisheries, apiculture, sericulture etc) are the backbone of the global economy and produce various feed materials which are life sustaining and health concern. Extreme weathers sign the emergence of various infectious disease and insect pest which reduce the overall productivity and nutrient concentration in food. Changing temperature and moisture in the soil can disturbs the availability of quality foods by disturbing the population and health of soil-inhabiting organism which plays an inevitable role in decomposing and releasing of essential nutrients on which the plant ecosystems depends for their life. Moreover, various studies have been conducted on climatic impacts on agriculture and allied sectors and modified by the availability of water (water shortage problem) and essential nutrient which affects the overall food security programme (Zhang, 2011; Lovejoy et al., 2012; Lovejoy, 2013;

Lyall, 2013). Therefore, the climatic impacts on food production and its availability are reported by both direct (altering in the growth pattern of plant and tree species) and indirect ways (disturbing the water availability, SOM, emergence of insect pest and disease) (Gregory et al., 2005; Newton et al., 2011; Miao et al., 2011).

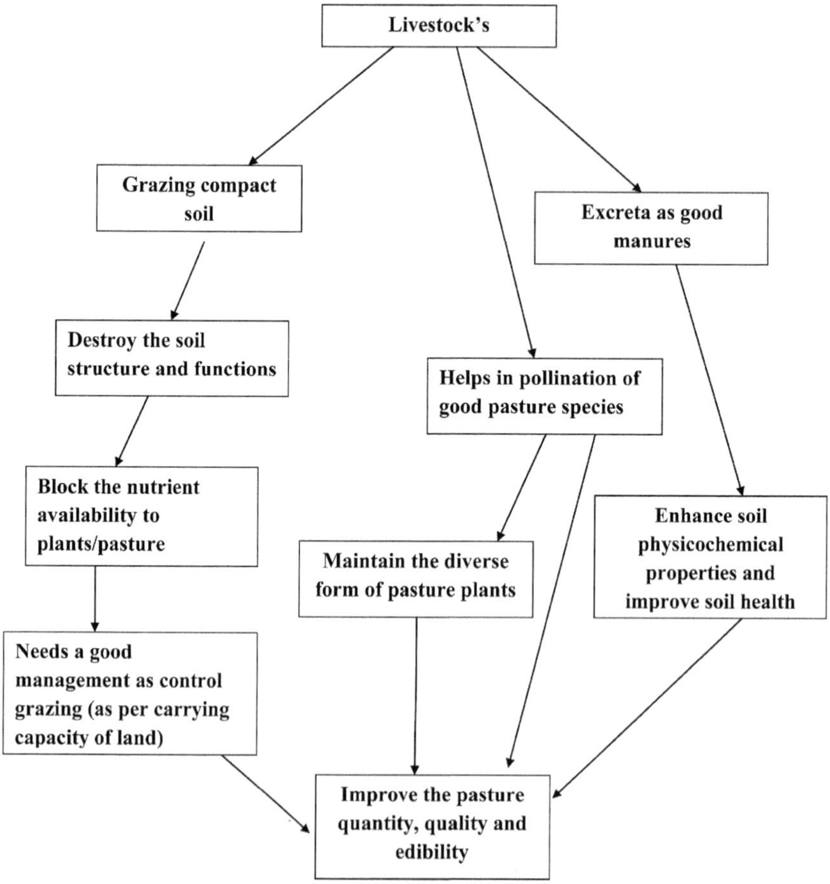

FIGURE 6.11 Linking concept among livestock's, soil fertility and pasture productions.

6.5 MITIGATION STRATEGIES

As we discussed earlier, CC is globally challenging issues today for all earth's organism which disturbs our ecosystem processes and deprive the health and economy of the world. In this context, the practices of farming systems in a climate-smart way will give surety in the mitigation of changing the climate and helps in improving the productivity and profitability of farmers. Involvement of tree improvement program, resistant tree and plant species verity, appropriate breeding and biotechnology program, etc. are some alternative practices to minimize the impact of extreme weather and species can withstand in that harsh and adverse conditions. Moreover, the adoption of climate-smart agriculture comprises agroforestry, conservation agriculture, no-till farming practices, IFS, organic farming systems, etc. have greatest potential to resolve the problem of CC and helpful for the strengthen the farmer's socioeconomic conditions. However, agroforestry is very good and new concept of primitive farming practices which is extensively spread to almost all tropics and is a good strategy for overcoming these variant climate situations by protecting soil, making the availability of food and nutrition and maintains the health of animal and humans along with economic security to farmers.

Therefore, various successful case studies are found in developing countries and spread almost in Asian and African countries where the climate is very harsh and adverse (Colin, 2013). As per one estimates, the combined practices of afforestation and agroforestry (for carbon sequestration (CS)), alternative for fossil fuels, minimizing the application of chemical fertilizers in agriculture land and checking deforestation activity will check the uncontrolled emission of GHGs in the atmosphere and these practices contribute up to 60% of the mitigation effort for global warming and CC phenomenon (Ramachandran et al., 2007). Also, the internationally recognized Kyoto protocol program works in this direction of controlling excessive emissions of carbon in the atmosphere and its mitigation by the sequestration process (Harper et al., 2007; UNFCCC, 2009).

Agroforestry works very effectively in the context of sink capacity of the atmospheric carbon in the woody parts on the long-term basis which helps in mitigating CC and global warming issues along with providing tangible and intangible benefits as multipurpose benefits for both national security and profitability (Sudha et al., 2007). The potential for storage and sequestration of carbon through the adoption of agroforestry are varies between 0.3–15.2 Mg C $ha^{-1}yr^{-1}$ in the tropics of the world (Nair et al., 2011). Therefore, the practices of forestation and cultivation of land can enhance the storage and sequestration of carbon in both above ground vegetation and belowground soil which increase the soil fertility through addition of SOM and helps in making quality soils. Thus, the practices of agroforestry, cover crop system, no-till farming system, conservation agriculture, application of mulch and biochar on cultivated field, application of breeding and biotechnological tools, residue management practices, introduction of high yield and improved resist variety will work in the context of higher storage and sequestration of carbon along with mitigating extreme climate with environmental security (CS in agricultural soils, 2012).

6.6 NEXUS BETWEEN SOIL SUSTAINABILITY AND FOOD SECURITY UNDER CHANGING CLIMATE

Today, extreme weather affects the global food production and its availability due to dryness and heat storm conditions in several Asian and African continents. Due to the higher temperature and fewer rainfall results water shortage in soil and high evapotranspiration conditions which affects global food production system. As we know, sustainable soil definitely makes the food availability through proper management and healthy farming practices in dry region. Moreover, a strong linkage exists between healthy soils and quality and nutritious food under the situation of CC. However, CC encourages the problem of Food and nutritional insecurity which affects all human health and environmental quality (Figure 6.12). Thus, climate-smart soil management practices are very important in the context of full utilization of soil as the provision of ecosystem services and availability of food to burgeoning people under changing climate along with the environmental security.

6.7 NEXUS BETWEEN SOIL, FOOD, AND HUMAN HEALTH

As we know, soil sustains the lives all plants, animals, and human through the provision of greatest ecosystem services. Soil stores essential elements; inhabit important decomposing microorganism, farmer's friendly earth-worms, protozoan, and harbors of various beneficial and harmful organisms which affect our health and lifestyle. Without soil, we can't imagine for food and life on the earth. Decompose live and dead organic matters and release an essential micro and macronutrients which are available to plants through soil solution and uptake by extensive root systems. Plants stores these essential nutrients as food material and optimum and efficient nutrients make plant healthy and strong. Herbivores consume these plants as a food material. Live-stock's feeds some edible grasses and pastures. Humans (omnivores) and other carnivores depend on these plants and animals for the food and affect health systems. Therefore, there is the great nexus among soil-food-health in our ecosystems. Healthy soils perform healthy ecosystem services which are prerequisites for human health along with environmental sustainability (Figure 6.13).

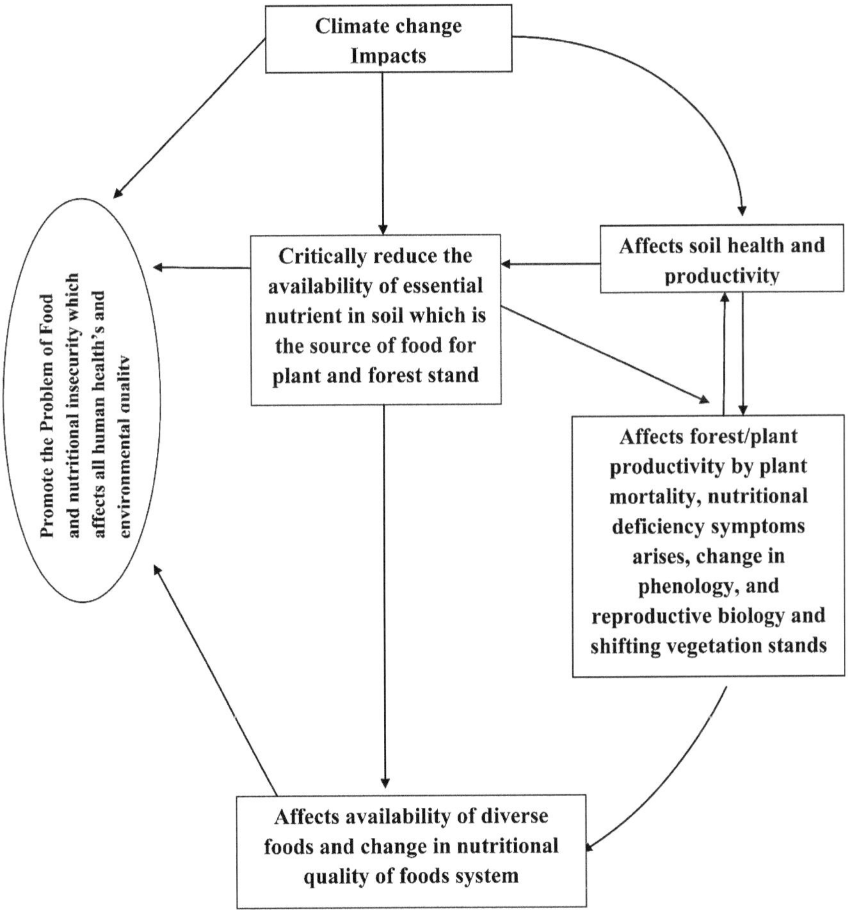

FIGURE 6.12 Nexus between soil and food system under changing climate.

6.8 ROADMAP AND FUTURE STRATEGIES FOR CLIMATE CHANGE (CC) MITIGATION AND ADAPTATION

The impacts of extreme weather situations are crystal clear for all as promoting loss of productivity, food unavailability, declining health and threatening the biodiversity along with economic loss and overall national

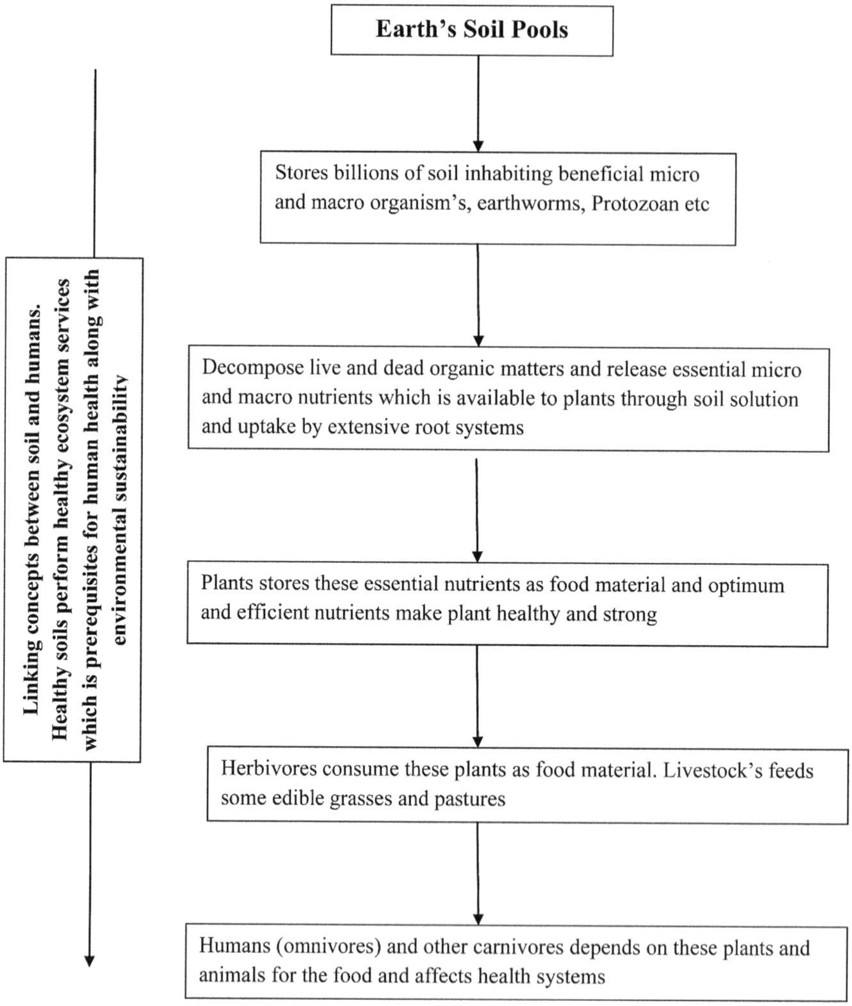

FIGURE 6.13 Links between soil, food, and human health.

security. In this context, there should be to develop a roadmap for reducing carbon pollution by minimizing continuous emissions of GHGs which can be possible through the involvement of all the government, non-government organizations and private sectors and other developmental programs. Creating awareness through government and non-government institution and its strong communications and linkage with people are some effective measures and strategies for solving the CC issue. However, a number of researches which is based on basic and applied science should be used for the better agriculture and healthy and quality soils under extreme weather situation which can also maintain food and nutritional security (FNS) program. R&D should be undertaken in the context of climate-smart soil fertility management for raising the higher fertility of soil, which in turn gives better productivity (minimize food insecurity issues) and helps in shortening the problems of land degradation.

6.9 CONCLUSION

Soil deterioration and degradation and FNS under the changing climate are a major issue and global concern today. Soil health and quality are the prime factor and prerequisites for the overall food production and the people's health. Due to anthropogenic and natural cause such as industrial development, mining, deforestation, heavy application of chemical fertilizers and faulty land conversion system that leads to GHGs emissions into the atmosphere which are affecting health and productivity of all biomes and disturbs ecosystem structure and function. Therefore, soil links to all development and services in various ways and soil management in a climate-smart way is prerequisite for global development in term of soil security, livestock, and human health and security along with environmental security. Also, science-based approach is very essential in the direction of climate educations in every institution for better health and quality of soil which directly links with food, environment, and national security.

KEYWORDS

- **agroforestry**
- **climate change**
- **farming system**
- **food security**
- **greenhouse gases (GHGs)**
- **organic matter**
- **food and nutritional security**

REFERENCES

Abhilash, P. C., Tripathi, V., Dubey, R. K., & Edrisial, S. A., (2015). Coping with changes: Adaptation of trees in a changing environment. *Trends Plant Sci.*, *20*(3), 137–138.

Allen, D. J., (2003). *Spring Dieback of Yellow Birch in North America: Historical Examination of Weather and Frost Hardiness*. Thesis BSc(F), University of New Brunswick.

Andreae, M. O., & Merlet, P., (2001). 'Emission of trace gases and aerosols from biomass burning.' *Global Biogeochemical Cycles*, *15*, 955–966.

Andreae, M. O., (1991). 'Biomass burning: Its history, use, and distribution and its impact on environmental quality and global climate.' In: Levine, J. S., (ed.), *Global Biomass Burning: Atmospheric, Climatic, and Biospheric Implications* (pp. 3–21). MIT Press, Cambridge, MA.

Arora, V. P. S., Bargali, S. S., & Rawat, J. S., (2011). Climate change: Challenges, impacts, and role of biotechnology in mitigation and adaptation. *Progressive Agriculture, 11*, 8–15.

Bodirsky, B. L., Popp, A., Lotze-Campen, A., Dietrich, J. P., Rolinski, S., Weindl, I., Schmitz, C., Müller, C., Bonsch, M., Humpenöder, F., Biewald, A., & Stevanovic, M., (2014). Reactive nitrogen requirements to feed the world in 2050 and potential to mitigate nitrogen pollution. *Nat. Commun.*, *5*, 38–58.

Bodirsky, B. L., Popp, A., Weindl, I., Dietrich, J. P., Rolinski, S., Scheiffele, L., Schmitz, C., & Lotze-Campen, H., (2012). N_2O emissions from the global agricultural nitrogen cycle—current state and future scenarios. *Biogeosciences, 9*, 4169–4197.

Bourque, C. P. A., Cox, R. M., Allen, D. J., Arp, P. A., & Meng, F. R., (2005). Spatial extent of winter thaw events in eastern North America: Historical weather records in relation to yellow birch decline. *Climate Change Biology*, *11*, 1477–1492.

Bouwman, L., Goldewijk, K. K., Hoek, K. W. V. D., Beusen, A. H. W., Vuuren, D. P. V., Willems, J., Rufino, M. C., & Stehfest, E., (2013). Exploring global changes in nitrogen and phosphorus cycles in agriculture induced by livestock production over the 1900–2050 periods. *Proc. Natl. Acad. Sci. USA, 110*, 20882–7.

Carbon Sequestration in Agricultural Soils, (2012). *Agriculture and Rural Development* (pp. 43–50). Report no. 67395-GLB, The World Bank.

Coble, A. P., Vadeboncoeur, M. A., Berry, Z. C., Jennings, K. A., McIntire, C. D., Campbell, J. L., Rustad, L. E., Templer, P. H., & Asbjornsen, H., (2017). Are Northeastern U. S. forests vulnerable to extreme drought? *Ecol. Process, 6*, 34. https://doi.org/10.1186/s13717–017–0100-x (Accessed on 9 October 2019).

Colin, M., (2013). "Agroforestry and smallholder farmers: Climate change adaptation through sustainable land use." *Capstone Collection.* Paper 2612.

Cook, E. R., Johnson, A. H., & Blasing, T. J., (1987). Forest decline: Modeling the effect of climate in tree rings. *Tree Physiol.*, *3*(1), 27–40.

Crutzen, P. J., & Andreae, M. O., (1990). Biomass burning in the tropics: Impact on atmospheric chemistry and biogeochemical cycles. *Science*, *250*, 1669–1678.

Decker, K. L. M., Wang, D., Waite, C., & Scherbatskoy, T., (2003). Snow removal and ambient air temperature effects on forest soil temperatures in northern Vermont. *Soil Science Society of America Journal*, *67*, 1234–1242.

FAO, (2007). *State of the World's Forests*. FAO of United Nations, Rome.

Ghahramani, A., & Moore, A. D., (2013). Climate change and broadacre livestock production across southern Australia: II. Adaptation options via grassland management *Crop Pasture Sci., 64*, 615–630.

Gregory, P. J., Ingram, J. S. I., & Brklacich, M., (2005). Climate change and food security. *Phil. Trans. Roy. Soc. London, 360*, 2139–2148.

Harper, R. J., Beck, A. C., Ritson, P., Hill, M. J., Mitchell, C. D., Barrett, D. J., Smettem, K. R. J., & Mann, S. S., (2007). The potential of greenhouse sinks to underwrite improved land management. *Ecological Engineering, 29*(4), 329–341.

Havlík, P., Leclère, D., Valin, H., Herrero, M., Schmid, E., Soussana, J. F., Müller, C., & Obersteiner, M., (2015). Global climate change, food supply and livestock production systems: A bioeconomic analysis. In: Elbehri, A., (ed.), *Climate Change and Food Systems: Global Assessments and Implications for Food Security and Trade*. Rome, Italy: Food Agriculture Organization of the United Nations (FAO).

Hein, R., Crutzen, P. J., & Heimann, M., (1997). 'An inverse modeling approach to investigate the global atmospheric methane cycle.' *Global Biogeochemical Cycles, 11*, 43–76.

Hoelzemann, J. J., Schultz, M. G., Brasseur, G. P., & Granier, C., (2004). Global wild land emission model (GWEM): Evaluating the use of global area burnt satellite data. *Journal of Geophysical Research, 109*, D14S04, doi: 10.1029/2003JD0036666.

Houston, D. R., (1999). In: Horsley, S. B., & Long, R. P., (eds.), *Sugar Maple Ecology and Health: Proceedings of an International Symposium* (pp. 19–26). Warren, PA. Gen. Tech. Rep. NE-261. Radnor, PA: U. S. Department of Agriculture, Forest Service, Northeastern Research Station.

Ito, A., & Penner, J. E., (2004). 'Global estimates of biomass burning emissions based on satellite imagery for the year 2000.' *Journal of Geophysical Research, 109*, D14S05. doi: 10.1029/2003JD004423.

Jhariya, M. K., Banerjee, A., Yadav, D. K., & Raj, A., (2018). Leguminous trees an innovative tool for soil sustainability. In: Meena, R. S., Das, A., Yadav, G. S., & Lal, R., (eds.), *Legumes for Soil Health and Sustainable Management* (pp. 315–345). Springer, ISBN 978–981–13–0253–4 (eBook), ISBN: 978–981–13–0252–7 (Hardcover). https://doi.org/10.1007/978–981–13–0253–4_10 (Accessed on 9 October 2019).

Jhariya, M. K., Bargali, S. S., & Raj, A., (2015). In: Miodrag, Z., (ed.), *Possibilities and Perspectives of Agroforestry in Chhattisgarh, Precious Forests - Precious Earth* (pp. 237–257). ISBN:978–953–51–2175–6, In-Tech, DOI: 10.5772/60841.

Keyser, T. L., & Brown, P. M., (2016). Drought response of upland oak (*Quercus L.)* species in Appalachian hardwood forests of the southeastern USA. *Annals of Forest Science, 73*, 971–986.

Kimura, M., Murase, J., & Lu, Y. H., (2004). Carbon cycling in rice field ecosystems in the context of input, decomposition and translocation of organic materials and the fates of their end products (CO_2 and CH_4). *Soil Biology and Biochemistry, 36*, 1399–1416.

Krausmann, F., Erb, K. H., Gingrich, S., Lauk, C., & Haberl, H., (2008). Global patterns of socioeconomic biomass flows in the year 2000: A comprehensive assessment of supply, consumption and constraints. *Ecol. Econ., 65*, 471–487.

Lara, L. J., & Rostagno, M. H., (2013). Impact of heat stress on poultry production. *Animals, 3*, 356–369.

Levine, J. S., Cofer, W. R., & Pinto, J. P., (2000). 'Biomass burning.' In: Khalil, M. A. K., (ed.), *Atmospheric Methane: Its Role in the Global Environment* (pp.190–201). Springer Verlag, New York.

Lovejoy, S., (2013). What is climate? *Eos Trans. AGU, 94*(1), 1–2.

Lovejoy, S., Schertzer, D., & Varon, D., (2012). Do GCM's predict the climate or macroweather? *Earth Syst. Dyn. Discuss., 3*, 1259–1286.

Lyall, S., (2013). *Heat, Flood or Icy Cold, Extreme Weather Rages Worldwide*. The New York Times: New York, NY, USA, 2013.

Mekonnen, M. M., & Hoekstra, A. Y., (2010). *The Green, Blue and Grey Water Footprint of Farm Animals and Animal Products* UNESCO-IHE, Delft, the Netherlands.

Miao, Y., Stewart, R. A., & Zhang, F., (2011). Long-term experiments for sustainable nutrient management in China: A review. *Agron. Sust. Dev.*, *31*, 397–414.

Nair, P. K. R., Nair, V. D., Kumar, B. M., & Showalter, J. M., (2011). Carbon sequestration in agroforestry systems. *Advances in Agronomy*, *108*, 237–307.

Newton, A. C., Johnson, S. N., & Gregory, P. J., (2011). Implications of climate change for diseases, crop yields and food security. *Euphytica*, *179*, 3–18.

Parihaar, R. S., (2016). Carbon stock and carbon sequestration potential of different land-use systems in hills and Bhabhar belt of Kumaun himalaya. *PhD Thesis* (p. 330). Kumaun University, Nainital.

Peden, D., Tadesse, G., & Misra, A., (2007). Water and livestock for human development. In: Molden, D., (ed.), *Water for Food, Water for Life: A Comprehensive Assessment of Water Management in Agriculture* (pp. 485–514). Oxford: Oxford University Press.

Poland, T. M., & McCullough, D. G., (2006). Emerald ash borer: Invasion of the urban forest and the threat to North America's ash resource. *Journal of Forestry*, *104*, 118–124.

Raj, A., & Jhariya, M. K., (2017). Sustainable agriculture with agroforestry: Adoption to climate change. In: Kumar, P. S., Kanwat, M., Meena, P. D., Kumar, V., & Alone, R. A., (ed.), *Climate Change and Sustainable Agriculture* (pp. 287–293). ISBN: 9789–3855–1672–6. New India Publishing Agency (NIPA), New Delhi, India.

Raj, A., Jhariya, M. K., & Bargali, S. S., (2018). Climate smart agriculture and carbon sequestration. In: Pandey, C. B., Gaur, M. K., & Goyal, R. K., (ed.), *Climate Change and Agroforestry: Adaptation Mitigation and Livelihood Security* (pp. 1–19). ISBN: 9789–386546067. New India Publishing Agency (NIPA), New Delhi, India.

Ramachandran, A., Jayakumar, S., Haroon, R. M., Bhaskaran, A., & Arockiasamy, D. I., (2007). Carbon sequestration: Estimation of carbon stock in natural forests using geospatial technology in the Eastern Ghats of Tamil Nadu, India. *Current Science*, *92*(3), 323–331.

Rathore, A. K., Jhariya, M. K., Jain, R., & Kumar, S., (2014). Agriculture: Cause, victim as well as mitigator of climate change. *Eco. Env. & Cons.*, *20*(3), 995–1000.

Rawat, V. R. S., (2010). Reducing emissions from deforestation in developing countries (REDD) and REDD plus under the UNFCCC negotiations, Research note. *Indian Forester*, *136*(1), 129–133.

Ribbons, R. R., (2014). Disturbance and climatic effects on red spruce community dynamics at its southern continuous range margin. *PeerJ.*, *2*, e293. http://doi.org/10.7717/peerj.293 (Accessed on 9 October 2019).

Root, T. L., Price, J. T., Hall, K. R., Schneider, S. H., Rosenzweig, C., & Pounds, J. A., (2003). Fingerprints of global warming on wild animals and plants. *Nature*, *421*, 57–60.

Singh, N. R., & Jhariya, M. K., (2016). Agroforestry and agrihorticulture for higher income and resource conservation. In: Narain, S., & Rawat, S. K., (ed.), *Innovative Technology for Sustainable Agriculture Development* (pp. 125–145). ISBN: 978–81–7622–375–1. Biotech Books, New Delhi, India.

Steinfeld, H., Gerber, P., Wassenaar, T., Castel, V., Rosales, M., & Haan, C. D., (2006). *Livestock's Long Shadow: Environmental Issues and Options*. Rome: Food and Agriculture Organization of the United Nations (FAO).

Sudha, P., Ramprasad, V., Nagendra, M. D. V., Kulkarni, H. D., & Ravindranath, N. H., (2007). Development of an agroforestry carbon sequestration project in Khammam district, India. *Mitigation and Adaptation Strategies for Climate Change, 12*, 1131–1152.

Thornton, P. K., & Gerber, P. J., (2010). Climate change and the growth of the livestock sector in developing countries. *Mitig. Adapt. Strateg. Glob. Change, 15*, 169–184.

Thornton, P. K., Van De Steeg, J., Notenbaert, A., & Herrero, M., (2009). The impacts of climate change on livestock and livestock systems in developing countries: A review of what we know and what we need to know. *Agricultural Systems, 101*, 113–127.

UNFCCC, (2009). *Land Use, Land-Use Change and Forestry*. United Nations. Available from: http://unfccc.int/2860.php (Accessed on 9 October 2019).

Vermeulen, S. J., Challinor, A. J., Thornton, P. K., Campbell, B. M., Eriyagama, N., Vervoort, J. M., Kinyangi, J., Jarvis, A., Laderach, P., Ramirez-Villegas, J., Nicklin, K. J., Hawkins, E., & Smith, D. R., (2013). Addressing uncertainty in adaptation planning for agriculture. *PNAS, 110*(21), 8357–8362.

Watson, R. T., Noble, I. R., Bolin, B., Ravindranath, N. H., Verardo, D. J., & Dokken, D. J., (2000). *Land Use, Land-Use Change, and Forestry: Intergovernmental Panel on Climate Change*. Cambridge University Press, Cambridge.

Weindl, I., Lotze-Campen, I., Popp, A., Müller, H. P., Herrero, M., Schmitz, C., & Rolinski, S., (2015). Livestock in a changing climate: Production system transitions as an adaptation strategy for agriculture. *Environmental Research Letters, 10*(9), 1–12.

Westhoek, H., Rood, T., Berg, M., Janse, J., Nijdam, D., Reudink, M., Stehfest, E., Lesschen, J. P., Oenema, O., & Woltjer, G. B., (2011). *The Protein Puzzle: The Consumption and Production of Meat, Dairy and Fish in the European Union.* PBPBL Netherlands Environmental Assessment Agency, The Hague.

Wuebbles, D. J., & Hayhoe, K., (2002). Atmospheric methane and global change. *Earth Science Reviews, 57*, 177–210.

Zhang, J., (2011). China's success in increasing per capita food production. *J. Exp. Bot., 62*, 3707–3711.

CHAPTER 7

Linking Social Dimensions of Climate Change: Transforming Vulnerable Smallholder Producers for Empowering and Resiliency

SUMIT CHAKRAVARTY,[1] ANJU PURI,[2] K. ABHA MANOHAR,[3] PRAKASH RAI,[4] UBALT LEPCHA,[3] VINEETA,[5] NAZIR A. PALA,[5] and GOPAL SHUKLA[5]

[1]*Professor, Uttar Banga Krishi Viswavidyalaya, Pundibari–736165, West Bengal, India, Mobile: 9434082687, E-mail: c_drsumit@yahoo.com*

[2]*Assistant Professor, Baring Union Christian College, Batala–143505, Punjab, India*

[3]*PhD Scholar, Uttar Banga Krishi Viswavidyalaya, Pundibari–736165, West Bengal, India*

[4]*Research Scholar, Dept. of Forestry, Uttar Banga Krishi Viswavidyalaya, Pundibari–736165, West Bengal, India*

[5]*Assistant Professor, Uttar Banga Krishi Viswavidyalaya, Pundibari–736165, West Bengal, India*

ABSTRACT

Climate change (CC) is an universal fact now. Its impact is experienced across the globe; however, tropical developing countries are more vulnerable. The impacts on different sectors are discussed. Majority of the poor population in these countries are rural and smallholders with agriculture as primary occupation. These countries depend more on natural resources and weather dependent agriculture from marginal land holding. The smallholders have low adaptive capacity due to poverty, isolation, marginality, inaccessible to information, illiteracy, lack of institutions, investment, and services, neglect by policymakers and lower human development. Rising temperature, changes in duration and intensity of precipitation associated with CC will significantly reduce farm productivity. This will lead to decline in net revenues from smallholder farming systems adversely affecting economies and livelihoods of the communities in terms of food security and poverty. Unfortunately, agricultural policy is still not well defined and socially integrated with respect to CC mitigation in the developing countries. As a result implementation of these policies is fraught with socio-economic and political difficulties. Social dimensions relate climate related policy and society. Policies must be developed and adopted considering social dimensions as adoption depend on people. This is possible by good governance that transforms and integrates social and economic interventions by empowering people through participation and decision making throughout the development process while equally sharing the benefits. Institutions should be responsible to empower people through holistic approaches so that people can transform themselves for overall sustainable development.

7.1 INTRODUCTION

Climate change (CC) is already a reality and is one of the toughest challenges the humanity is facing today. This is because of drastic rise in global temperature in the last few decades. Consequent of this drastic rise in temperature, extremities in weather are now more frequent and intense leading to vulnerable living conditions, insecure food supply and forcing displacement of people. The warming experienced in the past century is unprecedented in the past thousand years but felt all around the globe only during the second half of the twentieth century with warming in the last few decades mostly is anthropogenic. The impact of CC is global but is posing a serious threat to poorer or developing or less developed countries as it will compound existing poverty. The poorer countries are most vulnerable because most of them are tropically located with its citizens subsistently agrarian depending mainly on their marginal land holdings and natural resources along with their limited capacity to adapt to CC (DID, 2004). Globally, over a billion people live in extreme poverty on US$ 1–2 a day (Baulch, 2011; Chandy and Gertz, 2011; Chen and Ravallion, 2012; Summer, 2012a, b). Poverty leads to vulnerability because of less accessibility to resources resulting inability to adapt to climatic changes (IPCC, 2001). Consequently, the livelihood of these nations are increasingly threatened and further widening the gap between the developing and developed worlds. The risk unfortunately is also collateral because current development strategies are mostly ignorant of CC risks.

7.2 IMPACTS OF CLIMATE CHANGE (CC)

CC will influence sustainable development and effect livelihoods of people negatively, particularly the rural poor farmers living in the tropical countries. More than 200 million people are affected by climatic disasters every year (WFP, 2011). Majority of the tropical population lives in hazard prone areas. The continent Asia suffered most from natural disasters with 3.76 billion people from 1970 to 1999. Followed by Asia, Africa is next to suffer from natural disasters especially drought. The most damaging is flood followed by windstorms, earthquakes, and droughts in Latin America and the Caribbean (Charvériat, 2000). Annually, on an average from 2000 to 2004, climate disaster hit one out of 19 people in the developing world, with floods alone affecting more than 100 million people in Asia during 2007, displacing more than 21 million and killing more than 1000 people in south Asia (Ninan and Bedamatta, 2012).

CC impacts on human systems and ecosystems is diverse and adverse, negatively affecting livelihoods causing severe socio-economic stresses through decrease in farm productivity and yield, employment reduction and migration (Carey et al., 2012; Ericksen et al., 2012; Gray and Mueller, 2012; Handmer et al., 2012; Roy, 2012; Verner, 2012; Chakravarty et al., 2015; Kaushik and Sharma, 2015). CC impacts both direct and indirect will vary with geographic location and resilience and thus developing countries with complete dependency on agriculture and natural resources are more vulnerable than the developed countries (Hertel et al., 2010; World Bank, 2010). Studies indicate that European countries will be less affected (Tol et al., 2004) while Africa and Asia will be more vulnerable to precipitation changes and temperature increase than any other continents because of their sensitive water and agriculture sectors (World Bank, 2010; Müller et al., 2011). Australia in particular is reported sensitive to droughts (Nelson et al., 2007). Higher temperatures and changes in precipitation patterns will accelerate the process of land degradation, decreasing the agricultural production in Latin America and the Caribbean and indigenous and rural populations in rural North America and Europe (Tol et al., 2004).

Impacts of CC depend on demography, economics, and governance (Dasgupta et al., 2014). Detection and attribution of impacts is challenging and involves issues using traditional knowledge and local people's perceptions through comparing with global climatic changes (Udmale et al., 2014; Shukla et al., 2015; Dey et al., 2017a, b). The CC impacts on settlement, livelihoods, and incomes in rural areas have two categories of multi-step causal chains of impact (Dasgupta et al., 2014). The first is extreme climatic events as they cause direct loss of human property and life. The second one is impact on dependency of rural people through affecting agriculture and other natural systems which in turn affect livelihood.

7.2.1 AGRICULTURE, FORESTRY, AND BIODIVERSITY

CC impacts on agriculture, forestry, and biodiversity are direct that lowers productivity which is due to rising temperature and changes in duration and intensity of precipitation and its associated extreme events (Poudel and Kotani, 2013; Chakravarty et al., 2015). There will be northward shifting of pests and pathogens due to CC towards pole, while the response of crops and weeds to the intensity of carbon-dioxide fertilization will vary with carbon fixation pathway along with supply of water and nutrients (Black et al., 2011a). The impact of CC on agricultural systems, forestry, and biodiversity affecting production due to its enormous nature and weather dependence particularly in the developing countries will have significant economic implications in terms of income and well-being of people, specifically the rural and poor inhabitants (IPCC, 2013; Chakravarty et al., 2015).

Studies have indicated that future agricultural productivity and production stability will depend on the vulnerability of the area to CC, i.e., in already food insecure areas, productivity will significantly reduce further (FAO, 2010; UNEP, 2012). Indirect impacts include increase in food prices, and poverty rise, which will render the rural areas and rural communities more vulnerable to CC (Hertel et al., 2010). Poverty especially in rural areas will manifest through decreased profitability and livelihood loss resulting from changes in crop and land values, grassland or rangeland, fisheries, and tourism values (Dasgupta et al., 2014). Impacts in rural communities will also be visible through changes in accessibility or communication, availability of freshwater, health status and nutrition and changes in job availability and urbanization. Fluctuating crop prices due to CC will enforce the poor farmers to expand his farm land by clearing margin vegetation/deforestation which will modify natural habitat, resulting loss of topsoil and hence productivity (Lambin and Meyfroidt, 2011). Worldwide about 1–8% of the land gets degraded annually (Nellemann et al., 2009) with 13% reduction in agricultural productivity (Wood et al., 2000), which adversely influence the smallholder activity and livelihood (Neely and Fynn, 2011).

7.2.2 LIVESTOCK AND FISHERIES

CC impact on livestock will be influenced by change in herd dynamics and carrying capacity of grazing lands through drought or heat stress, flood, and disease; changes in rangeland composition, decrease in fodder production or loss of rangeland productivity, demand for meat products, heterogeneity, and inequality between livestock keepers (Franco et al., 2011; Ericksen et al., 2012; Verner, 2012) ultimately affecting the livelihood of pastoralists, ranchers, and meat farmers. It was also reported that due to increased probability of drought in Africa and Asia, farmers might shift from cropping to livestock (Jones and Thornton, 2009). About 36 million fisher folk in Asia, Africa, and Latin America will experience CC impact on fisheries which will affect their livelihoods and food security through changes in social system, fish stock distribution, abundance, and catch, destruction of fishing gear and infrastructure (Mills et al., 2011; Srikanthan, 2013). The impact on inland or marine fisheries in rural coastal areas will include loss or reduction of breeding habitats and mangroves (Hall, 2011).

7.2.3 WATER RESOURCES

The effect of CC on water resources will be through the size of the population dependent on these resources and will be universally experienced (Hoekstra and Mekonnen, 2012). Changes in water bodies either through drought/lower flow or inundation through flooding of water bodies and sea-level rise will cause temporary loss of land, land activities, sediment transport, damage to transportation, communication, and other infrastructures causing disruption in operations of water infrastructures, disruption in communication especially increasing the vulnerability of rural areas, change in storage capacity and failures of water allocation systems and conflicts among regions and nations (Meza et al., 2012; Das, 2015). Similarly, in temperate and arctic regions water infrastructure will deteriorate with warming leading to increased cost of construction and maintenance (Furgal and Prowse, 2008; Larsen et al., 2008; Lemmen et al., 2008).

Agriculture uses 70% of the available water by irrigation and is increasing due to increasing population (Davies et al., 2011; Fader et al., 2011; Konar et al., 2011; Bocchiola et al., 2013; Nana et al., 2014). There will be shortage of water supply in the event of CC in the continents of Asia, Africa, and Europe due to decrease in precipitation and melt water from glacial ice and snow, failure of ground water recharge and withering of water resources affecting more than 200 million people (Agarwal et al., 2014; Palazzoli et al., 2015). Amount and distribution of water available for agriculture will be reduced with increasing population due to its competing use in other sectors (Hanafi et al., 2012; Rochdane et al., 2012; Verner, 2012). Water shortages and pollution reduced agricultural production in the Indian Punjab (De Janvry, 2010). Water shortage in sub-Saharan Africa is becoming more and more critical day by day (Power, 2010). More than 650 million people in Africa are dependent on rain-fed agriculture which is already stressed with water scarcity and land degradation. Two-thirds of Africa's arable land will be lost by 2025 due to water scarcity or drought (FAO, 2009). Lesser the water available to cropland, greater is competing use of water between agriculture and other sectors (Vörösmarty et al., 2010). Water shortage in these regions is more accelerated due to CC now and will further worsen with time.

7.2.4 RECREATION AND TOURISM

Recreation and tourism is also reported sensitive to CC affecting livelihood of rural communities close to forest and involved in tourism (Lal et al., 2011; Nyaupane and Pouldel, 2011).

7.2.5 MINING, TRADE, AND INVESTMENT

Mining enterprise may not remain economically viable due to CC, thus people dependent will become vulnerable (Damigos, 2012; Backus et al., 2013). Apart from other economic and political factors, climatic condition like drought caused volatility and unpredictability in the trading environment (Anderson and Nelgen, 2012; Nazlioglu et al., 2013; Nelson et al., 2013). The areas with high probability of climate extremes will repel investment.

7.2.6 KNOWLEDGE

CC will influence knowledge transfers (Dasgupta et al., 2014), for example, use, dissemination, and transfer of traditional knowledge (Ettenger, 2012) are believed to be threatened by CC-induced migration (Gilles et al., 2013).

7.2.7 ECONOMIC LOSS

Valuation of economic loss due to CC is done both at individual and community levels (Farber et al., 2006) but is freight with difficulty (Aldred, 2012; Hoekstra and Mekonnen, 2012; Dasgupta et al., 2014). Valuation of losses due to CC is estimated as cost of losses/damage and adaptation. Some losses can be monetarily valued (productivity, infrastructure) while some like value of lives lost and ecosystem services losses are not (Handmer et al., 2012). The value of losses due to CC is equivalent to at least 5–7% of global yearly GDP (Stern, 2007).

Farm sector will incur losses due to heat waves, droughts, storms, inundation, and flooding (Handmer et al., 2012; Dasgupta et al., 2014). CC is expected to decline 50% agricultural output by 2020 (Manyeruke et al., 2013). South Asian countries by 2080 will incur GDP reduction of 1.4 and 1.7%, respectively due to loss in agricultural productivity and welfare (Zhai and Zhuang, 2009). Ethiopia was reported to incur a GDP loss of 10% from agriculture and linked sectors due to CC (Mideksa, 2010). Asia will be annually spending US$ 4.2–5 billions more for mitigating the adverse effect of CC on agriculture (Asian Development Bank and International Food Policy Research, 2009). Food grain production will reduce by 18% between 2030 and 2050 (Dasgupta et al., 2013) and up to 40% between 2080 and 2100 (Ninan and Bedamatta, 2012) in India, while Malaysia will loss US$ 54.17 million annually in rice production (Vaghefi et al., 2011). Some African countries by 2100, according to a report will have net farm revenues decline up to 25% amounting to a total loss of US$ 48.2 billion (Dinar et al., 2008). There will be 13% reduction of cropland values in the USA (Mendelsohn et al., 2007), and in California, only annual agricultural losses was estimated up to US$ 3 billion (Franco et al., 2011). Wittrock et al., (2011) estimated crop losses of CAN$ 7–171 per hectare in Canada due to drought. Annual reduction of farm property values in Brazil was estimated as 39% (Sanghi and Mendelsohn, 2008). Income of South Americans will reduce by 14–20% from 2020 to 2060 (Seo and Mendelsohn, 2008). In Mexico each degree of temperature rise will reduce the farmland values equivalent to 4–6000 pesos (Mendelsohn et al., 2010). United Nations Economic Commission for Latin America and the Caribbean (2010a, b) estimated reduction up to 25% in gross value of production in Guatemala, Belize, Costa Rica and Honduras. A World Bank study estimated a loss of US$ 7.6 billion by 2050 for Mozambique due to infrastructure and agriculture damages (World Bank, 2010).

7.3 VULNERABILITY

Vulnerability and CC impacts are related by exposure, sensitivity, and resiliency of people (Intergovernmental Panel on CC, 2001; Regmi and Adhikari, 2007). Agriculture particularly the rainfed is more vulnerable to CC (Ahmed et al., 2011; Bellon et al., 2011). Assessing vulnerability of farming communities or in rural areas requires taking into consideration the competing conceptualizations and terminologies of vulnerability using livelihood conditions and their adaptive capacity (O'Brien et al., 2007). Factors affecting Vulnerability or resilience of in rural farmers depend on farm diversification, availability of irrigation facilities, farm management and market accessibility (Dasgupta et al., 2014). Vulnerability is location specific but its causes and solutions are not but have different social, geographic, and temporal occurrence (Ribot, 2010). Non-climate factors affecting vulnerability were outlined by many workers (Nelson et al., 2002; Horton et al., 2010; Ahmed et al., 2011; Larson et al., 2011; Mougou et al., 2011; Seto, 2011; Sietz et al., 2011; Wutich et al., 2012) and referred by Dasgupta et al., (2014) is as follows:

- Location in terms of accessibility, site factors and dependency on climate.
- Economic constraints and poverty.
- Gender inequalities.
- Social, economic, and institutional impacts/trends.
- Role of institutions particularly the local ones in facilitating access to the resources.
- Culture and tradition.
- People's perceptions of CC.

Absence or inaccessibility to institutions render the rural and farming community remain unaware of interventions and incentive structure of adaptations and unable to empower themselves making them vulnerable to CC (Ribot, 2010; Romsdahl et al., 2013). Lack of access to assets mainly due to breakdown of traditional land tenure systems renders the small farmers and particularly the landless more vulnerable (Fraser et al., 2011; McSweeney and Coomes, 2011). Small farmers and poor people also become vulnerable when market is opened to international trade as it reduces their socioeconomic stratification and encourages monocropping which reducing crop diversity (Fraser et al., 2011; McSweeney and Coomes, 2011; Sietz et al., 2011; Rivera-Ferre et al., 2013a). This is because small-scale farming increases the vulnerability of the farmers (Gbetibouo et al., 2010; Bellon et al., 2011) as they are less accessible to technology, extension services and market (Brondizio and Moran, 2008). However, smaller farmers are in advantageous position as compared to their larger monocrop counterparts due to their diversified cropping practice particularly when climatic variations render a particular crop unfit for cultivation (Brondizio and Moran, 2008).

7.4 LIVELIHOOD

Livelihood in the developing countries depends on farming (Dev, 2012; Manyeruke et al., 2013). About three-quarter of poor population is rural with agriculture as primary occupation for livelihood and income (Ravallion et al., 2007; International Fund for Agricultural Development, 2010). Rural areas contribute about half of the total global population and majority of them live in less developed countries (UN-DESA, 2013). Rural areas have different patterns of settlement, infrastructure, and livelihoods and thus are exposed to specific vulnerabilities. Agriculture creates employment opportunities for 1.3 billion smallholders (0.2–2 ha) and landless workers with low resource availability that too is ecologically vulnerable (IFAD, 2012; GHI, 2014; GoI, 2014; Lal, 2016). Consequent of less available and vulnerable resource, smallholders are entrapped into a vicious cycle of poverty, hunger, and degradation (Lal, 2016).

Drastic reductions in farm productions over a past few decades indicate the vulnerability of the farming systems to CC. Increasing temperatures along with changes in precipitation pattern and prolificacy of weeds and pests are not only reducing crop yields but also increasing the likelihood of crop failures (Black et al., 2011a; Chakravarty et al., 2015). Unfortunately, CC impact in rural areas is more acute due to deficiency in inputs and infrastructure (Nelson et al., 2009). Strange but is a fact that developed and industrial nations are more responsible for CC in a way that they are significantly releasing more greenhouse gases (GHGs) but the poor developing nations will suffer more because they cannot afford to meet the cost of CC mitigation. This is because rural areas are dependent on subsistence agriculture and natural resources; high prevalence of poverty, illiteracy, isolation, sensitivity of their geographical locations, marginality, lack of access to information, limited capacity to adopt new livelihood strategies, investment, services, policy neglect and lower human development (Dasgupta et al., 2014). Smallholders adapted to climate variability due to their traditional knowledge developed from hit and trials of past experiences of their forefathers (Pettengell, 2010).

The access and availability of natural resources to these people will also be influenced by CC. Overall this will impact security and wellbeing of the rural people (Kumssa and Jones, 2010). Moreover, impacts on rural livelihoods in countries of Asia, Africa, and Latin America will be compounded owing to their small landholding and subsistence living through detrimental effect of CC on environmental and physical processes affecting farm and natural ecosystem production along with human and domestic animal health, non-agricultural livelihoods, economic policy, trade, globalization, and food prices (Oluoko-Odingo, 2011; Hamisi et al., 2012; Lerner et al., 2013). Rising food prices render rural people inaccessible to decent nutritious diet (Ruel et al., 2010). It was reported that moderate declines of agricultural production intensified food crises in Africa (Devereux, 2009). Post-harvest or value chain of agriculture is also affected by changing climate which ultimately also decide food prices (Tefera, 2012; Stathers et al., 2013).

CC affects rural people through influencing their capabilities or farm heterogeneity (in terms of land rights/ownership, education, food price affecting decent diet, infrastructure) and social relationship (within and outside household); assets (stores, resources, claims, and access), health, and occupation (Ruel et al., 2010; Claessens et al., 2012) that affect farm production. Several studies reported decline in net revenues from crop in rainfed and smallholder farming systems due to changing climate which was significant to economies and livelihoods of the communities in terms of food security and poverty (Kotir, 2011; Müller et al., 2011; Tscharntke et al., 2012).

CC is regarded as 'hunger risk multiplier' (World Food Programme, 2011). Majority of people in the developing countries are inadequately accessed to food which gets more acute with increasing food prices due to frequent droughts resulting more and more people pushed to poverty and hunger (Feyissa, 2007; Nelson et al., 2009). More than 800 million people are food insecure and many more suffering from 'hidden hunger' (Keatinge et al., 2011; FAO, 2011; Khush et al., 2012). Moreover, climate extreme events result in price hikes of food grains and volatility which further exacerbates hunger and malnutrition among rural households (Swinnen and Squicciarini, 2012; FAO, 2013). The effects of CC have been grossly felt by the developing countries that are already grappling with scarce food and drinking water reserves and poverty (Mugandani et al., 2012; The Zimbabwean, 2012). Rural poverty and rural extreme poverty along with hunger and malnutrition is rising in South Asia and sub-Saharan Africa (IFAD, 2010; Dasgupta et al., 2014). By 2050, hunger will increase by 10–20% solely due to productivity losses. This will significantly reduce calorie availability throughout the developing world causing 24 million more malnourished children (21%

more than 2000 AD), majority of which will be from sub-Saharan Africa (WFP, 2009). Moreover, prices of staple crop will also increase up to 150% by 2060 (Bailey, 2011; OECD and the FAO, 2011).

Growing demand for food will lead to instability in governance instigating instability, conflict, and displacement. People displaced by climate impacts will be inaccessible to health care services. Large areas in these countries have limited choices of livelihood with decreasing crop yields is also threatening famines and forcing internal migration within the country. Moreover, inundation of coastal areas in these countries will also result in large-scale migration (Chakravarty and Mallick, 2003). Droughts, storms, floods, and sea level rise along with socio-economic factors will induce migration (Black et al., 2011a, b; Gray and Mueller, 2012), which may cause abandonment of settlements (McLeman, 2011). During 2010–2011 about 42 million residents of Asia and Pacific were displaced from their home due to extreme weather events (Asian Development Bank, 2012). As a result, more, and more rural people are forced to abandon agriculture (Dasgupta et al., 2014). It was reported that migration remains the only option for the rural communities when saturation to agricultural adaptation are reached (Mertz et al., 2011).

CC impact on human health, nutrition, and ability to work or learn are significant. These impacts influence the accessibility to clean air, safe drinking water, and nutritious food and secure shelter. Extreme climate events may also damage health infrastructure, jeopardizing the health services (McMichael et al., 2003) and can negatively change natural, economic, and social systems that sustain health (Christensen et al., 2007; WHO, 2009). Africa, Asia, and Latin America are vulnerable to climate sensitive diseases like tuberculosis, diarrhea, malaria, dengue, and other vector- and rodent-borne diseases (Ahern et al., 2005; Boko et al., 2007). Modest warming since 1970s is causing over 140000 additional deaths annually (McMichael et al., 2004). Increased frequency and duration of severe heat waves had increased the risk of mortality and morbidity (Epstein et al., 1995). High temperatures with poor air quality will increase incidence of heat stress and smog-induced illnesses (Cruz et al., 2007). Increased forest fire due to dry climate and deforestation will affect human health due to biomass burning smoke (Haines and Patz, 2004; Patz, 2004).

It is believed that CC may have spread HIV due to escalating poverty, population displacement, and poor care systems (Harrus and Baneth, 2005; UNEP and JUNP on HIV/AIDS, 2008). Studies concluded that infants born during drought are more likely to be malnourished than infants born at other times (UNDP, 2007a–c). The rate of wasting and stunting among preschool children were reported increased due to deficient food supply, inadequate health care and increased exposure to contaminants following post-flood years (NRC and IDMC, 2011). Under- and malnourished child has less ability to grow and develop (Martinez and Fernandez, 2007). Declining water resources will result in 5–8% increase of arid and semi-arid lands by 2080s. This will further reduce food production and thus worsening chronic hunger (IPCC, 2007a–c).

Climatic disasters destroy and damage the service-delivery infrastructures affecting health and well-being. Climate extremes also restrict social protection systems and safety nets. Billions of people are still inaccessible to social security coverage (UNEP, 2008). The impacts of climate will vary among communities because of varying levels of development and differences in social structure. Age, gender, ethnicity, social class and caste are strongly associated with social vulnerability. For example, gender norms, roles, and relations already determine different impacts on women and men, including in relation to health (WHO, 2011). Unfortunately, existing policies and social protection systems are miserably inadequate to enhance resilience and adaptive capacity to mitigate CC impacts on employment.

7.5 FOOD SECURITY

World Food Summit (1996) has defined food security as "Food security exists when all people at all times have physical and economic access to safe and nutritious food which meets their dietary needs and food preferences for an active and healthy life" (Manyeruke et al., 2013). Following issues on food security due to CC was reported by IASC (2009):

- CC will act as a multiplier of existing threats to food security.
- There will be difficulties in achieving food security due to CC.
- Community-based development processes should be promoted.
- Strengthening crisis response and prevention strategies.

Food to be secured needs both physical and economic access to food meeting dietary needs food preferences of the people. Food can be secured on four pillars (FAO, 2006):

- Food availability, i.e., sufficient quantities of food are available.
- Food access, i.e., sufficient resources to access nutritious food.
- Food use, i.e., proper utilization of the available food.
- Stability in supply and utilization of food.

The aspects and impacts of food security vary regionally over time and strongly influenced by overall socio-economic conditions of a country (Schmidhuber and Tubiello, 2007). CC on agriculture is mostly adverse causing yield reduction and thus decline food security (Bhatt et al., 2014; Bocchiola, 2017). Recognizing food security is respecting human right, which is strongly related to issues of agricultural policy, economic development and trade (Garwe, 2008). Sufficient food stock in a country does not indicate that every individual has access to sufficient food. The prime cause of under- and malnourishment in developing countries is lack of access to food and poor distribution. In addition to climatic factors economic, political, and social conditions also influence food security in the developing countries (Ninan and Bedamatta, 2012). An individual is food secure only when he has enough income to procure the food he needs or is producing enough food to meet the household requirement at least through subsistence farming. There is a strong link between poverty and food security as poor faces hunger. Thus ensuring food security needs sufficient income.

There will be 10–20% more hungry people by 2050 with 20% more malnutrition in children (IASC, 2009) which will be 5–200 million more by 2100 (Wheeler and Von, 2013). Africa will have majority of hungry people by the 2080s (Fischer et al., 2002). India with 360 million undernourished and 300 million poor people has the largest number of hungry and deprived people in the world (Ninan and Bedamatta, 2012). The per capita availability of food grain in India declined from 510 grams in 1991 to 443 grams in 2007 (Ninan and Bedamatta, 2012).

Food security is governed by many issues like food production and purchasing power. Agriculture is the most important sector in the developing countries because influence of agricultural growth on poverty reduction exceeds the influence of growth in other sectors. Hunger will increase with CC due to decrease in food production and decreased purchasing power. Changes in temperature, precipitation, and climatic extremes will further degrade the land resource. This will decrease agricultural production thus increasing poverty in the developing countries. Declining and migration of fish stocks due to CC will impact the local food security in areas where fish is the prime source of protein for poor people. Unfortunately, agricultural policy on food security is still not well defined with respect to CC and drought mitigation in developing countries. In many of these countries implementation of policies for food security is hampered by socio-economic and political challenges creating inconsistencies in policy implementation in terms of many contradicting strategies. In addition to increasing food production, accessibility to proper food and resiliency to CC is required in these countries (IASC, 2009).

7.6 SOCIAL DIMENSIONS

People are both victims and drivers of CC. Policy interventions for successful adoption of adaptations depend on people. Social dimensions of CC relate climate, related policy and society. Effective interventions rely on the transformation of socio-economic factors aiding vulnerability and empowering the victims for actions to adopt measures not only for climate-resiliency and sustainable economic future but also the essence of just and equitable societies. Recognizing social dimensions of CC in true sense is justified on these conditions:

- Social dimensions though recognized by climate agreements in vogue but unfortunately in most elemental sense and practice.
- Linking social dimensions with climate policy is fundamental respect human rights.
- CC policies needs to be linked with social dimensions holistically in the strategies of action.
- Synergies between objectives and plan of action of CC strategies, sustainable development and human rights.
- Social dimensions can easily be linked with CC policy:
- Assessing the issues, i.e., assessing initial impacts on social consequences.
- Processing policies for development, i.e., empowering decision making through transparency and involvement of stakeholders throughout the policy cycle.
- Monitoring and evaluation of results through assessing and restructuring the specific policies and strategies making it unbiased and efficient.
- Knowledge gaps are associated with social dimensions of CC. Following approaches are recommended in response to knowledge gaps:
- Complementing scientific knowledge with traditional knowledge.
- More cooperation on institutional collaboration for research on social and climate issues.
- Strengthen climate science downscaling in health, disaster management and food and water security sectors.
- Integrating social dimensions in National Adaptation Programmes.

7.6.1 INTEGRATING SOCIAL DIMENSION WITH CLIMATE CHANGE (CC) POLICIES AND PROGRAMMES

Policies must be developed and adopted considering social dimensions. Social policies and institutions should be holistic, responsible, and accountable to empower people so that they are transformed into duty bound citizens and become resilient to live in a sustainably developed society in true sense (Mearns and Norton, 2010). Integrating good governance and social principles with climate policy design will build up confidence in public spending and investments. Such holistic development can be achieved through the following principles:

7.6.1.1 PARTICIPATION

Participation in policy formulation and implementation ensures opportunity of sustainable development in true sense to all stakeholders creating equality and resiliency.

7.6.1.2 ACCOUNTABILITY

Local, regional or national governments should be accountable and committed for efficient redressal and delivery system for ensuring human development.

7.6.1.3 NON-DISCRIMINATION AND EQUITY

Status of marginalized, discriminated, and vulnerable groups should be focused during CC mitigation and adaptation policy-making in terms of their participation for involving them in decision-making process to ensure equitable outcomes.

7.6.1.4 EMPOWERMENT

Empowering local stakeholders in decision making right from policy planning to execution and then evaluation will avoid conflict of interest safeguarding entitled claims and rights. Empowering is involving the stakeholder in their own development process, i.e., right to decide for themselves. Empowerment is thus not only a stepping stone for good CC policies but forbearer of overall sustainable development.

7.6.1.5 TRANSPARENCY

The success of CC policies depends on how transparently it is prepared by giving honest access of information to all stakeholders and following recommendations are suggested:

- Climate analysis should be complemented with social impact assessments.
- Social impact assessments should be exercised frequently or conducted at every stage of programme and policy development.
- There should be regular meetings or dialogue involving all the stakeholders.
- Rights and interests of the most vulnerable should be ensured while designing and implementing socially inclusive climate solutions.
- Investments in human capital should be promoted and prioritize.
- Large infrastructures should be for low-carbon growth, contribute to livelihood opportunities and balances societal equality.
- Bringing in social dimensions through responsive budgeting in climate finance i.e., children, health, and gender-responsive budgeting.
- Climate funding should be additional to current official development assistance to prevent fund diversion from essential development goals.
- Research gaps needs to be identified and prioritizing areas to bolster research.

Success of policy interventions for CC adaptations is based on the developed resilience of people, their livelihoods, health, and well-being. Such success requires proactive stakeholder from all sections of the society along with transformed social relations which integrates equity and empowerment with CC policy processes. Such transformative social policy effectively increase resiliency to climate by change by addressing equality, social justice, increasing productive capacity, harmony, and tuning institutions for good governance, equity, and empowerment (UNRISD, 2000).

CC policies make the institutions to recognize traditional in a system approach thus converting the institutions into delivery system of economic growth and sustainable development. Social mobilization and cooperation in response to CC is resulting inclusive and transformative social policies for good governance. These inclusive and transformed social policies along with realizing climate objectives can also achieve sustainable development benefits. Recent renewed development programs based on effective and holistic sustainable development policies integrating social, environmental, and economic aspects efficiently have achieved their objectives of sustainable development.

7.6.2 TRANSFORMED POLICIES

Pro people CC policies will give following desirable outcomes:

- Energy use efficiency and sustainable carbon release.
- Integrates livelihood skills-building with formal education.
- Gender unbiased strategies and recommendations.
- Responsible government delivering inclusive growth and development.
- Climate smart communities balancing rights and duties.
- Holistic sustainable climate smart development and growth.
- Social benefits in terms of health, food, and nutrition security, energy, migration as adaptation, employment, and social cohesion, democracy, and human rights.

7.6.2.1 GENERAL WELL-BEING

CC adaptation policies should have strategies that consider all round personal development including health care and well-being against changing climate.

7.6.2.2 BALANCED DIET

Strategies should encourage fund support for rural development and agriculture for adequate production, access to food, and balanced diet.

7.6.2.3 CARBON EFFICIENT ENERGY GENERATION AND ACCESS

Strategies should encourage development of carbon-efficient energy generation system like small-scale renewable energy units with guaranteed access.

7.6.2.4 SYNERGY WITH CLIMATE-INDUCED DISPLACEMENT

Strategies should be pro-poor and pro-development to prevent large scale climate-induced population displacement or develop community friendly small urban centers.

7.6.2.5 LIVELIHOOD OPPORTUNITIES

Creating opportunities for diversified income for sustainable livelihoods through encouraging green investments.

7.6.2.6 INVOLVEMENT AND PARTICIPATION

Policies should be efficient enough to address and deliver through a system of transparency with involvement and participation of all stakeholders in decision making with equitable sharing of benefits thus avoiding conflict of interests

7.6.3 TRANSFORMING SMALLHOLDERS

The way forward is now is transforming smallholders farming system into climate-smart farming system (Bhavnani et al., 2008; FAO, 2008; IFAD, 2013) through:

- Practicing the sustainable agricultural management practices in reality by involving the small and marginal producers to manage trade-offs between farm productivity and overall sustainability.
- Enough with policy barriers.
- Accessing smallholders with information.
- Strengthening research and development with adequate flow of funds to identify research gaps and solution.

7.6.4 RESEARCH GAPS

Following research gaps were identified by IFAD (2013) needs to be filled while moving forward:

- Understanding of biological communities i.e., their structure and ecosystem services they provide.
- Relationship between below- and above-ground services with crop growth and development.
- Understanding on land use management in relation to its environment that affects crop production.
- More information on how climate affect agriculture production and management.
- Critical understanding of valuation of ecosystem services associated with farming.
- Diversification of farming methods, practices, and management.
- Investments needed to achieve the proposed transformation of smallholders, and the benefits from these investments should be quantified.

7.7 ADAPTATION

Adaptation is progressive, developed with experience on past stresses based on local and indigenous knowledge can reduces vulnerability to CC but in many cases was induced by non-climatic factors (Berrang-Ford et al., 2011; Newsham and Thomas, 2011; Nakashima et al., 2012; Vincent et al., 2013; Rivera-Ferre et al., 2013b). In developing countries, it is reported that adaptation integrated with development process can reduce poverty and improve livelihood in rural areas (Sharma and Dahal, 2011; Nielsen et al., 2012). Farming initiatives for growth in the developing countries will develop resiliency in smallholders farming systems (Kotir, 2011). CC mitigation in rural areas of developed countries also needs adaptation (Kiem and Austin, 2013).

Adaptation should be continuous with varied interventions to address the causes of vulnerability with specific targets (McGray et al., 2007). Therefore, a holistic approach is required with a focus on socio-economic and environmental dimensions of adaptive capacity. CC vulnerability is significantly influence by poverty and poor economic development as poverty is both a cause and effect of vulnerability (Hammill et al., 2008). Adaptation to be effective and sustainable requires broad-based economic development converting subsistence smallholders to commercial farmers with higher-return livelihood activities (Frank and Buckley, 2012). It is believed that poor households may adopt following three livelihood strategies to realize this transition with effective development interventions:

- **Hanging In:** Activities that maintain livelihood.
- **Stepping Up:** Investment in existing activities for benefits
- **Stepping Out:** Investment in existing activities for assets.

Such development interventions will not only make small farmers resilience to CC but also will broaden their asset base and strengthen their capabilities critical to economic development. This is because asset and livelihood activity diversification enable smallholder accumulate more asset and innovate more ways to fight poverty reducing risk. Increasing engagement of these farmers to market will further aid this transition. Unfortunately as markets are not always stable and effective, encouraging strategies linked with market for diversification with higher risk and for higher returns often causes failure. The solution lies in correct identification and promotion of institutions and mechanisms by the smallholders to manage market risk and failures.

Farming societies respond to CC impacts by adjusting/modifying/changing their farming/management practices like in planting, harvesting, irrigation, manuring/fertilizing, crop/fodder varieties, herd size and composition, grazing, and feeding patterns, diversifying crops and livelihoods and practicing rain water harvesting and conservation agriculture (De Sherbinin et al., 2011; Manandhar et al., 2011; Pretty et al., 2011; Sowers et al., 2011; Chhetri et al., 2012; Erenstein et al., 2012; Huntjens et al., 2012; Rivera-Ferre and López-i-Gelats, 2012; Rivera-Ferre et al., 2013b; Speranza, 2013). Adopting adaptations make the farming community resilient to climatic changes, for example, farmers practicing diversified farming or generating income from diversified sources is more resilient than their specialized counterparts (Seo, 2010; Gachathi and Eriksen, 2011). Agriculture in developing countries must transform to become 'climate-smart' to prevent further deterioration of natural resources ensuring food security amongst the rural poor. For this to achieve, smallholders should have access to either improved or existing technologies or extension services. The more the farmers/growers adopt or develop sustainable agricultural techniques, the more they become resilient to CC.

Sustainable management of forest and fishery resources through sustainable management (like managing stand structure and composition, harvesting, forest fires, conservation of forest genetic resources or ecological restoration, selective use of fishing gear) involving the community along institutions with benefit sharing, insurance, adequate compensation and recognition of rights can play a key role to enhance biodiversity and environmental services (like water regulation and soil protection) ultimately will support livelihood adaptation in developing countries (Gentle and Maraseni,

2012; Porter-Bolland et al., 2012; Pramova et al., 2012; Girardin et al., 2013; Terrier et al., 2013). Studies across the globe have documented many local adaptations in agriculture, water, forestry, fisheries, and biodiversity (Amede et al., 2011; Bell et al., 2011; Biemans et al., 2011; Burte et al., 2011; Connell and Grafton, 2011; Lin, 2011; Tischbein et al., 2011; Klemm et al., 2012; Marshall, 2012; Rivera-Ferre and López-i-Gelats, 2012; Dey et al., 2017a).

Adaptations applicable to biophysical systems are permanent or remain unchanged over time and are defined as hard, while some applicable to social systems are temporary or soft (Dow et al., 2013). Physical (like land, water, asset base, technology, market, institutions); socio-economic (like tenure security, rights, off-farm income, employment opportunities, credit, culture, governance, policy interventions, gender, age) and information/knowledge (like farming experience, education, technological knowledge, extension services) constraints are major hindrance for adaptation or limit adaptive capacity of the farming and rural communities in the developing countries (Cunguara and Darnhofer, 2011; Jones and Boyd, 2011; Moumouni and Idrissou, 2013).

Resiliency towards CC along with climatic factors is also socio-economically influenced (Dorte, 2008; FAO, 2008). These include:

- Human capital like skill, knowledge, and education.
- Social status like rights and privileges.
- Physical assets like personal and community infrastructures and facilities.
- Natural resources
- Financial assets.

In turn sustainable exploitation of these resources by the people depends on functioning of and accessibility to local or higher institutions which empowers people for their own livelihood and well-being through following strategies (Adger, 2006):

- Transparency and accountability in governance which respect human rights and allows participation.
- Social and cultural norms deciding access, political power and equality.
- Social policies and services like social protection, health care and education.
- Area specific policies targeting multiple sectors.
- Support system in place for efficient execution of mitigation strategies for adaptation.
- Markets, finance, and credit institutions.

Policy recommendations to improve adaptive capacity of farming and rural communities recommended by many studies against changing climatic variables (Schroth et al., 2009; Laderach et al., 2010; Schepp, 2010; Eakin et al., 2011) and referred by Dasgupta et al., (2014) are as follows:

- Climate risks and opportunities needs to be analyzed for communities.
- Climate information should be available and accessible.
- Sustainable production techniques.
- Adaptable varieties to wider climatic conditions.
- Financial support and investment.
- Accessibility to knowledge and financial support and coordinated with organized small producers.
- Farmer/grower friendly carbon markets and simplified environmental service payments system in place.
- Development of value chain strategies.

Adaptation strategies for resilience can be effectively implemented through holistic approach of linking and managing the assets like governance, human resources, institutional structures, public finance, and natural resource management (NRM). Effective adaptation strategies are implemented on principle to support existing livelihoods complimenting folk or traditional knowledge by micro-insurance, infrastructure design and investments. Thus progress will require the following strategies:

- Efficient governance with open, transparent, and accountable policy and involvement of stakeholders in decision-making processes.
- Mainstreaming planning processes.
- Responsible nodal ministry.
- Bottom-up approaches for designing, planning, and implementation.
- Empowering stakeholders for participation in assessments and feedback.
- Holistic vulnerability assessments.
- Updated information and advanced forecasting system with effective dissemination to stakeholders.
- Integration of macroeconomic impacts with the national budget.
- Considering livelihood improvement and infrastructure development as critical for an effective poverty reduction strategy.

Agricultural adaptation is possible at farmers and other stakeholders (Ninan and Bedamatta, 2012); however, relative speed of adoption varies (Reilly and Schimmelpfennig, 1999; Table 7.1).

TABLE 7.1 Time Taken to Adopt an Adaptation Strategy

Adaptation Measure	Average Time Taken to Adopt (Years)
Improved cultivar	9
Irrigation from created infrastructure	75
Releasing improved cultivar	12
Plowing/tilling	11
Shifting to another land use	7
Machinery for irrigation	23
Fertigation	10

(Modified from Reilly and Schimmelpfennig, 1999).

Adaptations are autonomous and policy-driven for short and long time frame (Stern, 2007; Table 7.2).

TABLE 7.2 Adaptation in Practice

Response	Autonomous	Policy-Driven
Short	– Short duration like changing interculture operation practices – Minimizing risk through compensation or insuring	– Research for understanding cause, precaution, and minimizing the risk – Efficiency in addressing during the event of risk
Long	– Investment in climate resilience	– Efficient funding for building risk-mitigating facilities – Avoiding the impacts, e.g., land use planning

(Modified from Reilly and Schimmelpfennig, 1999).

Developing countries are trying different crop combinations and promoting integrated production management. Changing food habits also help to implement agricultural adaptation strategies (Ninan and Bedamatta, 2012). According to Ninan and Bedamatta (2012), the following broad categories of responses are beneficial:

- Regular and efficient extension services to farming community.
- Enlisting farming constraints of farming systems.
- Development of climate-smart crop and animal varieties.

- Effective social and farm insurance mechanism to protect the farmers against climate-related losses.
- Linking of service delivery systems for efficient support system to farmers.
- Removing subsidies to ensure speedy adoption of adaptation process by the farmers.

The developing countries should adopt sector-specific CC adaptation strategies. These sector-specific strategies recommended by UNFCCC (2007) are listed in Table 7.3.

A recent strategy to make smallholder farming climate-smart, while achieving food security is "4 per Thousand" goal (Lal, 2014, 2016; Le Foll, 2015; Chambers et al., 2016; Minasny et al., 2017, 2018) through targeting soil organic carbon (SOC) management and thus improving soil fertility, productivity, and mitigating CC (Lal, 2016; Chabbi et al., 2017; Minasny et al., 2017; Soussana et al., 2017; Van Groenigen et al., 2017; Baveye et al., 2018; De Vries, 2018; Nath et al., 2018; White et al., 2018). The responsibility of achieving this aspiring goal depends on efficiently empowering smallholders through inclusive, responsive, and accountable institutions and good governance integrating holistic social policies and climate policy designs to mobilize and transform smallholder producer responsive and resilient for sustainable developed society.

7.8 CONCLUSION

Global CC is today's greatest challenges and it is crucial to act now and take decisive and immediate action against CC. The solution is adaptation and mitigation and its fundamental is that it should be at the top of the political agenda in every country. Policies of developing countries to eradicate poverty, ensure food security, as well as to provide education and health services need to include adaptation strategies and implement it effectively considering social dimensions. Climate risk analysis can help to recommend appropriate adaptation strategy. Climate as resource needs efficient management or otherwise a risk to be faced which requires a portfolio of assets to be prepared. The assets are natural, physical, and biological including human, man-made, intellectual, policy, legislative, and service resources. Each asset should be value assessed for their integration into a holistic approach, transforming the actions into sustainable development for today and tomorrow.

TABLE 7.3 Adaptation Measures for Tropical Areas

Vulnerable Sectors	Reactive Adaptation	Anticipatory Adaptation
Irrigation Facilities	– Conserving groundwater and surface waterbody catchment – Increasing efficiency of irrigation facilities and its accessibility – Conserving rainwater for use – Using saline water after desalination	– Efficient water management through recycling, policy reform, and conservation – Vulnerability assessment and efficient response systems for water-related hazards
Farming and Livelihood	– Soil and water conservation – Ensuring productivity through improved varieties, sustainable agronomic practices, timely extension services, and capacity building	– Efficient crop/animal improvement program – Integrated nutrient and pest management – Multiple farming practices – Institutional back up: financial and market – Improved weather forecasting
Farm Family Well-Being	– Improved living infrastructure, hygiene, and sanitation – Efficient disaster management	– Efficient disaster forecasting system – Better surveillance and monitoring – Creating green surrounding and efficient pollution management systems
Land and Coastal Ecosystems	– Conserving natural ecosystems forest and increasing vegetation cover – Increasing trees in agricultural land use – Construction and maintenance of beach protection and other structures – Public participation and capacity building	– Creation of green and conserved – Developing climate-smart species/varieties and gene conservation – Environmental monitoring and assessment – Initiating area development programs integrating socioeconomic factors with effective research and monitoring base – Efficient environmental legislation

(Modified from UNFCCC, 2007).

KEYWORDS

- **adaptation**
- **climate change**
- **food security**
- **social dimension**
- **vulnerability**

REFERENCES

ADB & IFPR, (2009). *Building Climate Resilience in the Agriculture Sector in Asia and the Pacific* (p. 304). ADB and IFPRI, Mandaluyong City, Metro Manila, Philippines.

ADB, (2012). *Addressing Climate Change and Migration in Asia and the Pacific* (p. 82). ADB, Mandaluyong City, Metro Manila, Philippines.

Adger, N., (2006). Vulnerability. *Global Environmental Change, 16*, 268–281.

Agarwal, A., Babel, M. S., & Maskey, S., (2014). Analysis of future precipitation in the Koshi river basin, Nepal. *Journal of Hydrology, 513*, 422–434.

Ahern, M. J., Kovats, R. S., Wilkinson, P., Few, R., & Matthies, F., (2005). Global health impacts of floods, epidemiological evidence. *Epidemiological Reviews, 27*, 36–45.

Ahmed, S. A., Diffenbaugh, N. S., Hertel, T. W., Lobell, D. B., Ramankutty, N., Rios, A. R., & Rowhani, P., (2011). Climate volatility and poverty vulnerability in Tanzania. *Global Environmental Change, 21*, 46–55.

Aldred, J., (2012). Climate change uncertainty, irreversibility and the precautionary principle. *Cambridge Journal of Economics, 36*, 1051–1072.

Amede, T., Menza, M., & Awlachew, S. B., (2011). Zai improves nutrient and water productivity in the Ethiopian highlands. *Experimental Agriculture, 47*(S1), 7–20.

Anderson, K., & Nelgen, S., (2012). Trade barrier volatility and agricultural price stabilization. *World Development, 40*, 36–48.

Backus, G. A., Lowry, T. S., & Warren, D. E., (2013). The near-term risk of climate uncertainty among the U. S. states. *Climatic Change, 116*, 495–522.

Bailey, R., (2011). *Growing a Better Future: Food Justice in a Resource-Constrained World.* Oxford, Oxfam, Oxford.

Baulch, B. (ed.)., (2011). *Why Poverty Persists, Poverty Dynamics in Asia and Africa.* Cheltenham, UK.

Baveye, P. C., Berthelin, J., Tessier, D., & Lemaire, G., (2018). The "4 per 1000" initiative, a credibility issue for the soil science community? *Geoderma, 309*, 118–123.

Bell, A. R., Engle, N. L., & Lemos, M. C., (2011). How does diversity matter? The case of Brazilian river basin councils. *Ecology and Society, 16*, 42.

Bellon, M. R., Hodson, D., & Hellin, J., (2011). Assessing the vulnerability of traditional maize seed systems in Mexico to climate change. *Proceedings of the National Academy of Sciences of the United States of America, 108*, 13432–13437.

Berrang-Ford, L., Ford, J. D., & Paterson, J., (2011). Are we adapting to climate change? *Global Environmental Change, Human and Policy Dimensions, 21*, 25–33.

Bhatt, D., Maskey, S., Babel, M. S., Uhlenbrook, S., & Prasad, K. P., (2014). Climate trends and impacts on crop production in the Koshi River basin of Nepal. *Regional Environment Change, 14*, 1291–1301.

Bhavnani, A., Chiu, R. W. W., Janakiram, S., Silarszky, P., & Bhatia, D., (2008). *The Role of Mobile Phones in Sustainable Rural Poverty Reduction.* Washington, DC, ICT Policy Division, Global Information and Communications Department (GICT), World Bank.

Biemans, H., Haddeland, I., Kabat, P., Ludwig, F., Hutjes, R. W. A., Heinke, J., Von Bloh, W., & Gerten, D., (2011). Impact of reservoirs on river discharge and irrigation water supply during the 20th century. *Water Resources Research, 47*, W03509, doi: 10.1029/2009WR008929.

Black, R., Adger, W. N., Arnell, N. W., Dercon, S., Geddes, A., & Thomas, D., (2011b). The effect of environmental change on human migration. *Global Environmental Change, 21*(1), S3–S11.

Black, R., Kniveton, D., & Schmidt-Verkerk, K., (2011a). Migration and climate change, towards an integrated assessment of sensitivity. *Environment and Planning, 43*, 431–450.

Bocchiola, D., (2017). Agriculture and food security under climate change in Nepal. *Advances in Plants and Agriculture Research, 6*, 00237. doi: 10.15406/apar.2017.06.00237.

Bocchiola, D., Nana, E., & Soncini, A., (2013). Impact of climate change scenarios on crop yield and water footprint of maize in the Po valley of Italy. *Agricultural Water Management, 116*, 50–61.

Boko, M., Niang, I., Nyong, A., Vogel, C., Githeko, A., Medany, M., Osman-Elasha, B., Tabo, R., & Yanda, P., (2007). Africa. In: Parry, M. L., Canziani, O. F., Palutikof, J. P., Van Der Linden, P. J., & Hanson, C. E., (eds.), *Climate Change 2007, Impacts, Adaptation and Vulnerability. Contribution of Working Group II to the Fourth Assessment Report of the IPCC* (pp. 433–467). Cambridge University Press, Cambridge.

Brondizio, E. S., & Moran, E. F., (2008). Human dimensions of climate change, the vulnerability of small farmers in the Amazon. *Philosophical Transactions of the Royal Society, B., 363*, 1803–1809.

Burte, J. D. P., Coudrain, A., & Marlet, S., (2011). Use of water from small alluvial aquifers for irrigation in semi-arid regions. *Revista Ciência Agronômica, 42*, 635–643.

Carey, M., Huggel, C., Bury, J., Portocarrero, C., & Haeberli, W., (2012). An integrated socio-environmental framework for glacier hazard management and climate change adaptation, lessons from Lake 513, Cordillera Blanca, Peru. *Climatic Change, 112*, 733–767.

Chabbi, A., Lehmann, J., Ciais, P., Loescher, H. W., Cotrufo, M. F., Don, A., San, C. M., Schipper, L., Six, J., Smith, P., & Rumpel, C., (2017). Aligning agriculture and climate policy. *Nature Climate Change, 7*, 307–309.

Chakravarty, S., & Mallick, K., (2003). Agriculture in a green house world, what really will be? In: Kumar, A., (ed.), *Environmental Challenges of 21st Century* (pp. 633–652). APH Publishing House, New Delhi.

Chakravarty, S., Puri, A., & Shukla, G., (2015). Climate change vis-à-vis agriculture, Indian and global view- implications, abatement, adaptation and tradeoff. In: Sengar, R. S., & Sengar, K., (ed.), *Climate Change Effect on Crop Productivity* (pp. 1–88). CRC Press.

Chambers, A., Lal, R., & Paustian, K., (2016). Soil carbon sequestration potential of US croplands and grasslands, implementing the 4 per thousand initiative. *Journal of Soil Water Conservation,* https://doi.org/10.2489/jswc.71.3.68A (Accessed on 9 October 2019).

Chandy, L., & Gertz, G., (2011). *Poverty in Numbers, the Changing State of Global Poverty from 2005 to 2015*. Policy Brief 2011–01, Global Economy and Development at Brookings, the Brookings Institution Washington, DC.

Charvériat, C., (2000). *Natural Disasters in Latin America and the Caribbean, an Overview of Risk*. Inter-American Development Bank, Research Department–Working Paper no. 434.

Chen, S., & Ravallion, M., (2012). *An Update to the World Bank's Estimates of Consumption Poverty in the Developing World.* World, Bank, Washington, DC.

Chhetri, N., Chaudhary, P., Tiwari, P. R., & Yadaw, R. B., (2012). Institutional and technological innovation, Understanding agricultural adaptation to climate change in Nepal. *Applied Geography, 33*, 142–150.

Christensen, J. H., Hewitson, B., Busuioc, A., Chen, A., Gao, X., Held, I., et al., (2007). Regional climate projections. In: Solomon, S., Qin, D., Manning, M., Chen, Z., Marquis, M., Averyt, K. B., Tignor, M., & Miller, H. L., (eds.), *Climate Change 2007- The Physical Science Basis: Contribution of Working Group I to the Fourth Assessment Report of the Intergovernmental Panel on Climate Change*. Cambridge University Press, Cambridge, United Kingdom and New York, NY, USA.

Claessens, L., Antle, J. M., Stoorvogel, J. J., Valdivia, R. O., Thornton, P. K., & Herrero, M., (2012). A method for evaluating climate change adaptation strategies for smallscale farmers using survey, experimental and modeled data. *Agricultural Systems, 111*, 85–95.

Connell, D., & Grafton, Q., (2011). *Basin Futures: Water Reform in the Murray-Darling Basin* (p. 500). Australia National University (ANU), ANU Press, Canberra, Australia.

Cruz, R. V., Harasawa, H., Lal. M., Wu, S., Anokhin, Y., Punsalmaa, B., Honda, Y., Jafari, M., Li, C., & Hu Ninh, N., (2007). Asia. In: Parry, M. L., Canziani, O. F., Palutikof, J. P., Van Der Linden, P. J., & Hanson, C. E., (eds.), *Climate Change 2007, Impacts, Adaptation and Vulnerability. Contribution of Working Group II to the Fourth Assessment Report of the IPCC* (pp. 469–506). Cambridge University Press, Cambridge.

Cunguara, B., & Darnhofer, I., (2011). Assessing the impact of improved agricultural technologies on household income in rural Mozambique. *Food Policy, 36*, 378–390.

Damigos, D., (2012). Monetizing the impacts of climate change on the Greek mining sector. *Mitigation and Adaptation Strategies for Global Change, 17*, 865–878.

Das, D., (2015). Changing climate and its impacts on Assam, Northeast India. Bandung, *Journal of the Global South, 2*, 26 doi: 10.1186/s40728–015–0028–4.

Dasgupta, P., Bhattacharjee, D., & Kumari, A., (2013). Socio-economic analysis of climate change impacts on food grain production in Indian states. *Environmental Development, 8*, 5–21.

Dasgupta, P., Morton, J. F., Dodman, D., Karapinar, B., Meza, F., Rivera-Ferre, M. G., Toure Sarr, A., & Vincent, K. E., (2014). Rural areas. In: Field, C. B., Barros, V. R., Dokken, D. J., Mach, K. J., Mastrandrea, M. D., Bilir, T. E., et al., (eds.), *Climate Change 2014, Impacts, Adaptation, and Vulnerability. Part A, Global and Sectoral Aspects, Contribution of Working Group II to the Fifth Assessment Report of the Intergovernmental Panel on Climate Change* (pp. 613–657). Cambridge University Press, Cambridge, United Kingdom and New York, NY, USA.

Davies, W. J., Zhang, J., Yang, J., & Dodd, I. C., (2011). Novel crop science to improve yield and resource use efficiency in water-limited agriculture. *Journal of Agricultural Science, 149*, 123–131.

De Janvry, A., (2010). Agriculture for development, new paradigm and options for success. *Ecological Economics, 41*, 17–36.

De Sherbinin, A., Warner, K., & Ehrhart, C., (2011). Casualties of climate change. *Scientific American, 304*, 64–71.

De Vries, W., (2018). Soil carbon 4 per mille, a good initiative but let's manage not only the soil but also the expectations. *Geoderma, 309*. https://doi.org/10.1016/j.geoderma.207.05.023 (Accessed on 9 October 2019).

Dev, M. S., (2012). *Small Farmers in India, Challenges and Opportunities*. WP-2012-014. http://www.igidr.ac.in/pdf/publication/WP-2012–014.pdf (Accessed on 9 October 2019).

Devereux, S., (2009). Why does famine persist in Africa? *Food Security, 1*, 25–35.

Dey, T., Pala, N. A., Shukla, G., Pal, P. K., & Chakravarty, S., (2017b). Perception on impact of climate change on forest ecosystem in protected area of West Bengal, India. *Journal of Forest and Environmental Science, 33*, 1–7. https://doi.org/10.7747/JFES.2017.33.1.1 (Accessed on 9 October 2019).

Dey, T., Pala, N. A., Shukla, G., Pal, P. K., Das, G., & Chakravarty, S., (2017a). Climate change perceptions and response strategies of forest fringe communities in Indian Eastern Himalaya. *Environment, Development and Sustainability*. doi: 10.1007/s10668–017–9920–1.

DID, (2004). *The Impact of Climate Change on the Vulnerability of the Poor (03)*. Global and Local Environment Team, Policy Division, Department of International Development, London, UK.

Dinar, A., Hassan, R., Mendelsohn, R., & Benhin, J., (2008). *Climate Change and Agriculture in Africa, Impact Assessment and Adaptation Strategies* (p. 189). Earthscan, London, UK and Sterling, VA, USA.

Dorte, V., (2008). *Reducing Poverty, Protecting Livelihoods, and Building Assets in a Changing Climate, Social Implications of Climate Change in Latin America and the Caribbean*. The World Bank, Washington DC.

Dow, K., Berkhout, F., & Preston, B. L., (2013). Limits to adaptation to climate change, a risk approach. *Current Opinion in Environmental Sustainability, 5*, 384–391.

Eakin, H., Bojorquez-Tapia, L. A., Monterde, D. R., Castellanos, E., & Haggar, J., (2011). Adaptive capacity and social-environmental change, theoretical and operational modeling of smallholder coffee systems response in Mesoamerican Pacific Rim. *Environmental Management, 47*, 352–367.

Epstein, Y., Sohar, E., & Shapiro, Y., (1995). Exceptional heatstroke, a preventable condition. *Israel Journal of Medical Science, 31*, 454–462.

Erenstein, O., Sayre, K., Wall, P., Hellin, J., & Dixon, J., (2012). Conservation agriculture in maize- and wheat-based systems in the (sub) tropics, lessons from adaptation initiatives in South Asia, Mexico, and Southern Africa. *Journal of Sustainable Agriculture, 36*, 180–206.

Ericksen, P., De Leeuw, J., Thornton, P., Said, M., Herrero, M., & Notenbaert, A., (2012). Climate change in sub-Saharan Africa, what consequences for pastoralism? In: Catley, A., Lind, J., & Scoones, I., (eds.), *Pastoralism and Development in Africa, Dynamic Change at the Margins* (pp. 71–82). Routledge, London, UK, and New York, NY, USA.

Ettenger, K., (2012). Aapuupayuu (the weather warms up), climate change and the Eeyouch (Cree) of Northern Quebec. In: Castro, A. P., Taylor, D., & Brokensha, D. W., (eds.), *Climate Change and Threatened Communities, Vulnerability, Capacity and Action* (pp. 107–117). Practical Action Publishing, Rugby, UK.

Fader, M., Gerten, D., Thammer, M., Lotze-Campen, H., & Lucht, W., (2011). Internal and external green-blue agricultural water footprints of nations, and related water and land savings through trade. *Hydrology and Earth System Science, 15*, 1641–1660.

FAO, (2006). *The State of Food Insecurity in the World 2005*. FAO, Rome.

FAO, (2008). *Water and the Rural Poor, Interventions for Improving Livelihoods in Sub-Saharan Africa.* FAO, Rome.

FAO, (2009). *How to Feed the World in 2050.* FAO, Rome.

FAO, (2010). *Climate-Smart Agriculture, Policies, Practices and Financing for Food Security, Adaptation and Mitigation.* FAO, Rome.

FAO, (2011). *The State of the World's Land and Water Resources for Food and Agriculture (SOLAW) – Managing Systems at Risk.* Rome, FAO, Rome and Earthscan, London.

FAO, (2013). *FAOSTAT Database.* FAO, Rome. Available at: http://www.fao.org/faostat/en/#data/RF.

Farber, S., Costanza, R., Childers, D. L., Erickson, J., Gross, K., Grove, M., Hopkinson, C. S., Kahn, J., Pincetl, S., Troy, A., Warren, P., & Wilson, M., (2006). Linking ecology and economics for ecosystem management. *Bioscience, 56,* 121–133.

Feyissa, R., (2007). *The Sub-Saharan African Agriculture, Potential, Challenges and Opportunities.* Paper written for the 1stconference, Can Africa Feed Itself, Oslo, Norway.

Fischer, G., Shah, M., & Van Velthuizen, H., (2002). *Climate Change and Agricultural Variability, a Special Report, on Climate Change and Agricultural Vulnerability.* Contribution to the World Summit on Sustainable Development. Johannesburg (Global, agriculture).

Franco, G., Cayan, D. R., Moser, S., Hanemann, M., & Jones, M., (2011). Second California Assessment, integrated climate change impacts assessment of natural and managed systems. *Climatic Change, 109*(1), 1–19.

Frank, J., & Buckley, P., (2012). *Small-Scale Farmers and Climate Change* (p. 31). How can farmer organizations and Fairtrade build the adaptive capacity of smallholders? IIED, London.

Fraser, E. D. G., Dougill, A. J., Hubacek, K., Quinn, C. H., Sendzimir, J., & Termansen, M., (2011). Assessing vulnerability to climate change in dry land livelihood systems, conceptual challenges and interdisciplinary solutions. *Ecology and Society, 16,* 3. www.ecologyandsociety.org/vol16/iss3/art3/ (Accessed on 9 October 2019).

Furgal, C., & Prowse, T., (2008). Northern Canada. In: Lemmen, D. S., Warren, F. J., Lacroix, J., & Bush, E., (eds.), *From Impacts to Adaptation, Canada in a Changing Climate 2007* (pp. 61–118). Government of Canada, Ottawa, ON, Canada.

Gachathi, F. N., & Eriksen, S., (2011). Gums and resins, the potential for supporting sustainable adaptation in Kenya's dry lands. *Climate and Development, 3,* 59–70.

Garwe, D., (2008). *Intellectual Property Rights and Food Security* (pp. 26–50). Regional status report on trade and development, agro-biodiversity and food sovereignty. CTDT Publication. Harare.

Gbetibouo, G. A., Ringler, C., & Hassan, R., (2010). Vulnerability of the South African farming sector to climate change and variability, an indicator approach. *Natural Resources Forum, 34,* 175–187.

Gentle, P., & Maraseni, T. N., (2012). Climate change, poverty and livelihoods, adaptation practices by rural mountain communities in Nepal. *Environmental Science and Policy, 21,* 24–34.

GHI, (2014). *Global Agricultural Productivity Report: Global Revolutions in Agriculture, the Challenge and Promise of 2050.* GAP 801 17th Street, NW, Suite 200, Washington, D. C. 20006, USA.

Gilles, J. L., Thomas, J. L., Valdivia, C., & Yucra, E. S., (2013). Laggards or leaders, conservers of traditional agricultural knowledge in Bolivia. *Rural Sociology, 78,* 51–74.

Girardin, M. P., Ali, A. A., Carcaillet, C., Blarquez, O., Hely, C., Terrier, A., Genries, A., & Bergeron, Y., (2013). Vegetation limits the impact of a warm climate on boreal wildfires. *New Phytologist, 199,* 1001–1011.

GoI., (2014). *All India Reports on Number and Area of Agricultural Holdings from 1976–1977, 1980–1981, 1990–1991, 2000–2001, 2005–2006 and 2010–2011*. Agricultural Census Division, Department of Agriculture and Co-operation, GoI, New Delhi.
Gray, C., & Mueller, V., (2012). Drought and population mobility in rural Ethiopia. *World Development, 40*, 134–145.
Haines, A., & Patz, J. A., (2004). Health effects of climate change. *JAMA, 291*, 99–103.
Hall, S. J., (2011). Climate change and other external drivers in small-scale fisheries, practical steps for responding. In: Pomeroy, R. S., & Andrew, N., (eds.), *Small-Scale Fisheries Management, Frameworks and Approaches for the Developing World* (pp. 132–159). CABI Publishing, Wallingford, UK and Cambridge, MA, USA.
Hamisi, H. I., Tumbo, M., Kalumanga, E., & Yanda, P., (2012). Crisis in the wetlands, combined stresses in a changing climate–experience from Tanzania. *Climate and Development, 4*, 5–15.
Hammill, A., Mathew, R., & McCarter, E., (2008). Microfinance and climate change adaptation. *IDS Bulletin, 39*, 113–122. Available from: http://www.iadb.org/intal/intalcdi/PE/2008/01609.pdf (Accessed on 9 October 2019).
Hanafi, S., Mailhol, J. C., Poussin, J. C., & Zairi, A., (2012). Estimating water demand at irrigation scheme scales using various levels of knowledge, applications in northern Tunisia. *Irrigation and Drainage, 61*, 341–347.
Handmer, J., Honda, Y., Kundzewicz, Z. W., Arnell, N., Benito, G., Hatfield, J., Mohamed, I. F., Peduzzi, P., Wu, S., Sherstyukov, B., Takahashi, K., & Yan, Z., (2012). Changes in impacts of climate extremes, human systems and ecosystems. In: Field, C. B., Barros, V., Stocker, T. F., Qin, D., Dokken, D. J., Ebi, K. L., et al., (eds.), *Managing the Risks of Extreme Events and Disasters to Advance Climate Change Adaptation. A Special Report of Working Groups I and II of the Intergovernmental Panel on Climate Change* (pp. 231–290). Cambridge University Press, Cambridge, UK, and New York, NY, USA.
Harrus, S., & Baneth, G., (2005). Drivers for the emergence and re-emergence of vector-borne protozoal and rickettsial organisms. *International Journal for Parasitology, 35*, 1309–1318.
Hertel, T. W., Burke, M. B., & Lobell, D. B., (2010). The poverty implications of climate induced crop yield changes by 2030. *Global Environmental Change, 20*, 577–585.
Hoekstra, A. Y., & Mekonnen, M. M., (2012). The water footprint of humanity. *Proceedings of the National Academy of Sciences of the United States of America, 109*, 3232–3237.
Horton, G., Hanna, L., & Kelly, B., (2010). Drought, drying and climate change, emerging health issues for aging Australians in rural areas. *Australasian Journal on Ageing, 29*, 2–7.
Huntjens, P., Lebel, L., Pahl-Wostl, C., Camkin, J., Schulze, R., & Kranz, N., (2012). Institutional design propositions for the governance of adaptation to climate change in the water sector. *Global Environmental Change, 22*, 67–88.
IASC, (2009). *Climate Change, Food Insecurity and Hunger*. Technical paper for IASC task force on climate change. Submitted by WFP, FAO, IFRC and OXFAM as well as WHO, WVI, CARE, CARITAS and Save the Children.
IFAD, (2010). *Rural Poverty Report 2011: New Realities, New Challenges, New Opportunities for Tomorrow's Generation* (p. 319). IFAD, IFAD, Rome, Italy.
IFAD, (2012). *Sustainable Smallholder Agriculture, Feeding the World, Protecting the Planet* (p. 12). IFAD, Rome, Italy.
IFAD, (2013). *Smallholders, Food Security and the Environment* (p. 52). IFAD, Rome.
IPCC, (2001). In: McCarthy, J. J., Canziani, O. F., Leary, N. A., Dokken, D. J., & White, K. S., (eds.), *Climate Change 2001, Impacts, Adaptation and Vulnerability. A Contribution of the Working Group II to the Third Assessment Report of the Intergovernmental Panel on Climate Change*. Cambridge University Press, UK.

IPCC, (2007a). *Climate Change 2007, The Physical Science Basis*. Contribution of Working Group I to the Fourth Assessment Report of the Intergovernmental Panel on Climate Change. Cambridge University Press, Cambridge.

IPCC, (2007b). *Climate Change 2007, Impacts, Adaptation and Vulnerability*. Contribution of Working Group II to the Fourth Assessment Report of the Intergovernmental Panel on Climate Change. Cambridge University Press, Cambridge.

IPCC, (2007c). *Climate Change 2007, Mitigating the Effects of Climate Change*. Contribution of Working Group III to the Fourth Assessment Report of the Intergovernmental Panel on Climate Change. Cambridge University Press, Cambridge.

IPCC, (2013). Summary for policymakers. In: Stocker, T. F., Qin, D., Plattner, G. K., Tignor, M., Allen, S. K., Boschung, J., et al., (eds.), *Climate Change 2013, The Physical Science Basis. Contribution of Working Group I to the Fifth Assessment Report of the Intergovernmental Panel on Climate Change*. Cambridge University Press, Cambridge, United Kingdom and New York, NY, USA.

Jones, L., & Boyd, E., (2011). Exploring social barriers to adaptation, insights from Western Nepal. *Global Environmental Change, Human and Policy Dimensions, 21*, 1262–1274.

Jones, P. G., & Thornton, P. K., (2009). Croppers to livestock keepers, livelihood transitions to 2050 in Africa due to climate change. *Environmental Science and Policy, 12*, 427–437.

Kaushik, G., & Sharma, K. C., (2015). Climate change and rural livelihoods-adaptation and vulnerability in Rajasthan. *Global NEST Journal, 1*, 41–49.

Keatinge, J. D. H., Yang, R. Y., Hughes, J. D., Easdown, W. J., & Holmer, R., (2011). The importance of vegetables in ensuring both food and nutritional security in attainment of the millennium development goals. *Food Security, 3*, 491–501.

Khush, G., Lee, S., Cho, J. I., & Jeon, J. S., (2012). Biofortification of crops for reducing malnutrition. *Plant Biotechnology Reports, 6*, 195–202.

Kiem, A., & Austin, E., (2013). Drought and the future of rural communities, opportunities and challenges for climate change adaptation in regional Victoria, Australia. *Global Environmental Change, 23*, 1307–1316.

Klemm, O., Schemenauer, R. S., Lummerich, A., Cereceda, P., Marzol, V., Corell, D., et al., (2012). Fog as a fresh-water resource, overview and perspectives. *Ambio, 41*, 221–234.

Konar, M., Dalin, C., Suweis, S., Hanasaki, N., Rinaldo, A., & Rodriguez-Iturbe, I., (2011). Water for food, The global virtual water trade network. *Water Resources Research, 47*, W05520.

Kotir, J. H., (2011). Climate change and variability in sub-Saharan Africa, a review of current and future trends and impacts on agriculture and food security. *Environment, Development and Sustainability, 13*, 587–605.

Kumssa, A., & Jones, J. F., (2010). Climate change and human security in Africa. *International Journal of Sustainable Development and World Ecology, 17*, 453–461.

Laderach, P., Haggar, J., Lau, C., Eitzinger, A., Ovalle, O., Baca, M., Jarvis, A., & Lundy, M., (2010). *Mesoamerican Coffee, Building a Climate Change Adaptation Strategy* (p. 4). CIAT Policy Brief, International Center for Tropical Agriculture (CIAT), CIAT, Cali, Colombia.

Lal, P., Alavalapati, J., & Mercer, E., (2011). Socio-economic impacts of climate change on rural United States. *Mitigation and Adaptation Strategies for Global Change, 7*, 1381–2386.

Lal, R., (2014). Small landholder farming and global food security. In: Lal, R., & Stewart, B. A., (eds.), *Advances in Soil Science, Soil Management of Smallholder Agriculture*. CRC Press, Taylor & Francis, Group 6000 Broken Sound Parkway NW.

Lal, R., (2016). Potential and challenges of conservation agriculture in sequestration of atmospheric CO_2 for enhancing climate-resilience and improving productivity of soil of small landholder farms. *CAB Reviews, 11*, 1–16.

Lambin, E. F., & Meyfroidt, P., (2011). Global land use change, economic globalization, and the looming land scarcity. *Proceedings of the National Academy of Sciences of the United States of America, 108*, 3465–3472.

Larsen, P., Goldsmith, S., Smith, O., Wilson, M., Strzepek, K., Chinowsky, P., & Saylor, B., (2008). Estimating future costs for Alaska public infrastructure at risk from climate change. *Global Environmental Change, 18*, 442–457.

Larson, K., Ibes, D. C., & White, D. D., (2011). Gendered perspectives about water risks and policy strategies, a tripartite conceptual approach. *Environment and Behavior, 43*, 415–438.

Le Foll, S., (2015). *4 Per 1000, a New Program for Carbon Sequestration in Agriculture.* French Minister of Agriculture, Agrifood and Forestry, Paris, France.

Lemmen, D. S., Warren, F. J., Lacroix, J., & Bush, E., (2008). *From Impacts to Adaptation, Canada in a Changing Climate 2007*. Government of Canada, Ottawa, ON, Canada.448p.

Lerner, A. M., Eakin, H., & Sweeney, S., (2013). Understanding pen-urban maize production through an examination of household livelihoods in the Toluca Metropolitan Area, Mexico. *Journal of Rural Studies, 30*, 52–63.

Lin, B. B., (2011). Resilience in agriculture through crop diversification, adaptive management for environmental. *Bioscience, 61*, 183–193.

Manandhar, S., Vogt, D. S., Perret, S. R., & Kazama, F., (2011). Adapting cropping systems to climate change in Nepal, a cross-regional study of farmers' perception and practices. *Regional Environment Change, 11*, 335–348.

Manyeruke, C., Hamauswa, S., & Mhandara, L., (2013). The effects of climate change and variability on food security in Zimbabwe, a socio-economic and political analysis. *International Journal of Humanities and Social Science, 3*, 270–286.

Marshall, A., (2012). Existing agbiotech traits continue global march. *Nature Biotechnology, 30*, 207, doi: 10.1038/nbt.2154.

Martinez, R., & Fernandez, A., (2007). *The Cost of Hunger, Social and Economic Impact of Child Undernutrition in Central America and the Dominican Republic*. Economic Commission for Latin America and the Caribbean and the World Food Programme, Santiago de Chile.

McGray, H., Hammill, A., & Bradley, R., (2007). *Weathering the Storm, Options for Framing Adaptation and Development.* World Resources Institute, Washington DC.

McLeman, R. A., (2011). Settlement abandonment in the context of global environmental change. Global environmental change, *Human and Policy Dimensions, 21*(1), S108–S120.

McMichael, A. J., Campbell-Lendrum, D. H., Corvalán, C. F., Ebi, K. L., Githeko, A. K., Scheraga, J. D., & Woodward, A., (2003). *Climate Change and Human Health – Risks and Responses* (p. 322). World Health Organization, Geneva.

McMichael, A. J., Campbell-Lendrum, D., Kovats, R. S., Edwards, S., Wilkinson, P., Edmonds, N., Nicholls, N., Hales, S., Tanser, F. C., Le Sueur, D., Schlesinger, M., & Andronova, N., (2004). Climate change. In: Ezzati, M., Lopez, A. D., Rodgers, A., & Murray, C. J. L., (eds.), *Comparative Quantification of Health Risks, Global and Regional Burden of Disease Due to Selected Major Risk Factors* (pp. 1543–1649). World Health Organization, Geneva.

McSweeney, K., & Coomes, O. T., (2011). Climate-related disaster opens a window of opportunity for rural poor in northeastern Honduras. *Proceedings of the National Academy of Sciences of the United States of America, 108*, 5203–5208.

Mearns, R., & Norton, A., (2010). *Social Dimensions of Climate Change: Equity and Vulnerability in a Warming World.*, The World Bank, Washington DC.

Mendelsohn, R., Basist, A., Kurukulasuriya, P., & Dinar, A., (2007). Climate and rural income. *Climatic Change, 81*, 101–118.

Mendelsohn, R., Christensen, P., & Arellano-Gonzalez, J., (2010). A Ricardian analysis of Mexican farms. *Environment and Development Economics, 15*, 153–171.

Mertz, O., Mbow, C., Reenberg, A., Genesio, L., Lambin, E. F., D'haen, S., Zorom, M., Rasmussen, K., Diallo, D., Barbier, B., Bouzou, M. I., Diouf, A., Nielsen, J. Ø., & Sandholt, I., (2011). Adaptation strategies and climate vulnerability in the Sudano-Sahelian region of West Africa. *Atmospheric Science Letters, 12*, 104–108.

Meza, F., Wilks, D., Gurovich, L., & Bambach, N., (2012). Impacts of climate change on irrigated agriculture in the Maipo Basin, Chile, reliability of water rights and changes in the demand for irrigation. *Journal of Water Resources Planning and Management, 138*, 421–430.

Mideksa, T. K., (2010). Economic and distributional impacts of climate change, the case of Ethiopia. *Global Environmental Change, Human and Policy Dimensions, 20*, 278–286.

Mills, D. J., Westlund, L., De Graaf, G., Kura, Y., Willman, R., & Kelleher, K., (2011). Underreported and undervalued, small-scale fisheries in the developing world. In: Pomeroy, R. S., & Andrew, N., (eds.), *Small-Scale Fisheries Management, Frameworks and Approaches for the Developing World* (pp. 1–15). CABI, Wallingford, UK and Cambridge, MA, USA.

Minasny, B., Arrouays, D., McBratney, A. B., Angers, D. A., Chambers, A., Chaplot, V., et al., (2018). Rejoinder to comments on Minasny et al., 2017 soil carbon 4 per mille. *Geoderma, 292 & 309*, pp. 59–86, 124–129.

Minasny, B., Malone, B. P., McBratney, A. B., Angers, D. A., Arrouays, D., Chambers, A., et al., (2017). Soil carbon 4 per mille. *Geoderma, 292*, 59–86.

Mougou, R., Mansour, M., Iglesias, A., Chebbi, R. Z., & Battaglini, A., (2011). Climate change and agricultural vulnerability, a case study of rain-fed wheat in Kairouan, Central Tunisia. *Regional Environmental Change, 11*(1), S137–S142.

Moumouni, I., & Idrissou, L., (2013). Innovation systems for agriculture and climate in Benin, an inventory. In: Morton, J., (ed.), *Climate Learning for African Agriculture Working Paper No. 3, (trans.)* (p. 24). African Forum for Agricultural Advisory Services (AFAAS), Forum for Agricultural Research in Africa (FARA), and Natural Resources Institute (NRI), University of Greenwich, NRI, Kent, UK.

Mugandani, R., Wuta, M., Makarau, A., & Chipindu, B., (2012). Re-classification of agro-ecological regions of Zimbabwe inconformity with climate variability and change. *African Crop Science Journal, 20*, 361–369.

Müller, C., Cramer, W., Hare, W. L., & Lotze-Campen, H., (2011). Climate change risks for African agriculture. *Proceedings of the National Academy of Sciences of the United States of America, 108*, 4313–4315.

Nakashima, D. J., McLean, K. G., Thulstrup, H. D., Castillo, A. R., & Rubis, J. T., (2012). *Weathering Uncertainty, Traditional Knowledge for Climate Change Assessment and Adaptation* (p. 120). United Nations Educational, Scientific and Cultural Organization (UNESCO), Paris, France and United Nations University (UNU), Darwin, Australia.

Nana, E., Corbari, C., & Bocchiola, D., (2014). A hydrologically based model for crop yield and water footprint assessment, study of maize in the Po valley. *Agricultural Systems, 127*, 139–149.

Nath, A. J., Lal, R., Sileshi, G. W., & Das, A. K., (2018). Managing India's small landholder farms for food security and achieving the "4 per Thousand" target. *Science of the Total Environment, 634*, 1024–1033.

Nazlioglu, S., Erdem, C., & Soytas, U., (2013). Volatility spillover between oil and agricultural commodity markets. *Energy Economics, 36*, 658–665.

Neely, C., & Fynn, A., (2011). *Critical Choices for Crop and Livestock Production Systems That Enhance Productivity and Build Ecosystem Resilience*. SOLAW Background Thematic Report TR11. FAO, Rome.

Nellemann, C., MacDevette, M., Manders, T., Eickhout, B., Svihus, B., Prins, A. G., & Kaltenborn, B. P. (eds.)., (2009). *The Environmental Food Crisis, The Environment's Role in Averting Future Food Crises.* A UNEP Rapid Response Assessment. GRID-Arendal, United Nations Environment Programme. Norway.

Nelson, G. C., Rosegrant, M. W., Koo, J., Robertson, R., Sulser, T., Zhu, T., et al., (2009). *Climate Change, Impact on Agriculture and Costs of Adaptation* (p. 19). International Food Policy Research Institute (IFPRI), Washington, DC, USA.

Nelson, G. C., Valin, H., Sands, R. D., Havlik, P., Ahammad, H., Deryng, D., et al., (2013). Climate change effects on agriculture, economic responses to biophysical shocks. *Proceedings of the National Academy of Sciences of the United States of America, D.* doi: 10.1073/pnas.1222465110.

Nelson, R., Kokic, P., & Meinke, H., (2007). From rainfall to farm incomes–transforming advice for Australian drought policy. II. Forecasting farm incomes. *Australian Journal of Agricultural Research, 58*, 1004–1012.

Nelson, V., Meadows, K., Cannon, T., Morton, J., & Martin, A., (2002). Uncertain predictions, invisible impacts, and the need to mainstream gender in climate change adaptations. *Gender and Development, 10*, 51–59.

Newsham, A. J., & Thomas, D. S. G., (2011). Knowing, farming and climate change adaptation in North-Central Namibia. *Global Environmental Change, 21*, 761–770.

Nielsen, J. Ø., D'haen, S., & Reenberg, A., (2012). Adaptation to climate change as a development project, a case study from northern Burkina Faso. *Climate and Development, 4*, 16–25.

Ninan, K. N., & Bedamatta, S., (2012). *Climate Change, Agriculture, Poverty and Livelihoods, a Status Report* (p. 35). Working Paper 277. The Institute for Social and Economic Change, Bangalore.

NRC & IDMC, (2011). *Displacement Due to Natural Hazard-Induced Disasters: Global Estimates for 2009 and 2010.* Norwegian Refugee Council and Internal Displacement Monitoring Centre, Oslo.

Nyaupane, G. P., & Poudel, S., (2011). Linkages among biodiversity, livelihood, and tourism. *Annals of Tourism Research, 38*, 1344–1366.

O'Brien, K., Eriksen, S., Nygaard, L. P., & Schjolden, A., (2007). Why different interpretations of vulnerability matter in climate change discourses. *Climate Policy, 7*, 73–88.

OECD-FAO, (2011). *OECD-FAO Agricultural Outlook 2011–2020.* OECD and FAO, Paris.

Oluoko-Odingo, A. A., (2011). Vulnerability and adaptation to food insecurity and poverty in Kenya. *Annals of the Association of American Geographers, 101*, 1–20.

Palazzoli, I., Maskey, S., Uhlenbrook, S., Nana, E., & Bocchiola, D., (2015). Impact of prospective climate change upon water resources and crop yield in the Indrawati basin, Nepal. *Agricultural Systems, 133*, 143–157.

Patz, J. A., (2004). Global warming. Health impacts may be abrupt as well as long term. *BMJ, 328*, 1269–1270.

Pettengell, C., (2010). *Climate Change Adaptation, Enabling People Living in Poverty to Adapt.* Oxfam International Research Report. Oxfam International, Oxford, UK. Available at: http://

www.oxfam.org.uk/resources/policy/climate_change/downloads/rr_climate_change_adaptation_full_290410.pdf (Accessed on 10 October 2019).

Porter-Bolland, L., Ellis, E. A., Guariguata, M. R., Ruiz-Mallén, I., Negrete-Yankelevich, S., & Reyes-García, V., (2012). Community managed forests and forest protected areas, an assessment of their conservation effectiveness across the tropics. *Forest Ecology and Management, 268*, 6–17.

Poudel, S., & Kotani, K., (2013). Climatic impacts on crop yield and its variability in Nepal, do they vary across seasons and altitudes? *Climatic Change, 116*, 327–355.

Power, A., (2010). Ecosystem services and agriculture, Tradeoffs and synergies. *Philosophical Transactions of the Royal Society, B, Biological Sciences, 365*, 2959–2971.

Pramova, E., Locatelli, B., Djoudi, H., & Somorin, O. A., (2012). Forests and trees for social adaptation to climate variability and change. *Wiley Interdisciplinary Reviews, Climate Change*, *3*, 581–596.

Pretty, J., Toulmin, C., & Williams, S., (2011). Sustainable intensification in African agriculture. *International Journal of Agricultural Sustainability, 9*, 5–24.

Ravallion, M., Chen, S., & Sangraula, P., (2007). New evidence on the urbanization of global poverty. *Population and Development Review, 33*, 667–701.

Regmi, B. R., & Adhikari, A., (2007). *Human Development Report (2007)*. Climate change and human development-risk and vulnerability in a warming world. Country Case Study-Nepal. UNDP.

Reilly, J. M., & Schimmelpfennig, D., (1999). Agricultural impact assessment, vulnerability, and the scope for adaptation. *Climatic Change, 43*, 745–488.

Ribot, J., (2010). Vulnerability does not fall from the sky, towards multi-scale pro-poor climate policy. In: Mearns, R., & Norton, N., (eds.), *Social Dimensions of Climate Change, Equity and Vulnerability in a Warming World* (pp. 47–57). The World Bank, Washington, DC, USA.

Rivera-Ferre, M. G. Ortega-Cerdà, M., & Baumgärtner, J., (2013a). Rethinking study and management of agricultural systems for policy design. *Sustainability Science, 5*, 3858–3875.

Rivera-Ferre, M. G., & López-i-Gelats, F., (2012). *The Role of Small Scale Livestock Farming in Climate Change and Food Security* (p. 146). Vétérinaires Sans Frontières Europa (VSF-Europe), Brussels, Belgium.

Rivera-Ferre, M. G., Di Masso, M., Mailhost, M., López-i-Gelats, F., Gallar, D., Vara, I., & Cuellar, M. (2013b). *Understanding the Role of Local and Traditional Agricultural Knowledge in a Changing World Climate, The Case of the Indo-Gangetic Plains* (p. 98). CGIAR Research Program on Climate Change, Agriculture and Food Security (CCAFS), a collaboration of the Consultative Group on International Agricultural Research (CGIAR) and the Earth System Science Partnership (ESSP), Frederiksberg, Denmark.

Rochdane, S., Reichert, B., Messouli, M., Babqiqi, A., & Khebiza, M. Y., (2012). Climate change impacts on water supply and demand in Rheraya Watershed (Morocco) with potential adaptation strategies. *Water, 4*, 28–44.

Romsdahl, R. J., Atkinson, L., & Schultz, J., (2013). Planning for climate change across the US great plains, concerns and insights from government decision-makers. *Journal of Environmental Studies and Sciences, 3*, 1–14.

Roy, T. N., (2012). Economic analysis of producer's perceptions about impact of climate change on fisheries in West Bengal. *Agricultural Economics Research Review, 25*, 161–166.

Ruel, M. T., Garrett, J. L., Hawkes, C. R., & Cohen, M. C., (2010). The food, fuel, and financial crises affect the urban and rural poor disproportionately, a review of the evidence. *The Journal of Nutrition, 140*(1), 170S–176S.

Sanghi, A., & Mendelsohn, R., (2008). The impacts of global warming on farmers in Brazil and India. *Global Environmental Change, 18*, 655–665.

Schepp, K., (2010). *How Can Small-Scale Coffee and Tea Producers Adapt to Climate Change. AdapCC Final Report – Results & Lessons Learnt* (p. 37). Cafédirect and Deutsche Gesellschaft für Internationale Zusammenarbeit (GIZ) GmbH, GTZ, Eschborn, Germany.

Schmidhuber, J., & Tubiello, F. N., (2007). Global food security under climate change. *Proceedings of the National Academy of Sciences of the United States of America, 104*, 19703–19708.

Schroth, G., Laderach, P., Dempewolf, J., Philpott, S., Haggar, J., Eakin, H., Castillejos, T., Moreno, J. G., Pinto, L. S., Hernandez, R., Eitzinger, A., & Ramirez-Villegas, J., (2009). Towards a climate change adaptation strategy for coffee communities and ecosystems in the Sierra Madre de Chiapas, Mexico. *Mitigation and Adaptation Strategies for Global Change, 14*, 605–625.

Seo, S. N., & Mendelsohn, R., (2008). A Ricardian analysis of the impact of climate change on South American farms. *Chilean Journal of Agricultural Research, 68*, 69–79.

Seo, S. N., (2010). Is an integrated farm more resilient against climate change? A microeconometric analysis of portfolio diversification in African agriculture. *Food Policy, 35*, 32–40.

Seto, K. C., (2011). Exploring the dynamics of migration to mega-delta cities in Asia and Africa, contemporary drivers and future scenarios. *Global Environmental Change, 21*(1), S94–S107.

Sharma, M., & Dahal, S., (2011). *Assessment of Impacts of Climate Change and Local Adaptation Measures in Agriculture Sector and Livelihoods of Indigenous Community in High Hills of Sankhuwasabha District*. Rampur, Chitwan, Nepal. National Adaptation Program of Action, Ministry of Environment, Kathmandu, Nepal.

Shukla, G., Kumar, A., Pala, N. A., & Chakravarty, S., (2015). Farmer's perception and awareness of climate change a case study from Kanchandzonga Biosphere Reserve, India. *Environment, Development and Sustainability, 4*, doi: 10.1007/s10668–015–9694–2.

Sietz, D., Lüdeke, M. K. B., & Walther, C., (2011). Categorization of typical vulnerability patterns in global dry lands. *Global Environmental Change, 21*, 431–440.

Soussana, J. F., Lutfalla, S., Ehrhardt, F., Rosenstock, T., Lamanna, C., Havlik, P., Richards, M., Wollenberg, E., Chotte, J. E., Torquebiau, E., Ciais, P., Smith, P., & Lal, R., (2017). Matching policy and science, Rationale for the '4 per 1000 – soils for food security and climate' initiative. *Soil Tillage Research.* https://doi.org/10.1016/j.still.2017.12.002 (Accessed on 10 October 2019).

Sowers, J., Vengosh, A., & Weinthal, E., (2011). Climate change, water resources, and the politics of adaptation in the Middle East and North Africa. *Climatic Change, 104*, 599–627.

Speranza, C. I., (2013). Buffer capacity, capturing a dimension of resilience to climate change in African smallholder agriculture. *Regional Environmental Change, 13*, 521–535.

Srikanthan, S., (2013). Impact of climate change on the fishermen's livelihood development, a case study of village in coromandel coast. *IOSR Journal of Humanities and Social Science, 12*, 49–54.

Stathers, T., Lamboll, R., & Mvumi, B. M., (2013). Postharvest agriculture in a changing climate, its importance to African smallholder farmers. *Food Security, 5*, 361–392.

Stern, N., (2007). *The Economics of Climate Change, The Stern Review* (p. 712). Cambridge University Press, Cambridge, UK.

Summer, A., (2012b). *Where do the Worlds' Poor Live? A New Update*. IDS Working Paper Vol. 212 No. 393. Institute of Development Studies, London, UK.

Sumner, A., (2012a). Where do the poor live? *World Development, 405*, 865–877.
Swinnen, J., & Squicciarini, P., (2012). Mixed messages on prices and food security. *Science, 27*, 405–406.
Tefera, T., (2012). Post-harvest losses in African maize in the face of increasing food shortage. *Food Security, 4*, 267–277.
Terrier, A., Girardin, M. P., Perie, C., Legendre, P., & Bergeron, Y., (2013). Potential changes in forest composition could reduce impacts of climate change on boreal wildfires. *Ecological Applications, 23*, 21–35.
The Zimbabwean, (2012). *Climate Change to Impact Heavily on Food Security*. [Online]. Available at: http://www.thezimbabwean.co.uk/news/32154/climate-change-to-impact-heavily-on-food-security.html (Accessed on 10 October 2019).
Tischbein, B., Manschadi, A. M., Hornidge, A. K., Conrad, C., Lamers, J. P. A., Oberkircher, L., Schorcht, G., & Vlek, P. L. G., (2011). Proposals for the more efficient utilization of water resources in the province of Khorezm, Uzbekistan. *Hydrologie und Wasserbewirtschaftung, 55*, 116–125.
Tol, R. S. J., Downing, T. E., Kuik, O. J., & Smith, J. B., (2004). Distributional aspects of climate change impacts. *Global Environmental Change, Human and Policy Dimensions, 14*, 259–272.
Tscharntke, T., Clough, Y., Wanger, T., Jackson, L., Motzke, I., Perfecto, I., Vandermeer, J., & Whitbread, A., (2012). Global food security, biodiversity conservation and the future of agricultural intensification. *Biological Conservation, 151*, 53–59.
Udmale, P., Ichikawa, Y., Manandhar, S., Ishidaira, I., & Kiem, S. M., (2014). Farmers' perception of drought impacts, local adaptation and administrative mitigation measures in Maharashtra State, India. *International Journal of Disaster Risk Reduction, 10*, 250–269.
UN DESA, (2013). *World Population Prospects, The 2012 Revision, Highlights and Advance Tables* (p. 94). UN DESA Population Division Working Paper No. ESA/P/WP.228. UN DESA Population Division, New York, NY, USA.
UN ECLAC, (2010b). *Economics of Climate Change in Latin America and the Caribbean – Summary 2010* (p. 107). UN ECLAC, Santiago, Chile.
UNDP, (2007a). *Central Asia, Background Paper on Climate Change*. Human Development Report, 2007. UNDP, New York.
UNDP, (2007b). *Human Development Report, 2007: Fighting Climate Change, Human Diversity in a Divided World*. UNDP, New York.
UNDP, (2007c). *Human Development Report 2007/2008: Fighting Climate Change, Human Solidarity in a Divided World.* United Nations Development Programme, New York.
UNECLAC for Latin America and the Caribbean, (2010a). *The Economics of Climate Change in Central America–Summary 2010* (p. 144). UN ECLAC, Santiago, Chile.
UNEP, (2008). *Green Jobs, Towards Decent Work in a Sustainable, Low-Carbon World.* United Nations Environment Programme, Nairobi.
UNEP, (2012). *Avoiding Future Famines, Strengthening the Ecological Foundation of Food Security Through Sustainable Food Systems*. United Nations Environment Programme, Nairobi.
UNEP-JUNP on HIV/AIDS, (2008). *Climate Change and AIDS, a Joint Working Paper*. UNEP-JUNP on HIV/AIDS, Nairobi.
UNFCCC, (2007). *Impacts, Vulnerabilities and Adaptation in Developing Countries* (p. 64). UNFCCC Bonn, Germany.
UNRISD, (2000). *Combating Poverty and Inequality, Structural Change, Social Policy and Politics*. UNRISD Geneva.

Vaghefi, N., Nasir, S. M., Makmom, A., & Bagheri, M., (2011). The economic impact of climate change on the rice production in Malaysia. *International Journal of Agricultural Research, 6*, 67–74.
Van Groenigen, J. W., Van Kessel, C., Hungate, B. A., Oenema, O., Powlson, D. S., & Van Groenigen, K. J., (2017). Sequestering soil organic carbon, a nitrogen dilemma. *Environmental Science and Technology, 51*, 4738–4739.
Verner, D., (2012). *Adaptation to a Changing Climate in the Arab Countries, a Case for Adaptation Governance and Leadership in Building Climate Resilience* (p. 402). Mena Development Report, the World Bank, Washington, DC, USA.
Vincent, K., Cull, T., Chanika, D., Hamazakaza, P., Joubert, A., Macome, E., & Mutonhodza-Davies, C., (2013). Farmers' responses to climate variability and change in southern Africa, is it coping or adaptation. *Climate and Development, 5*, 194–205.
Vörösmarty, C. J., McIntyre, P. B., Gessner, M. O., Dudgeon, D., Prusevich, A., Green, P., Glidden, S., Bunn, S. E., Sullivan, C. A., Liermann, C. R., & Davies, P. M., (2010). Global threats to human water security and river biodiversity. *Nature, 467*, 555–561.
WB, (2010). *Economics of Adaptation to Climate Change, Synthesis Report* (p. 101). The International Bank for Reconstruction and Development / the World Bank, Washington, DC, USA.
WFP, (2009). *Climate Change and Hunger, Responding to the Challenge*. World Food Programme, Rome.
WFP, (2011). *Climate Change and Hunger, Towards a WFP Policy on Climate Change*. World Food Programme, Rome.
Wheeler, T., & Von Braun, J., (2013). Climate change impacts on global food security. *Science, 341*, 508–513.
White, R., Davidson, B., Lam, S., & Chen, D., (2018). A critique of the paper 'soil carbon 4 per mille' by Minasny et al. (2017). *Geoderma, 309*, 115–117.
WHO, (2009). *Protecting Health from Climate Change, Connecting Science, Policy and People*. WHO Geneva.
WHO, (2011). *Gender, Climate Change and Health*. WHO, Geneva.
Wittrock, V., Kulreshtha, S. N., & Wheaton, E., (2011). Canadian prairie rural communities, their vulnerabilities and adaptive capacities to drought. *Mitigation and Adaptation Strategies for Global Change, 16*, 267–290.
Wood, S., Sebastian, K., & Scherr, S. J., (2000). *Pilot Analysis of Global Ecosystems, Agro-ecosystems*. International Food Policy Research Institute and World Resources Institute, Washington, DC.
Wutich, A., York, A. B., Brewis, A., Stotts, R., & Roberts, C. M., (2012). Shared cultural norms for justice in water institutions, results from Fiji, Ecuador, Paraguay, New Zealand, and the U. S. *Journal of Environmental Management, 113*, 370–367.
Zhai, F., & Zhuang, J., (2009). *Agricultural Impact of Climate Change, a General Equilibrium Analysis with Special Reference to Southeast Asia* (p. 17). ADBI Working Paper 131, Asian Development Bank Institute, Tokyo, Japan. Available at: www.adbi.org/files/2009.02.23.wp131.agricultural.impact.climate.change.pdf (Accessed on 10 October 2019).

FIGURE 3.1 Urban greening in Nigeria (Akanwa, 2018).

FIGURE 3.2 Rural areas in Africa showing its rich agroforestry resources (Akanwa, 2016).

FIGURE 3.3 Unsustainable quarrying activities destroy lands, soils, trees, vegetation, landscapes, and displace animals from their habitat. The quarry sites are abandoned as artificial pools after closure without plans of re-afforestation (Akanwa, 2016).

FIGURE 3.4 All-season farming involves the combination of agricultural activities with tree planting and growth. It also employs the application of improved seeds, chemical fertilizers, and pesticides to promote growth, disease-resistant, and pest-free crops (Ogbuene, 2012; Akanwa, 2016).

FIGURE 3.5 Huge expanse of forest areas located in the rural areas (Akanwa, 2016).

(a) (b)

FIGURE 8.1 False smut of rice.

Source: https://commons.wikimedia.org/wiki/File:False_smut_of_rice.jpg; https://commons.wikimedia.org/wiki/File:Rice_false_smut.jpg

FIGURE 8.2 (a) Single wilted and dead *Plumeria* tree (b, c) alone and several dead trees in a row.

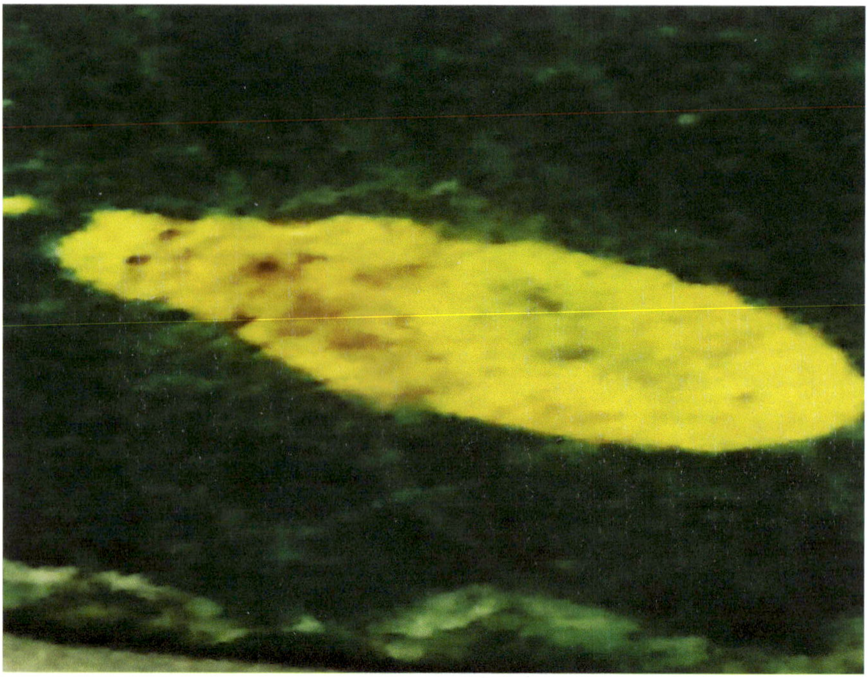

FIGURE 8.3 Close up of Rust infection on leaf of *Plumeria.*

a b

FIGURE 8.4 *Taphrina* sp. fungal infection on *Cordia dichotoma* leaves.

FIGURE 11.1 Study site-I Patelpara, Navapara.

FIGURE 11.2 Study site-II Godhanpur.

FIGURE 11.3 Study site-III New bus stand.

FIGURE 11.4 Study site-IV Darripara.

FIGURE 11.5 Study site-V D.C. road.

FIGURE 11.6 Study site-VI Khalpara.

FIGURE 11.7 Study site-VII Namnakala.

FIGURE 11.8 Study site-VIII Deviganj.

FIGURE 11.9 Study site-IX Thanganpara.

FIGURE 11.10 Study site-X Gangapur.

FIGURE 11.11 Study site-XI Old bus stand.

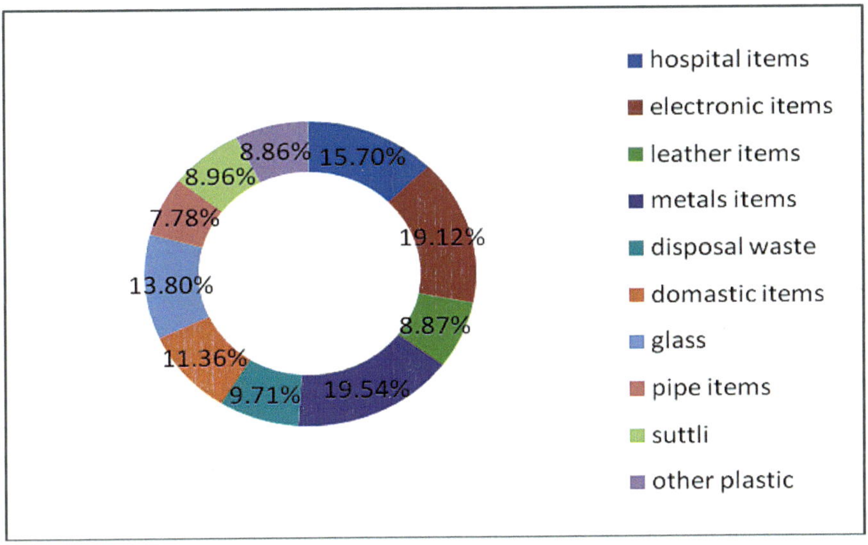

FIGURE 11.21 Item wise contribution of waste material to the total waste load of Ambikapur township.

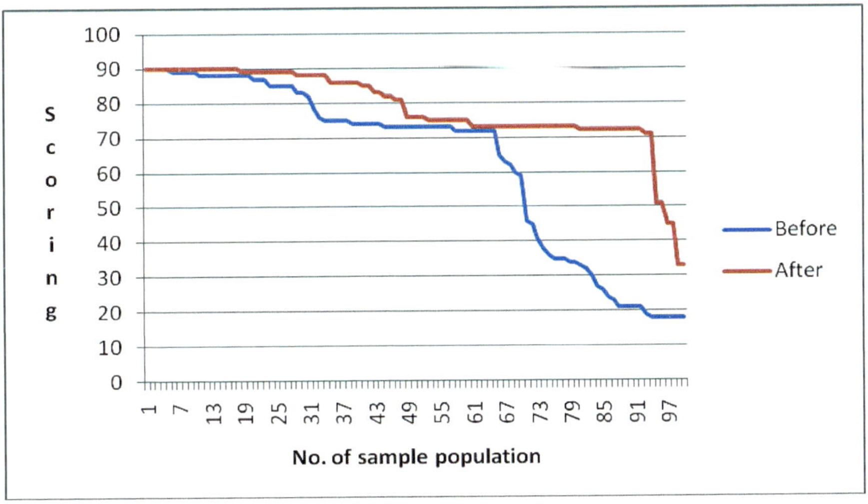

FIGURE 12.4 Scoring of sample population of media learning.

CHAPTER 8

Invasion of Major Fungal Diseases in Crop Plants and Forest Trees Due to Recent Climatic Fluctuations

NARENDRA KUMAR and S. M. PAUL KHURANA

Amity Institute of Biotechnology, Amity University Haryana, Manesar, Gurgaon–122413, India, E-mail: narendra.microbiology@rediffmail.com (N. Kumar); E-mail: smpaulkhurana@gmail.com (S. M. Paul Khurana)

ABSTRACT

The changing scenarios of climate on earth are basically due to modifications in the biosphere, hydrosphere, their interactions, and fluctuating different atmospheric conditions. This is the reason of emergence of new pathogen records in a vast number. The diversity of pathogens is an outcome of geographical situations like host availability, abundance of new climatic situations. Their distribution is changing very fast mainly because of climatic conditions. The global change in climate is due to inert greenhouse gas (GHGs) concentration of atmosphere and their impact either may be neutral, negative/positive having potential to either increase/decrease without any impact on region and period of disease. It is not easy to confirm impacts and needs help of specialists of different areas to be sure with discussion on broader angles. Climate affects various stages of life of the pathogens. The hosts also pose various challenges in different ecosystems. The chapter focuses on various types of climatic fluctuations on major fungal pathogenic attacks on the crops and forestry plants.

The historical context is summarized below regarding diseases of plants in respect to CO_2 concentration and other gases. The chapter describes how altered conditions will favor spread of diseases. What will be chemical, biological control in this regard. The chapter also discusses first reports of rust, leaf distortion and reddening of *Cordia dichotoma,* Rust disease of

Plumeria (Frangipani), leaf spot disease of *Calotropis procera* and *Acrophialophora levis* producing wilt in Plumeria. It needs more research on deviations of climatic conditions affecting plant health in the plains and hilly areas and dense forests.

8.1 INTRODUCTION

Climate change (CC) has an association with conditions of atmospheric weather. This happens due to alterations in average weather. The alterations in climate happen due to many reasons viz., biotic potential and radiations coming on earth, the movement of seven large plates on large-scale and volcanic eruptions. It can be clearly stated that there is an upset in equilibrium in diversity of several pathogens. The alterations in climate results additional pressure in spread and incidence of pest distribution, which is not normal to workout. Although, pest may causes large yet only partly predictable effects at a time.

The CCs have direct role on ozone depletion of stratosphere which in turn has direct impact on biodiversity loss, physiology, and morphology of plants and ecosystem balance and finally resulting in desertification and also effects soil fertility indirectly. Changing climatic conditions impact crop production through changes in precipitation patterns, invasive species, average temperatures and increase in pests which results more frequently due to fluctuations in weather situations and erosion of soil (Khurana and Kumar, 2016).

The ending years of twentieth and starting of the 21st century was recorded as most warm period. The 4th Assessment Report of Intergovernmental Panel on Climate Change (IPCC) records that this is because of increasing concentration of greenhouse gases (GHGs) as a main cause. The alterations in climatic conditions can result attack of pathogens in plants via anthropogenic processes such as fast urbanization, soil, air, and water pollution. Through this more exotic species comes (Regniere, 2012). Prospero et al., (2009) recorded that human made reaction with nature results diseases like sudden oak death.

The CC affects Natural plant population and agriculture ecosystems in entire world (Stern, 2007). The microbes directly impact crops and do interactions finally results diseases. What are the effects of CC on plants less studies have been done (Garrett et al., 2006). The change of weather may help development of severe plant epidemics (Bosch et al., 2007). This will create food problem if affecting food crops (Anderson et al., 2004). If valuable species are affected the landscapes will be damaged (Bergot et al., 2004).

Due to human activities in the atmosphere, concentration of GHG is increasing fast and finally causing the global CC. Besides, eighteenth century industrial revolution resulted in much more application of natural resources. It proceeds through use of fossil fuel, more deforestation along with other patterns of use of lands, etc. CO_2concentration reached significantly higher which was not there in past six hundred fifty thousand years (Siegenthaler et al., 2005). Now CO_2 concentration has increased rapidly in comparison to last decades (Canadell et al., 2007). It has also been recorded for CH_4 (methane), N_2O (nitrous oxide), along with GHGs (Spahni et al., 2005; IPCC, 2007).

The average temperature worldwide now increased by 0.2°C per decade in thirty years (Hansen et al., 2006). It will be continued if GHGs concentrations would not be stable. For getting equilibrium longer period would be needed (IPCC, 2007).

Development of plant diseases due to changing environmental set up has been in knowledge of common people for last two thousand years. It was seen in plants when humans were living like hunter-gatherers (Agrios, 2005). Scheffer (1997) mentioned that damages because of diseases are different and normally localized. There was extensive damage due to pests.

With increase in agriculture there was increase in incidence of pathogens attack in plants and sometimes it took the shape of famines due to reduced food production (Agrios, 2005; Scheffer, 1997). During start of 19th century and in 18th century nutritional factor along with air humidity were observed for disease occurrence (Colhoun, 1973). Disease is a phase of dynamic actions where pathogens and host interact and finally produce physiological and morphological alterations (Gaumann, 1950).

Strange and Scott (2005) mentioned that there are 10% losses in food production globally because of pathogenic agents which pose a threat to security of food. Annually losses by plant diseases cost up to US$ 220 billion (Agrios, 2004). It needs modifications of current phyto-sanitary scenario. For formation of resistant forms yet new control technology is needed so that serious losses may be handled (Ghini, 2005).

The crop diseases are mainly due to fungal, bacterial, or viral infections which spread rapidly through vectors. This produced spread of sudden oak death (phytopthora) (Prospero et al., 2009), soybean rust (Rosenzweig et al., 2001; Schneider et al., 2005). This also may be an addition in spread of crop diseases. The summer precipitation due to fluctuations in climate was found to increase needle blight in British Columbia (Woods et al., 2005). The increased incidence of fungal load was due to increasingCO_2concentration which decreased plants diversity and also increased nitrogen deposition (Mitchell et al., 2003).

There are many reports that the CC poses a big serious threat. Manning and Tiedemann (1995) investigated an upward direction in diseases based on potential effects in rising CO_2 concentration on plant infections. This was based on responses of plant in new environmental conditions. Plant biomass got increased due to increase in productivity of leaves, shoots, flowers, fruit, and finally forming more tissue which is infected through pathogens. The enhancement in carbohydrate parameters may enhance sugar dependent disease causing agents resulting powdery mildews and rusts. More canopy density and size of plant may stimulate higher growth and sporulation. This spreads leaf targeting fungal species having high humidity of air and no rains and resulting, leaf necrosis, powdery mildews and even rusts. The crop residues represent better situations of survival in case of necrotrophic pathogens. Chakraborty et al., (2000) found that increased concentration of CO_2influences both host and pathogen on the spread of anthracnose in plants of *Stylosanthes scabra* caused through *Colletotrichum gloeosporioides.*

But for crops few reports are available on what will be essential in protection of plant in coming future. This chapter aims to discuss and report effects of changing climatic scenario in plant health. It describes impacts of increasing CO_2 concentration in atmosphere and its results for control of diseases.

8.2 RECENT CLIMATE ALTERATIONS AND PREDICTIONS

Recently some researchers working on change of climate have reported their observations. For instance, Chakraborty et al., (2008) mentioned that present CO_2 concentration in atmospheric will exceed up to 400 ppm in next few years which will be much higher than normal concentrations of 180–300 ppm. The main problem of global CO_2 increase is because of land-use changes, burning of fossil fuels and even deforestation (Cerri et al., 2007; Paterson and Lima, 2010). Increase in the CO_2 concentration and various GHGs now have already increased average temperature at global level by 0.6–0.7°C in past century (Mann et al., 1998; Walther et al., 2002; Benvenuti, 2009).

This has resulted in shorter and warmer winters which in turn affected distribution phenology and abundance of many species (Körner and Basler, 2010; Matesanz et al., 2010). An estimate of +2–4°C average temperatures at global level during the 21st century further remarks for a shift (Milad et al., 2011). Due to CCs some eco-regions will be more affected (Engler et al., 2011; Heyder et al., 2011). The most bio-diverse eco-regions are at greater risk (Beaumont et al., 2011). In Europe, the Southern places will show more sensitivity to CC than the Northern ones mainly due to increased summer droughts (Kůdela, 2009). In boreal and mountainous regions of North America temperature will be more in comparison to average global temperature (Bentz et al., 2010). Tropical regions will experience a lower absolute temperature increase. The tropical metabolic rates will be affected more strongly in comparison to extratropical latitudes due to the nonlinear effects of temperature in metabolism (Dillon et al., 2010). Metabolic activity decides the activity of plant pathogens. The current temperatures are also near to the maximum physiological limit (Cerri et al., 2007). The increase of following parameters viz., severe wind to heatwaves, drought to floods, hail storms, and rain have direct effect on plant health (Boland et al., 2004; Hegerl et al., 2011; Peng et al., 2011). There will be more spread of water-borne diseases. The heatwaves and droughts will expose plants to infection, and storm will enhance wind-borne dispersal of pathogenic propagules.

"Global temperatures are already about 1°C (2°F) above those of pre-industrial times; The countries now agreed to limit that warming to 2°C above preindustrial levels at the end of the century" (Thompson, 2016). Jones (2017) mentioned that "reaching 400 ppm is a stark reminder that the world is still not on a track to limit CO_2 emissions and therefore climate impacts." The upgrading in CO_2 concentrations will pose increased growth in different plant species. This results 'greening' of vegetation normally. The slower-growing larger tree species may lose out in comparison to faster-developing competitors (Richard et al., 2018).

8.3 LOSSES CAUSED IN CROPS DUE TO FUNGI

Fungi cause many plant diseases and changed the scenario of human history. For example, ergot of rye occurs due to *Claviceps purpurea* that contains useful compounds for pharmaceuticals and also source of LSD. This also resulted bubonic plague (Jans, 2016). There was an Irish famine in year 1945/46 due to failure of potato production in Europe due to fungus *Phytophthora infestans.* There was death of one in eight of Irish population. The world agriculture is now facing significant losses because of plant diseases despite scientific advances in 150 years. Naturally, we have learned much to avoid massive problematic situations like the Irish famine, Bengal Famine. The statistical calculations soundly cleared that crop losses due to fungal infections are more significant. It also depends on whether parameters. Oerke (2006) mentioned that the world agriculture suffers plant disease losses up to 16% annually. The global potential loss for wheat is 50% which is more in cotton that is 80%. The losses in soybean, wheat, cotton are 26–29%. But it is 40, 31 and 37% for potato, maize, rice, respectively. The losses due to weed was 34% while pests caused half of that loss (Oerke, 2006).

Different groups of fungi may cause serious plant diseases. The rice blast occurs by *Magnaporthe grisea/Magnaporthe oryzae.* This is creating most of damage worldwide. The destruction is so high that each year 60 million people can feed on it (Knogge, 1998). The various species of *Armillaria* results disease in trees and shrubs that comes in category of aggressive pathogen. *A. ostoyae* present in Northeast Oregon Blue Mountains (USA) showed killing of 30% pines of the Ponderosa which is about 400–1000 years old (Ferguson et al., 2003). *Cercospora beticola* causing destructive leaf spot disease in sugar beet affects its yield and quality. It survives in infected crop residues. It produces spores which is the prime source of infective agent coming on leaves producing infection in the next crop (Khan et al., 2008).

Stemphylium blight of lentil produces severe leaf drop which results in complete defoliation of plants. It produces disease in large irregular patches. Sometimes it becomes severe in nitrogen stress conditions and it shows only terminal leaves. In Montana North Dakota and Saskatchewan, it becomes problematic when rains are not earlier in the season. In these areas, the disease spreads in the last third of the growing season, covering from late bloom to late pod fill (Wunsch, 2013).

8.4 CLIMATIC FLUCTUATIONS IN RESPECT TO DISEASES IN PLANTS

The climate affects both agriculture and forestry. On a global scale, the term agriculture and CC are interrelated. The changes are many viz., in average temperatures, rainfall, heatwaves, atmospheric CO_2, ground-level ozone, new diseases, new pests and nutritional quality in some foods, sea level, and much more we are witnessing from past few years. This produces an impact in agriculture where the results are given in terms of demand, farm profitability, supply, prices, and trade. Short-term alterations are very sensitive to agriculture for the long term and annual crops. Crop yield are dependent on many factors such as soil, seed, pest, diseases, fertilizers, agronomic practices (Howdy, 2018). The CC have various effects on forests which are multiple complexity forms, but rising CO_2 concentration, higher temperatures, changes of precipitation, flooding, drought duration and frequency will all have significant pose in tree growth. Forest ecosystems might be affected. This results an up-gradation of frequency of pests and disease outbreaks (biotic disturbances). This brings alterations in wild fire occurrence and wind storm intensity and frequency (abiotic disturbances).

Agrios (2005) mentioned that plant diseases show a major effect in agriculture. Due to CC this would be a long tenure shift from disease conducive to disease suppressive forms (Fuhrer, 2003; Perkins et al., 2011). So, new infections may be useful in terms of CC indicators (Garrett et al., 2009). No data on long-term datasets are available (Scherm, 2004) but have potential to reveal the role of change of climate in plant health (Fabre et al., 2011). There will be accelerated pathogen development due to alterations of abiotic stresses and extreme weather conditions (Sutherst et al., 2011). There will be an up in intensity of tree pathogens, due to drought affecting the cell physiology of host (Desprez-Loustau et al., 2006). The dry situation directly affects the pathogens, for example, *Heterobasidion irregulare* of central Italy. This shows better adaptation in the Mediterranean climate in comparison to native *H. annosum* species (Garbelotto et al., 2010).

8.5 EFFECTS OF INCREASING TEMPERATURES

A fluctuation in temperature favors development of various currently inactive microbes, which may cause an epidemic. Temperature and precipitation changes resulting change in climate will affect growth rate, physiology, pathogenicity, resistance in host plant (Charkraborty and Datta, 2003).

An up-gradation in temperatures will impact incidence of infections in agricultural crops. Swaminathan (1986) highlighted number of diseases in rice, maize, citrus, and tomato were much higher in tropics 500–600, 125, 248, 278 than under temperate conditions which were 54, 85, 50 and 32, respectively.

There is accumulation of protective pigments/phytoalexins in host cells due to direct sunlight. Temperature being a factor, affects the incidence of diseases (Kudela, 2009). With enhancement in temperature, winter duration and growth rate and pathogen's reproduction gets changed. The disease incidence of vector borne nature would also be enhanced/reduced. This influences development and distribution of vectors. This will influence phenology; interspecific interactions resultantly increased the risk of attack through migrant pests. Changing temperatures directly affects plants and the pathogens. The studies revealed that oats and wheat are more sensitive to rust because of up-gradation in temperature (Coakley et al., 1999). The parameters of pathogenic growth and infection in specified temperature are normally used in various mathematical models to forecast plant diseases.

The onset of earlier warm temperatures may result late blight producing epidemics and needs an increase of fungicides in disease control. The up temperature effect will vary in all months for plants. In colder regions warming will relieve cold stress while in hotter months there will be heat stress. For instance, rice yield was estimated in the Philippines to decline 10% for each 1°C up of minimum temperature during dry season (Peng et al., 2004).

The plant growth and development has dependence on surrounding temperature which is minimum, maximum, and optimum. These values have been recorded by Hatfield et al., (2008, 2011) for many species. The changes in 30–50 years will be 2–3°C (IPCC, 2007). The extreme temperature events/heatwaves will be more intense, more frequent, which may last longer in coming years (Meehl et al., 2007). Increase in temperature could cause yield declines between 2.5 to 10% in 21st century (Hatfield et al., 2011). Table 8.1, evidences that increasing temperatures have effect in seed germination, growth, development, photosynthetic alterations, physiological, dry matter alterations, water loss, crop quality, reduction in yield and even in plant disease development.

The following diseases spread due to abiotic factors, for example, red band needle blight (*Dothistroma septosporum*), sudden oak death (*Phytophthora ramorum*), Swiss needle disease (*Phaeocryptopus gaeumannii*). These were due to an up in temperature in winter (Stone et al., 2008). In Kashmir, the *Alternaria* epidemic of Apple caused by *Alternaria mali* was due to rains (Bhat et al., 2015).

TABLE 8.1 The Higher Temperature Impacts on Various Activities of Plants

Qualitative Effects of Up Temperature Conditions	
1. a. Inhibited seed germination	2. a. Reduction in plant growth
b. Improper development	b. Alteration in photosynthesis
c. Oxidative stress	c. Changes of dry matter partitioning
d. Poor crop quality	d. Poor yield
e. Water Loss	e. Interspecific interactions
3. Changes in Life cycle Patterns	4. Higher disease susceptibility

8.6 EFFECT OF INCREASING CO_2 CONCENTRATION

Manning and Tiedemann (1995) studied effect of increasing CO_2 concentration from 1930–1993 and they found an upward trend. There was a severe result of up CO_2 concentration on plants which is dependent on plant responses for new environment. Chakraborty and Pangga (2004) observed an increased severity of diseases in CO_2enriched environments.

The host plants get affected by up CO_2 faces alterations in host pathogen interactions. Pangga et al., (2004) stressed on induced resistance importance studies. They reported a check in *Phyllosticta minima* (red maple pathogen) incidence, severity of *Acer rubum* under CO_2-enriched atmosphere which was mainly due reduction in conduction in stomata, resulting smaller pores for germ tubes infection (Mcelrone et al., 2010). It showed reduced silicon content in leaves of rice which reduced respiration rates and caused an up in severity of rice blast (Kobayashi et al., 2006).

The effect of increased CO_2 may be observed at various pathogen host interaction cycles. Hibberd et al., (1996a, b) studied *Erysiphe graminis* (powdery mildew) cycle for barley. Chakraborty et al., (2000) studied effect of *Colletotrichum gloeosporioides* pathogen in *Stylosanthes scabra* which results anthracnoses in it. $CO_2$was found to influence pathogen and even host which resulted in differences in resistance of disease. Under controlled conditions the up in CO_2 concentration resulted an up in plant growth. Hibberd et al., (1996a, b) highlighted that the plus points of CO_2 on growth depends on plant resistance. The slow growth of the pathogen's germ-tube, even appressorium checked the conidial germination on leaves. This resulted in loss of severity of disease. The pathogen penetration occurs by stomata. The up in CO_2 produces reduction of stomatal density in leaf.

The pathogen alterations are due to an up in CO_2 concentration. To find out aggressiveness of *Colletotrichum* against *Stylosanthes scabra*, Chakraborty and Datta (2003) studied levels of resistance up to 25 infection cycles in two cultivars. The CO_2concentration was 350 and 700 ppm. There was increase in aggressiveness in resistant cultivar which was absent insusceptible cultivar. It also increased the pathogen isolate fecundity due to up environmental CO_2 concentration.

The up CO_2 concentration in surroundings of living may interact through nitrogen fertilization and even irrigation. Thompson et al., (1993) investigated interactions of powdery mildew in England in wheat caused by *Erysiphe graminis*. Thompson and Drake (1994) studied C_3 and C_4 plants for N and water contents for fungal and insect disease severity. The growth of pathogen may be affected by an up in CO_2 concentrations producing greater number of fungal spores. The increased CO_2 may result physiological alterations plant which increase resistance in host for pathogenic attacks (Coakley et al., 1999). Mitchell et al., (2003) studied the effects of an up in CO_2 concentration, input nitrogen and species diversity of ecosystem for foliar fungal attacks in

C_3, C_4 plants. Tiedemann and Firsching (2000) studied the effect of an up ozone (O_3) concentration wheat plants grown in spring when infected with *Puccinia recondita* f. sp. *Tritici* resulting rust of leaf. They noticed that rust problem was checked by O_3, but not affected by an up of CO_2. An up of CO_2 largely reduced rate of photosynthesis, growth, and yield parameters. It could not compensate the effects of infection caused by fungi. Percy et al., (2002) investigated on increased CO_2 and O_3 effects in *Melampsora medusae* which cause leaf rust. In this investigation CO_2 could not alter incidence of rust. This increased fourfold in enhanced O_3while CO_2exposure could not remove the negative effects of O3. The infection increased almost threefold than the control. A study conducted in Canada and United States and measured O_3 from 1999 to 2003 along with growth response. They developed exposure based regression models which predicted that growth alteration within the North American ambient air quality context for *Populus tremuloides* (Percy et al., 2007). Karnosky et al., (2002) investigated on leaf the impacts of O_3 which increased rust incidence on poplar. An up ofCO_2concentrations may accelerate plant biomass production while higher carbohydrate dose in the host tissue enhanced incidence of infection of bio-trophic rust fungi (Chakraborty et al., 2002).

The studies on effects of high atmospheric CO_2 have received little interest on plant pathogen interactions but there are reports with conflicting results. Increased level of CO_2 has potential to affect the pathogens. For instance, according to Chakraborty et al., (2002), the growth of appressorium, conidium, and germ tube in *C. gloeosporioides* fungi was slower on CO_2 (700 ppm). The rate of germination of conidia on leaves was lesser at CO_2 concentration 700 ppm in comparison to 350 ppm. At 700 ppm CO_2 concentration germination, sporulation was much. It increased the rust infection by 3 to 5 times on poplar due to ozone while the response was checked by increased CO_2. There was a high degree increase in potato plants when infected by *Phytophthora infestatans* due to increase of CO_2 (from 400 ppm to 700 ppm) (Karnosky et al., 2002).

The little studies were performed for verifying the impacts of CO_2 in disease prevalence. Runion et al., (1994) investigated upon soil borne pathogens in FACE experiment. There was increase in "damping off" through infection of *Rhizoctonia solani* on cotton crop. This was recorded during an up in CO_2. Jwa and Walling (2001) investigated relation between an up CO_2 concentration and root rot in tomato, due to *Phytophthora parasitica*. The pathogen's incidence in roots was lower in tomato plants growing at 700 ppm in comparison to 350 ppm CO_2. While the hyphal development of *P. parasítica in vitro* could not alter at both levels.

Osozawa et al., (1994) studied the impression of gaseous phase CO_2 on two soils in which one being suppressive but other conducive in cabbage club root disease resulted due to *Plasmodiophora brassicae.* They noticed an up in CO_2 concentration increased the formation of disease, at high levels of soil moisture.

There are relatively many studies on elevated concentration of atmospheric CO_2in terms of growth in plant. Recent investigations are on record (Loladze, 2002; Jans, 2016). Increased CO_2 concentration in general gives a positive effect on growth even though there might be some differences among species. CO_2 enrichment in atmosphere brings metabolism alterations in the form of growth rates. The photosynthetic rate increased but rate of transpiration decreased leaf area per unit but total transpiration rate increases which is due to larger area of leaf in plant (Jwa and Walling, 2001; Li et al., 2003). The changes are because of higher power in terms of water use/nitrogen (Thompson and Drake, 1994). The photosynthesis gets enhanced because of reduction of tension in atmospheric CO_2 and O_2 fixed by ribulose 1,5-bisphosphate carboxylase-oxygenase enzyme (RUBISCO). The O_2 concentration in atmosphere checks CO_2 absorption normally. This can trigger rate of photorespiration.

The diseases and pests presence in rice are direct connection with temperature and relative humidity. Higher CO_2 dose is a prime agent for changes to climate at global level which results an increased temperature. The IPCC broadcast that means temperature at global level would rise from 0.9°C to 3.5°C in 2100. Under such situations the precipitation frequency, drought intensity. The UV-B radiation would increase, which affects rice plant morphology and affect intensity of pests. There will be alteration in pest plant relations which can cause positive or negative impact on diseases and pests. *Fusarium moniliforme* causes disease in rice during Aus season and it was of no value for Boro crop. But in year 2000 it become a big problem for Boro rice areas of single Boro. This was due to an up in minimum temperature at winter time which might have accelerated and increased Bakanae in Boro season (Haq et al., 2008). *Rhizoctonia solanica using sheath blight* was a minor disease problem in early 1970s but presently it is most destructive problem for rice (Haq et al., 2010). The diseases in rice viz., false smut, sheath blight and sheath rot got increased (Figure 8.1.) (Biswas et al., 2011; Ghosh et al., 2012).

In spite of good effects of CO_2 on plant, it could not find out surely whether impacts would-be the same. Some studies were done in tropical conditions under controlled atmospheres may not reflect actual plant responses in field conditions because it may be constant and complex interactions of temperature, precipitation, and many others. The pathogen's response helps to produce new races fastly in atmosphere with an up temperature and elevated concentration CO_2concentration (Chakraborty, 2013).

(a) (b)

FIGURE 8.1 (See color insert.) False smut of rice.
Source: https://commons.wikimedia.org/wiki/File.

The observed effects of CO_2 in plant disease can counteract many others. Investigations evidenced that speedy growth rates. The upCO_2produced denser canopies having higher humidity which can help the pathogens to easily infect plants. However, at high CO_2 lower rate of decomposition is observed.

8.7 CLIMATE CHANGE (CC) V/S FOREST DISEASES

The climate has strong influence on forest tree diseases. The disease incidence and distribution are directly related with microorganism infections. The drought or typhoon conditions being extreme weather can destroy large number of forest trees directly through overwhelming their physiological capacity and structural power. Expected alternations in climate increase the pathogen stress and produced conditions favorable for several diseases in forest. The wood decay rates by forest fungal species also change and have potential of changing carbon cycles in the forest (Frankel, 2008).

In Western North America, conditions like changes in weather extremes, precipitation and warming are affecting forest trees diseases. The following instances cover entire range purely from abiotic factors drought related decline (decline of Alaska yellow cedar) and through *Dothistroma septosporum*-red band pine needle blight which is a weak pathogen. Also there is an aggressive but non-native pathogen as example of sudden death oak due to *Phytophthora ramorum*.

1. **Alaska Yellow-Cedar Decline:** Earlier melting of snow exposes fine roots in colder temperatures which produces freeze injury in Alaska destroying millions of *Chamaecyparis nootkatensis* (D. Don) Spach-yellow cedar (Hennon and D'Amore, 2007; Hennon and Shaw, 1997). The best instance of climatic impact on a forests tree is the severe mortality of yellow cedar in up to 1000 kilometers in the northern Pacific coast of North America. The reasons were reduced tree mortality basically due to two main causes such as soil drainage and snow depth. This is applied as a model for future and present suitability for cedar habitat (Hennon et al., 2012).
2. **Red Band Needle Blight (*Dothistroma septosporum*):** There was high mortality of British Columbia plantations and mature stands of lodge pole pine (*Pinus contorta* Dougl. ex Loud.) through needle blight. The mortality was adjacent to places covered by *Dendroctonus ponderosae* (mountain pine beetle). The increase of precipitation in summer was main cause in disease outbreak which is beyond the limit of other recorded weather patterns (Woods et al., 2005).
3. **Sudden Death of Oak (*Phytophthora ramorum*):** The extreme weather resulted sudden death related tree mortality of oak. The rate of sudden death of oak in Oregon and California increase abruptly and subside. This pattern was driven due to heavy rains which extended weather wet in warm periods, creating optimal conditions for the infection. Infected plants face a reduction in capacity in managing water. It survives till high temperatures exist and extended up to dry periods. This overwhelms vascular capability of the trees finally causing death. This pattern of two cycles was recorded in California: 1998–2001 and 2005–2008 (Frankel, 2007). The Bay Area place faced an all-time record up to 25 rainy days during March 2006. This is followed by rain in July that is above three months of longest string of hot weather ever recorded.

It was observed that *Plumeria acutifolia* (Apocynaceae family) was found wilting in Gurgaon. The incidence was alarming (10–15%) at IMT, Manesar. This caused wilting and even death of trees in a row (Figure 8.2a, b, c.). This wilting was observed in all age groups. This produced yellowing of leaves which starts from margins after this complete yellowing in old and young leaves results up to death of the plant in 3–4 years. This was first report of *Acrophialophora levis* causing wilt (Kumar et al., 2016).

Coleosporium sp. causing rust infection of young Plantations of *Plumeria obtuse* and *P. hybrid* at various locations of Gurgaon (March, 2014 to March, 2015). It resulted in leaf drying and death (Figure 8.3). The rust pustules diameter on leaves was between 0.5–3.0 mm. Both young and mature leaves showed susceptibility to the pathogen. This caused chlorosis, necrosis, and premature leaf fall (Kumar and Khurana, 2016a).

Aecidium sp. (Pucciniaceae) leaf spot in *Cordia dichotoma* G. Forest another important medicinal tree was noticed first time in Panchgaon, Gurgaon Haryana. It was seen in February-March 2014–2016 having oval to rounded, chlorotic lesions measuring 4–14 mm in diameter on living leaves (Kumar and Khurana, 2016b). The worst case is cankers disease due to *Geosmithia morbida* which is spread by walnut twig beetle. This destroyed the eastern black walnut (*Juglans nigra*) plantations in certain places of the United States (Jans, 2016).

We observed a serious leaf spot disease in *Calotropis procera* (Aiton) W. T. Aiton. (Asclepiadaceae) locally known as madar in Panchgaon, Manesar, Gurgaon (Haryana) in January 2014. This was alarming in January February 2017 and spot was present on every plant in this location. The Disease incidence was greater than 90%. This was yet a new record from Aravali Hills near AUH, Gurgaon (Kumar and Khurana, 2017a).

A serious leaf spot, initially white (Figure 8.4a) and then reddening and distortion (Figure 8.4b) was observed on naturally growing trees of glue berry – *Cordia dichotoma* at Panchgaon (Manesar), Gurgaon since March 2014. This was confirmed as *Taphrina* sp. based on morphological observations of ascospores and asci. This was the First record of *Taphrina sp.* causing leaf distortion, reddening in *C. dichotoma* (Kumar and Khurana, 2017b).

FIGURE 8.2 (See color insert.) (a) Single wilted and dead *Plumeria* tree (b, c) alone and several dead trees in a row.

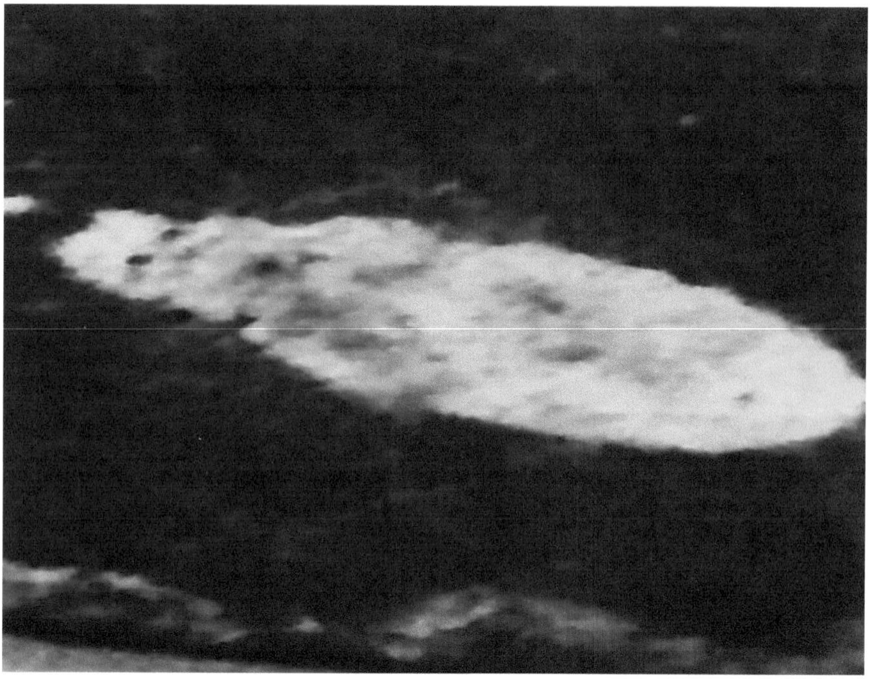

FIGURE 8.3 (See color insert.) Close up of Rust infection on leaf of *Plumeria.*

FIGURE 8.4 (See color insert.) *Taphrina* sp. fungal infection on *Cordia dichotoma* leaves.

4. **Drought Related Decline:** The forest trees stressed by abiotic factors, for example, drought, have least capacity to survive long (Winnett, 1998). There was worst drought in California. This resulted

over $2.5 billion damage (Keeley et al., 2004; Kliejunas et al., 2008). The drought of New Mexico and Arizona caused a severe loss of *Pinus edulis* Engelm. (Pinyon pine) around 12000 km^2 in the Southwest (Breshears et al., 2005).

8.8 FOREST DISEASE PREDICTIONS

i. There is a great uncertainty that how the forest tree diseases may respond to the CC. Changes may be different in comparison in relation to diseases of forest. Because of warming effects diseases may occur at higher elevations than in current climatic conditions.
ii. For newly developing forests the pathogens may play new roles against their health. Various new complex disease situations are expected.
iii. There will be many alterations in plant disease epidemiology. Prediction of disease outbreaks may not be typical during unstable weather and changing climate.
iv. In a rapidly fluctuating environment the resistance of host to pathogens could be overcome rapidly due to rapid development of aggressiveness.
v. Warmer winters could mostly lead for greater overwintering success in pathogens with increase in disease severity.
vi. The interactions can establish significant effects in carbon cycling of forest.

To study the effects many pathosystem have been studied (Elad and Pertot, 2014). Predictive models were searched. Phenological models of grapevine, phenological models of grape powdery mildew and the European grapevine moth models depict CC (Caffarra et al., 2012). They recorded that warmer region having profitable viticulture and increased temperatures results detrimental effect in yield. The models have been searched developed to predict forest pathogens in Australia such as dynamic simulation model known as CLIMEX was developed (Sutherst et al., 1999). This was used in telling impending geographical dispersion of forest tree species. It is useful in knowing potential dispersion of numerous pathogens (Venette and Cohen, 2006). In Portugal progress of modeling prediction in projecting pine wilt problem was done in Plant health risk and monitoring evaluation (PHRAME) project (Evans et al., 2008). The Manter et al., (2005) showed disease predictive model which is based on temperature and in combination of GIS (geographical information systems) and estimates disease levels in USA.

8.9 CLIMATIC PERTURBATION AND INHIBITION OF FUNGAL DISEASES

The management of crop disease plans depends much on the climatic status. CCs will bring about changes in geographical and temporal distributions of diseases so the control methods must be changed to be effective. Inspite of much importance a few discussions are available that how chemical methods of control may be affected by changes in climate. Changes can easily change fungicide residue dynamics of foliage and products degradation may get modified. The penetration and translocation mechanism of fungicides also gets altered in CO_2-enriched atmosphere (Chakraborty and Pangga, 2004; Pritchard and Amthor, 2005).

There may be change in fungicide market. Chen and McCarl (2001) conducted a study by regression analysis for pesticide usage and variations in climate on several US parts showing that the average cost of pesticide usage per acre of cotton, cotton, soybeans, wheat, and potatoes. It was found that it gets increased when precipitation increases. Accordingly average use of pesticide usage cost for cotton, corn, potato, and soybean increases when temperature gets increased but pesticide cost decreases in wheat due to temperature.

The main effect of chemical management of plant diseases depends on cultural realm. Everybody is facing the results of anthropogenic actions. Exploiting resources may create awareness to conduct this in sustainable manner. In the next few decades society will exert enough pressure to use only organic or non-chemical control of diseases.

The genetic resistance or susceptibility in plants in to diseases is affected by CC. The alterations in physiology may change resistance in cultivars obtained through genetic engineering and traditional techniques. Many studies have provided evidence on the alterations as increase in papillae production, photosynthetic rates, silicon accumulation and changes in enzymes which influence resistance (Osswald et al., 2006).

Paoletti and Lonardo (2001) observed increased CO_2 dose effects on disease. They reported genetic resistance in *Cupressus sempervirens* to canker due to *Seiridium cardinal.* They reported that it maintained resistance at genetic level during cultivation at a high CO_2 dose in atmosphere. There are many reports on impacts of temperature with other climatic parameters (Huang et al., 2006).

The plant disease resistance, in some cases, may be more affected in comparison to others. Significant alterations in physiology of host have potential to increase plant resistance. In this, the greatest problem for genetic resistance may be an up in cycles of pathogens. The investigations highlighted delay in the initial infection (Chakraborty et al., 2000).

Practically there is hardly any information on effect of climate alterations on biocontrol of plant diseases. CC influences dynamics of microbial population in phyllosphere. The microbial activity in soil may be changed by soil nutrients, soil water content and by increasing temperature. The nitrogen when mixed to natural soilalong with pollutants and fertilizers will produce serious effects on agricultural systems (Nosengo, 2003).

Rezácová et al., (2005) studied the effects on saprophytic fungi *Chlonostachys rosea* and reported it to serve as biocontrol for *Botrytis*. For insect pest management *Metarrhizium anisopliae* falls in the category of entomopathogens.

Warwick (2001) observed the biological control through use of *Acremonium vittelinum* and *A. persicinum* on coconut tar spot caused through *Catacaumator rendiella* and *C. palmicola.* But have little information on antagonists. This will be useful in power management of biocontrol agents. Inspite of this, it is essential to get response of pathogen to climate. This will allow in drawing conclusions as to what may happen on a large scale biocontrol under natural conditions and also upon introduction to bio-agents.

The predictions of results of CC and biocontrol of plants diseases is very complex and now depends on indirect observations. Nevertheless, this vulnerability of bio-control agents will alter surely due to CC (Garrett et al., 2006).

Ghini and Bettiol (2008) mentioned that change of climate is useful for both introduced (exotic) and natural (local) biological control. Adaptability of some specific alternative systems of agriculture could minimize negative impacts by adopting the new cultivars with other practices. The developing countries will face more difficulties due to change in climate, the reason being their poor technical competence. This is also because of scarce for resources the adoption of measures. Further society awareness for environmental issues will surely ask to reduce pollutant emissions. There will be an up in system complexity up to level of biological control.

8.10 FUTURE RESEARCH POTENTIAL

To overcome this goal, the specialists covering different agriculture fields need to jump their boundaries so the effect through CC may be investigated. CC shows positive/negative results on fungal infestation and there is limited work on fungal attacks. It needs modified chemical/biocontrol methods for implementation against diseases. It creates a new responsibility for a breeder. It needs development of models for Climate prediction against fungal infections. Since this is a burning issue in the present century, it needs collaboration from all disciplines to reach on single platform for this global issue. It needs more rational approaches for consideration to find out the actual mechanism of CC impact in fungal infections.

8.11 SUMMARY AND CONCLUSIONS

The effects of CC on diseases plants pests are a big issue to observe. CC affects quality and yield of crops and forms new diseases. It covers many factors which affect plant health. Due to climatic fluctuations there will be changes in physiology of plant along with change in host pathogen interactions, modification in resistance and development, spread of pathogens. The change in Climate can either have neutral, negative or positive impact in plants. This is because of pathogen's specific nature and interactions. Since CC functions globally, hence we need deep knowledge of epidemic progression. Many hitherto unknown plant diseases now recorded in Haryana, India viz., *Aecidium* sp. causing serious rust on *Cordia dichotoma* leaf. Rust disease of Plumeria due to *Coleosporium* spp., *Taphrina* sp. causing reddening and leaf distortion in *Cordia dichotoma* (glue berry). Serious leaf spot problem of *Calotropis procera* (Aiton) W.T. Aiton. By *Alternaria alternata* and *Acrophialophora levis* producing wilt in *Plumeria* at Gurgaon, Haryana, India. For disease control, information is generally required at the field scale for the devastating diseases. It needs the evaluation of potential effects of change in climate and detailed assessment to investigate mechanisms that are essential for an epidemic.

CONFLICT OF INTEREST STATEMENT

We declare that we have no conflict of interest.

ACKNOWLEDGMENT

Authors are thankful to the Amity University, Haryana authorities for the facilities and constant encouragement for research and innovation.

KEYWORDS

- **climate change**
- **diseases of crops**
- **fungal pathogens**
- **greenhouse gases**

REFERENCES

Agrios, G. N., (2004). *Plant Pathology* (5th edn., p. 922). Elsevier, USA.

Agrios, G. N., (2005). *Plant Pathology*. Burlington, MA: Elsevier Academic Press.

Anderson, P. K., Cunningham, A. A., Patel, N. G., Morales, F. J., Epstein, P. R., & Daszak, P., (2004). Emerging infectious diseases of plants: Pathogen, pollution, climate change and agrotechnology drivers. *Trends Ecol. Evol., 19*, 535–544.

Beaumont, L. J., Pitman, A., Perkins, S., Zimmermann, N. E., Yoccoz, N. G., & Thuiller, W., (2011). Impacts of climate change on the world's most exceptional ecoregions. *Proceedings of the National Academy of Sciences USA, 108*, 2306–2311.

Bentz, B. J., Régnière, J., Fettig, C. J., Hansen, E. M., Hayes, J. L., & Hicke, J. A., (2010). Climate change and bark beetles of the Western United States and Canada: Direct and indirect effects. *BioScience, 60*, 602–613.

Benvenuti, S., (2009). Potenzialeimpattodeicambiamenti climatic nell'evoluzione floristica di fitocenosi spontanee in agroecosistemi mediterranei. *RivistaItaliana di Agronomia, S1*, 45–67.

Bergot, M., Cloppet, E., Perarnaud, V., Deque, M., Marcais, B., & Desprez-Loustau, M. L., (2004). Simulation of potential range expansion of oak disease caused by *Phytophthora cinnamomi* under climate change. *Global Change Biology, 10*, 1539–1552.

Bhat, K. A., Peerzada, S. H., & Anwar, A., (2015). *Alternaria* epidemic of Apple in Kashmir. *African Journal of Microbiology Research, 9*, 831–837.

Biswas, B. K. Golam, F. Alam, N., Hossain, M. K., & Rahman, M. M., (2011). Disease index construction against prevalence of major diseases in fine rice. *Afr. J. Agric. Res., 6*(21), 4954–4959.

Boland, G. J., Melzer, M. S., Hopkin, A., Higgins, V., & Nassuth, A., (2004). Climate change and plant diseases in Ontario. *Canadian Journal of Plant Pathology, 26*(3), 335–350.

Bosch, J., Carrascal, L. M., Duran, L., Walker, S., & Fisher, M. C., (2007). Climate change and outbreaks of amphibian chytridiomycosis in a montane area of Central Spain; is there a link? *Proc. R. Soc., 274*, 253–260.

Breshears, D. D., Cobb, N. S., Rich, P. M., Price, K. P., Allen, C. D., Balice, R. G., et al., (2005). Regional vegetation die-off in response to global-change type drought. *Proceedings of the National Academy of Sciences of the United States of America, 102*, 15144–15148.

Caffarra, A., Rinaldi, M., Eccel, E., Rossi, V., & Pertot, I., (2012). Modeling the impact of climate change on the interaction between grapevine and its pests and pathogens: European grapevine moth and powdery mildew. *Agriculture, Ecosystems and Environment, 148*, 89–101.

Canadell, J. G., Le Quéré, C., Raupach, M. R., Field, C. B., Buitenhuis, E. T., Ciais, P., Conway, T. J., Gillett, N. P., Houghton, R. A., & Marland, G., (2007). Contributions to accelerating atmospheric CO_2 growth from economic activity, carbon intensity, and efficiency of natural sinks. *Proceedings of the National Academy of Sciences*, *104*, 1866–1870.

Cerri, C. E. P., Sparovek, G., Bernoux, M., Easterling, W. E., Melillo, J. M., & Cerri, C. C., (2007). Tropical agriculture and global warming: Impacts and mitigation options. *Scientia Agricola, 64*, 83–99.

Chakraborty, S., & Datta, S., (2003). How will plant pathogens adapt to host plant resistance at elevated CO2 under a changing climate? *New Phytol.*, *159*, 733–742.

Chakraborty, S., & Pangga, I. B., (2004). Plant disease and climate change. In: Gillings, M., & Holmes, A., (ed.), *Plant Microbiology* (pp.163–180). London: BIOS Scientific.

Chakraborty, S., (2013). Migrate or evolve: Options for plant pathogens under climate change. *Glob Chang Biol.*, *19*, 1985–2000.

Chakraborty, S., Luck, J., Hollaway, G., Freeman, A., Norton, R., & Garrett, K. A., (2008). Impacts of global change on diseases of agricultural crops and forest trees. *CAB Reviews*, *3*, 054.

Chakraborty, S., Murray, G., & White, N., (2002). *Impact of Climate Change on Important Plant Diseases in Australia.* A report for the Rural Industries Research and Development Corporation, RIRDC Publication No W02/010, RIRDC Project No CST-4A.

Chakraborty, S., Tiedemann, A. V., & Teng, P. S., (2000). Climate change: Potential impact on plant diseases. *Environmental Pollution, 108*, 17–326.

Chen, C. C., & McCarl, B. A., (2001). An investigation of the relationship between pesticide usage and climate change. *Climatic Change*, *50*, 475–487.

Coakley, S. M., Scherm, H., & Chakraborty, S., (1999). Climate change and plant disease management. *Annu. Rev. Phytopathol.*, *37*, 399–426.

Colhoun, J., (1973). Effects of environmental factor on plant disease. *Annual Review of Phytopathology, 11*, 343–364.

Desprez-Loustau, M. L., Marçais, B., Nageleisen, L. M., Piou, D., & Vannini, A., (2006). Interactive effects of drought and pathogens in forest trees. *Annals Forest Sci.*, *63*, 597–612.

Dillon, M. E., Wang, G., & Huey, R. B., (2010). Global metabolic impacts of recent climate warming. *Nature, 467*, 704–707.

Elad, Y., & Pertot, I., (2014). Climate change impacts on plant pathogens and plant diseases. *Journal of Crop Improvement*, *28*(1), 99–139.

Engler, R., Randin, C. F., Thuiller, W., Dullinger, S., Zimmermann, N. E., Araujo, M. B., et al., (2011). 21st century climate change threatens mountain flora unequally across Europe. *Global Change Biology*, *17*(7), 2330–2341.

Evans, S., Evans, H., & Ikegami, M., (2008). *Modeling PWN-Induced wilt Expression: A Mechanistic Approach, Pine Wilt Disease: A Worldwide Threat to Forest Ecosystems* (pp. 259–278). Springer.

Fabre, B., Piou, D., Desprez-Loustau, M. L., & Marçais, B., (2011). Can the emergence of pine Diplodia shoot blight in France be explained by changes in pathogen pressure linked to climate change? *Global Change Biol.*, *17*, 3218–3227.

Ferguson, B. A., Dreisbach, T. A., Parks, C. G., Filip, G. M., & Schmitt, C. L., (2003). Coarse-scale population structure of pathogenic *Armillaria* species in a mixed-conifer forest in the blue mountains of northeast Oregon. *Canadian Journal of Forest Research, 33*, 612–623.

Frankel, S. J., (2007). *Climate Change's Influence on Sudden Oak Death*. PACLIM 2007, 2007, Monterey, CA. http://www.fs.fed.us/psw/cirmount/meetings/paclim/pdf/frankel_talk_ PACLIM2007.pdf (Accessed on 10 October 2019).

Frankel, S. J., (2008). *Forest Plant Diseases and Climate Change*. U. S. Department of Agriculture, Forest Service, Climate Change Resource Center. http://www.fs.fed.us/ccrc/topics/plant-diseases.shtml (Accessed on 10 October 2019).

Fuhrer, J., (2003). Agroecosystem responses to combinations of elevated CO_2, ozone, and global climate change. *Agric. Ecosystems Environ.*, *97*, 1–20.

Garbelotto, M., Linzer, L., Nicolotti, G., & Gonthier, P., (2010). Comparing the influences of ecological and evolutionary factors on the successful invasion of a fungal forest pathogen. *Biol. Invasions*, *12*, 943–957.

Garrett, K. A., Dendy, S. P., Frank, E. E., Rouse, M. N., & Travers, S. E., (2006). Climate change effects on plant disease: Genomes to ecosystems. *Annu. Rev. Phytopathol.*, *44*, 489–509.

Garrett, K. A., Nita, M., DeWolf, E. D., Gomez, L., & Sparks, A. H., (2009). Plant pathogens as indicators of climate change. In: Letcher, T., (ed.), *Climate Change: Observed Impacts on Planet Earth* (pp. 425–437). Elsevier, Dordrecht.

Gaumann, E., (1950). *Principles of Plant Infection* (p. 543). London: Crosby Lockwood.

Ghini, R., & Bettiol, W., (2008). Impacto das mudanças climáticas globais sobre o controlebiológico de doenças de plantas. *Summa Phytopathologica*, *34*, 193–194.

Ghini, R., (2005). *Mudanças climáticas globais e Doenças de Plantas* (p. 104). Jaguariúna: EmbrapaMeioAmbiente.

Ghosh, N., Prakash, S., Das, C., Mandal, Gupta, S., Kingsuk, D., Dey, N., & Adak, M. K., (2012). Variations of antioxidative responses in two rice cultivars with polyamine treatment under salinity stress. *Physiology and Molecular Biology of Plants*, *18*(4), 301–313.

Hansen, J., Sato, M., Ruedy, R., Lo, K., Lea, D. W., & Medina-Elizade, M., (2006). Global temperature change. *Proceedings of the National Academy of Sciences*, *103*, 14288–14293.

Haq, M., Tahar Mia, M. A., Rabbi, M. F., & Ali, M. A. M., (2008). Incidence and severity of rice diseases and insect pests in relation to climate change. *International Symposium on Climate Change: Climate Change and Food Security in South Asia*. Dhaka Bangladesh.

Haq, M., Tahar Mia, M. A., Rabbi, M. F., & Ali, M. A. M., (2011). Incidence and severity of rice diseases and insect pests in relation to climate change. In: Lal, R., Sivakumar, M. V. K., Faiz, S. M. A., Rahman, A. H. M. M., & Islam, K. R., (eds.), *Climate Change and Food Security in South Asia*. Springer.

Hatfield, J. L., Boote, K. J., Kimball, B. A., Ziska, L. H., Izaurralde, R. C., Ort, D. R., & Thomson, A. M. (2011). Climate impacts on agriculture: Implications for crop production. *Agron. J.*, *103*, 351–370.

Hatfield, J., Boote, K., Fay, P., Hahn, L., Izaurralde, C., B. Kimball, B. A., Mader, T., Morgan, J., Ort, D., Polley, W., Thomson, A., & Wolfe, D., (2008). Agriculture. In: *The Effects of Climate Change on Agriculture, Land Resources, Water Resources, and Biodiversity in the United States* (p. 362). A Report by the U. S. Climate Change Science Program and the Subcommittee on Global Change Research. Washington, DC., USA.

Hegerl, G. C., Hanlon, H., & Beierkuhnlein, C., (2011). Climate science: Elusive extremes. *Nature Geoscience*, *4*, 142–143.

Hennon, P. E., David, V. D., Schaberg, P. G., Wittwer, D. T., & Shanley, C. S., (2012). Shifting climate, altered niche, and a dynamic conservation strategy for yellow-cedar in the North Pacific Coastal Rainforest. *BioScience*, *62*(2), 147–156.

Hennon, P., & D'Amore, D., (2007). *The Mysterious Demise of an Ice-Age Relic: Exposing the Cause of Yellow-Cedar Decline: Science Findings* (Issue 93, p. 6). Portland, OR: U. S. Department of Agriculture, Forest Service, Pacific Northwest Research Station.

Hennon, P., & Shaw, C. G., (1997). The enigma of yellow-cedar decline: What is killing these long lived, defensive trees? *Journal of Forestry, 95*(12), 4–10.

Heyder, U., Schaphoff, S., Gerten, D., & Lucht, W., (2011). Risk of severe climate change impact on the terrestrial biosphere. *Environmental Research Letters, 6*, 034036. doi: 10.1088/1748–9326/6/3/034036.

Hibberd, J. M., Whitbread, R., & Farrar, J. F., (1996a). Effect of 700 μmol mol^{-1} CO_2 and infection with powdery mildew on the growth and carbon partitioning of barley. *New Phytologist, 134*, 309–315.

Hibberd, J. M., Whitbread, R., & Farrar, J. F., (1996b). Effect of elevated concentrations of CO_2 on infection of barley by *Erysiphe graminis*. *Physiological and Molecular Plant Pathology, 48*, 37–53.

Howdy, (2018). *Climate Change Threats to Indian Agriculture*. http://www.iac2016.in/climate-change-threat-india (Accessed on 10 October 2019).

Huang, Y. J., Evans, N., Li, Z. Q., Eckert, M., Chèvre, A. M., Renard, M., & Fitt, B. D. L., (2006). Temperature and leaf wetness duration affect phenotypic expression of Rlm6-mediated resistance to *Leptosphaeria maculans* in *Brassica napus*. *New Phytologist, 170*, 129–141.

Intergovernmental Panel Climate Change (IPCC), (2007). *Climate Change 2007: Impacts, Adaptation and Vulnerability: Contribution of Working Group II to the Fourth Assessment Report of the Intergovernmental Panel on Climate Change*. Cambridge University Press, Cambridge, U. K., & New York, NY.

Jans, S. A., (2016). *The Silent Invasion of the Killer* fungus.elsevier.com/connect/the-silent-invasion-of-the-killer-fungus (assessed on 23/5/18).

Jones, N., (2017). *How the World Passed a Carbon Threshold and Why it Matters*. Yale Environment 360. https://e360.yale.edu/features/how-the-world-passed-a-carbon-threshold-400ppm-and-why-it-matters59 (Accessed on 10 October 2019).

Jwa, N. S., & Walling, L. L., (2001). Influence of elevated CO_2 concentration on disease development in tomato. *New Phytologist, 149*, 09–18.

Karnosky, D. F., Percy, K. E., Xiang, B. X., Callan, B., Noormets, A., Mankovska, B., Hopkin, A., Sober, J., Jones, W., Dickson, R. E., & Isebrands, J. G., (2002). Interacting elevated CO_2 and tropospheric O_3 predisposes aspen (*Populus tremuloides* Michx.) to infection by rust (*Melampsora medusae* f. sp *tremuloidae*). *Global Change Biology, 8*, 329–338.

Keeley, J. E., Fotheringham, C. J., & Moritz, M. A., (2004). Lessons from the October 2003 wildfires in southern California. *Journal of Forestry, 102*, 26–31.

Khan, J., Del Rio, L. E., Nelson, R., Rivera-Varas, V., Secor, G. A., & Khan, M. F. R., (2008). Survival, dispersal, and primary infection site for *Cercospora beticola* in sugar beet. *Plant Disease, 92*, 741–745.

Khurana, S. M. P., & Kumar, N., (2016). Climate change: Perspectives of challenges and options in plant disease/pest management. *International Journal of Innovative Horticulture, 5*(2), 59–74.

Kliejunas, J. T., Geils, B., Glaeser, J. M., Goheen, E. M., Hennon, P., Mee-Sook, K., Kope, H., Stone, J., Sturrock, R., & Frankel, S. J., (2008). *Climate and Forest Diseases of Western North America: A Literature Review* (p. 36). Albany, CA: U. S. Department of Agriculture, Forest Service, Pacific Southwest Research Station. http://www.fs.fed.us/psw/topics/climate_change/forest_disease (Accessed on 10 October 2019).

Knogge, W., (1998). Fungal pathogenicity. *Current Opinion in Plant Biology, 1*, 324–328.

Kobayashi, T., Ishiguro, K., Nakajima, T., Kim, H. Y., Okada, M., & Kobayashi, K., (2006). Effects of elevated atmospheric CO_2 concentration on the infection of rice blast and sheath blight. *Phytopathology, 96*, 425–431.

Körner, C., & Basler, D., (2010). Phenology under global warming. *Science, 327*, 1461–1462.

Kudela, V., (2009). Potential impact of climate change on geographic distribution of plant pathogenic bacteria in Central Europe. *Plant Protect. Sci., 45*, S27-S32.

Kumar, N., & Khurana, S. M. P., (2016a). "Rust disease of Frangipani (Plumeria) caused by *Coleosporium* sp. in Gurgaon, Haryana." *International Journal of Current Microbiology and Applied Sciences (IJCMAS), 5*(02), 590–597.

Kumar, N., & Khurana, S. M. P., (2016b). *Aecidium* sp. causing serious leaf rust on *Cordia dichotoma* G. Forst in Haryana-first record. *International Journal of Current Microbiology and Applied Sciences, 5*(8), 55–58.

Kumar, N., & Khurana, S. M. P., (2017a). Serious Leaf Spot Disease Problem of *Calotropis procera* (Aiton) W. T. Aiton. by *Alternaria alternata* in Gurgaon (Haryana), India. *Int. J. Curr. Microbiol. App. Sci., 6*(5), 403–407.

Kumar, N., & Khurana, S. M. P., (2017b). First report of *Taphrina* sp. causing leaf distortion and reddening in glue berry, *Cordia dichotoma. Medicinal Plants, 9*(3), 221–222.

Kumar, N., Kulshreshtha, D., Sharma, S., Aggarwal, R., & Khurana, S. M. P., (2016). First record of *Acrophialophora levis* causing wilt of Plumeria in Gurgaon. *Indian Phytopathology, 69*(4), 400–406.

Li, F., Kang, S., Zhang, J., & Cohen, S., (2003). Effects of atmospheric CO_2 enrichment, water status and applied nitrogen on water- and nitrogen-use efficiencies of wheat. *Plant and Soil, 25*(4), 279–289.

Loladze, L., (2002). Rising atmospheric CO_2 and human nutrition: Toward globally imbalanced plant stoichiometry? *TRENDS in Ecology and Evolution, 17*, 457–461.

Mann, M. E., Bradley, R. S., & Hughes, M. K., (1998). Global scale temperature patterns and climate forcing over the past six centuries. *Nature, 392*, 779–787.

Manning, W. J., & Tiedemann, A. V., (1995). Climate change: Potential effects of increased atmospheric carbon dioxide (CO_2), ozone (O_3), and ultraviolet-B (UV-B) radiation on plant diseases. *Environmental Pollution, 88*, 219–245.

Manter, D. K., Reeser, P. W., & Stone, J. K., (2005). A climate-based model for predicting geographic variation in Swiss needle cast severity in the Oregon Coast Range. *Phytopathology, 95*, 1256–1265.

Matesanz, S., Gianoli, E., & Valladares, F., (2010). Global change and the evolution of phenotypic plasticity in plants. *Annals of the New York Academy of Sciences, 1206*, 35–55.

Mcelrone, A. J., Hamilton, J. G., Krafnick, A. J., Aldea, M., Knepp, R. G., & De Lucia, E. H., (2010). Combined effects of elevated CO_2 and natural climatic variation on leaf spot diseases of red bud and sweetgum trees. *Environ. Pollut., 158*, 108–114.

Meehl, G. A., Stocker, W. D., Collins, P., Friedlingstein, A. T., Gaye, A. T., Gregory, A., et al., (2007). Global climate projections. In: Solomon, S., Qin, D., Manning, M., Chen, Z., Marquis, M., Averyt, K. B., Tignor, M., & Miller, H. L., (eds.), *Climate Change 2007: The Physical Science Basis*. Contribution of Working Group I to the Fourth Assessment Report of the Intergovernmental Panel on Climate Change. Cambridge University Press, Cambridge, United Kingdom and New York, NY, USA.

Milad, M., Schaich, H., Bürgi, M., & Konold, W., (2011). Climate change and nature conservation in Central European forests: A review of consequences, concepts and challenges. *Forest Ecology and Management, 261*, 829–843.

Mitchell, C. E., Reich, P. B., Tilman, D., & Groth, J. V., (2003). Effects of elevated CO_2, nitrogen deposition, and decreased species diversity on foliar fungal plant disease. *Global Change Biology*, *9*, 438–451.

Nosengo, N., (2003). Fertilized to death. *Nature, 425*, 894–895.

Oerke, E. C., (2006). Crop losses to pests. *The Journal of Agricultural Science*, *144*, 31–43.

Osozawa, S., Iwama, H., & Kubota, T., (1994). Effect of soil aeration on the occurrence of clubroot disease of crucifers. *Soil Science and Plant Nutrition*, *40*, 445–455.

Osswald, W. F., Fleischmann, F., & Heiser, I., (2006). Investigations on the effect of ozone, elevated CO_2 and nitrogen fertilization on host-parasite interactions. *Summa Phytopathologica*, *32S*, S111–S113.

Pangga, I. B., Chakraborty, S., & Yates, D., (2004). Canopy size and induced resistance in *Stylosanthes scabra* determine anthracnose severity at high CO_2. *Phytopathology*, *4*, 221–227.

Paoletti, E., & Lonardo, V. D., (2001). *Seiridium cardinale* cankers in a tolerant *Cupressus sempervirens* clone under naturally CO_2-enriched conditions. *Forest Pathology*, *31*, 307–311.

Paterson, R. R. M., & Lima, N., (2010). How will climate change affect mycotoxins in food? *Food Research International*, *43*, 1902–1914.

Peng, S., Huang, J., Sheehy, J. E., Laza, R. C., & Visperas, R. M., (2004). Rice yields decline with higher night temperature from global warming. *Proc. Natl. Acad. Sci. USA, 101*, 9971–9975.

Percy, K. E., Awmarck, C. S., Lindroth, R. L., Kubiske, M. E., Kopper, B. J., Isebrands, J. G., et al., (2002). Altered performance of forest pests under atmospheres enriched by CO_2 and O_3. *Nature*, *420*, 403–407.

Percy, K. E., Nosal, M., Heilman, W., Dann, T., Sober, J., Legge, A. H., & Karnosky, D. F., (2007). New exposure-based metric approach for evaluating O_3 risk to North American aspen forests. *Environmental Pollution*, *147*, 554–566.

Perkins, L. B., Leger, E. A., & Nowak, R. S., (2011). Invasion triangle: An organizational framework for species invasion. *Ecol. Evol.*, *1*, 610–625.

Pritchard, S. G., & Amthor, J. S., (2005). *Crops and Environmental Change* (p. 421). Binghamton: Food Products Press.

Prospero, S., Grunwald, N. J., Winton, L. M., & Hansen, E. D. M., (2009). Migration patterns of the emerging plant pathogen *Phytophthora ramorum* on the west coast of the United States of America. *Phytopathol.*, *99*, 739–749.

Quarles, W., (2007). Global warming means more pests. *The IPM Practitioner*, *29*(9/10), 1–8.

Regniere, J., (2012). Invasive species, climate change and forest health. In: Schlichter, T., & Montes, L., (eds.), *Forests in Development: A Vital Balance* (pp. 27–37). Springer, Berlin.

Rezácová, V., Blum, H., Hrselová, H., Gamper, H., & Gryndler, M., (2005). Saprobic microfungi under *Lolium perenne* and *Trifolium repens* at different fertilization intensities and elevated atmospheric CO_2 concentration. *Global Change Biology*, *11*, 224–230.

Richard, A., Betts, D., & Neall, M., (2018). How much CO_2 at 1.5°C and 2°C? *Nature Climate Change*, *8*(7), 546.

Rosenzweig, C., Iglesias, A. A., Yang, X. B., Epstein, P. R., & Chivian, E., (2001). *"Climate Change and Extreme Weather Events-Implications for Food Production, Plant Diseases, and Pests"* (p. 24) NASA Publications. http://digitalcommons.unl.edu/nasapub (Accessed on 10 October 2019).

Runion, G. B., Curl, E. A., Rogers, H. H., Backman, P. A., Rodríguez-Kábana, R., & Helms, B. E. (1994). Effects of free-air CO_2 enrichment on microbial populations in the rhizosphere and phyllosphere of cotton. *Agricultural and Forest Meteorology*, *70*, 117–130.

Scheffer, R. P., (1997). *The Nature of Disease in Plants*. New York, N. Y.: Cambridge University Press.

Scherm, H., (2004). Climate change: Can we predict the impacts on plant pathology and pest management? *Canadian J. Plant Pathol.*, *26*, 267–273.

Schneider, R. W., Hollier, C. A., & Whitam, H. K., (2005). "First report of soybean rust caused by *Phakopsora pachyrhizi* in the continental United States. *Plant Disease, 89*(7), 741–774.

Siegenthaler, U., Stocker, T. F., Monnin, E., Luthi, D., Schwander, J., Stauffer, B., Raynaud, D., Barnola, J. M., Fischer, H., Masson-Delmotte, V., & Jouzel, J., (2005). Stable carbon cycle-climate relationship during the late pleistocene. *Science*, *310*, 1313–1317.

Spahni, R., Chappellaz, J., Stocker, T. J., Loulergue, L., Hausammann, G., Kawamura, K., Fluckiger, J., Schwander, J., Raynaud, D., Masson-Delmotte, V., & Jouzel, J., (2005). Atmospheric methane and nitrous oxide of the late pleistocene from Antarctic ice cores. *Science*, *310*, 1317–1321.

Stern, N., (2007). The economics of climate change: The stern review. Cambridge University Press, Cambridge, UK. *Int. J. Modern Plant and Anim. Sci.*, *1*(3), 105–115.

Stone, J. K., Coop, L. B., & Manter, D. K., (2008). Predicting effects of climate change on Swiss needle cast disease severity in Pacific Northwest forests. *Can. J. Plant Pathol.*, *30*, 169–176.

Strange, R. N., & Scott, P. R., (2005). Plant disease: A threat to global food security. *Annual Review of Phytopathology*, *43*, 83–116.

Sutherst, R. W., Constable, F., Finlay, K. J., Harrington, R., Luck, J., & Zalucki, M. P., (2011). Adapting to crop pest and pathogen risks under a changing climate. *Wiley Interdisciplinary Reviews - Climate Change*, *2*, 220–237.

Sutherst, R. W., Maywald, G. F., Yonow, T., & Stevens, P. M., (1999). *CLIMEX-Predicting the Effects of Climate on Plants and Animals*. Melbourne, CSIRO publishing.

Swaminathan, M. S., (1986). Changing paradigms in Indian agriculture – the way ahead. *The Hindu*, pp. 22–46.

Thompson, A., (2016). *UK Forecasters: More Warming in Store Over Next 5 Years*. http://www.climatecentral.org/news/five-year (Accessed on 10 October 2019).

Thompson, G. B., & Drake, B. G., (1994). Insects and fungi on a C3 sedge and a C4 grass exposed to elevated atmospheric CO_2 concentrations in open-top chambers in the field. *Plant, Cell and Environment*, *17*, 1161–1167.

Thompson, G. B., Brown, J. K. M., & Woodward, F. I., (1993). The effects of host carbon dioxide, nitrogen and water supply on the infection of wheat by powdery mildew and aphids. *Plant, Cell and Environment*, *16*, 687–694.

Tiedemann, A. V., & Firsching, K. H., (2000). Interactive effects of elevated ozone and carbon dioxide on growth and yield of leaf rust-infected versus non-infected wheat. *Environmental Pollution*, *108*, 357–363.

Venette, R. C., & Cohen, S. D., (2006). Potential climatic suitability for establishment of Phytophthora ramorum within the contiguous United States. *Forest Ecology and Management*, *231*, 18–26.

Walther, G. R., Post, E., Convey, P., Menzel, A., Parmesan, C., & Beebee, T. J. C., (2002). Ecological responses to recent climate change. *Nature*, *416*, 389–395.

Warwick, D. R. N., (2001). Colonização de estromas de *Sphaerodothisacrocomiae* agente causal da lixagrande do coqueiropor *Acremoniumpersicinum*. *Fitopatologia Brasileira*, *26*, 220.

Winnett, S. M., (1998). Potential effects of climate change on U. S. forests: A review. *Climate Research*, *11*, 39–49.

Woods, A., Coates, K. D., & Hamann, A., (2005). "Is an unprecedented dothistroma needle blight epidemic related to climate change?" *BioScience*, *55*(9), 761–769.

Wunsch, M., (2013). *Management of Stemphylium Blight of Lentils*. North Dakota State University Camington Extension Research Center.

CHAPTER 9

Utilization of Boiling Water of Rice: A Case Study of Sustainable Water Management at Laboratory Scale, Ambikapur, Surguja, Chhattisgarh, India

VIJAY RAJWADE,[1] ARNAB BANERJEE,[2] MADHUR MOHAN RANGA,[3] MANOJ KUMAR JHARIYA,[4] DHIRAJ KUMAR YADAV,[4] and ABHISHEK RAJ[5]

[1]*Post Graduate Student, University Teaching Department, Department of Environmental Science, Sarguja Vishwavidyalaya, Ambikapur, Chhattisgarh–497001, India*

[2]*Assistant Professor, University Teaching Department, Department of Environmental Science, Sarguja Vishwavidyalaya, Ambikapur, Chhattisgarh–497001, India, Mobile: +919926470656, E-mail: arnabenvsc@yahoo.co.in*

[3]*Professor, University Teaching Department, Department of Environmental Science, Sarguja Vishwavidyalaya, Ambikapur, Chhattisgarh–497001, India*

[4]*Assistant Professor, University Teaching Department, Department of Farm Forestry, Sarguja Vishwavidyalaya, Ambikapur, Chhattisgarh–497001, India, E-mail: manu9589@gmail.com (M. K. Jhariya); E-mail: dheeraj_forestry@yahoo.com (D. K. Yadav)*

[5]*PhD Scholar, Indira Gandhi Krishi Vishwavidyalaya, Raipur, Chhattisgarh–492012, India, E-mail: ranger0392@gmail.com*

ABSTRACT

An experimental study at the laboratory scale was conducted to evaluate the influence of boiling water of rice on germination physiology of selected crop species. Seeds of gram and wheat undergo various treatments, i.e., 0%,

20%, 40%, 60%, and 100% dilution. The treatments were arranged in RBD (randomized block design) along with replication. Various data were recorded on germination physiology of wheat plants during the present investigation. High level of application of boiling water of rice revealed inhibitory effect on germination physiology of wheat seedlings. Rice boiling water at 20% dilution revealed positive impact over germination physiology of selected crop seedlings.

9.1 INTRODUCTION

Germination is a mechanism of growth of embryonic plant of seed. Male and female unite to form an embryo within seed produced inside a fruit, which germinates into a young plant. Seeds undergo a quiescence stage, which represents less growth or remains inactive. Therefore, mobilization takes place from a site to another site. They will grow when the growth condition becomes favorable. The architecture of seed comprises of an embryo contained within seed coat with stored substance. Favorable condition helps germination of seeds; growth takes place within the embryonic tissues and ultimately developing into a seedling (Chachalis and Reddy, 2000).

Germination occurs in the case of various plant species. Initiation of germination takes place through growth of embryonic tissue, developing towards seedling. Germination occurred in the case of monocotyledonous as well as plants with two-cotyledons (Chachalis and Reddy, 2000). In monocotyledonous plants, embryo's radical and cotyledon were covered through coleorhizae and coleoptiles. Coleorhiza is the initiator in the germination process to come out from the seed coat. Further, the radical emerges. Then the coleoptiles are pushed within ground and which ultimately reaches the surface layer. Thereafter the leaf emerges from an opening and further elongation ceases after certain time interval.

9.2 REQUIREMENTS FOR THE GERMINATION PROCESS

Seed germination is both internally and externally dependent. Factors such as oxygen, water, photoperiod, light intensity, temperature acts externally upon the process of seed germination. Different nature of propagules along with their germination depends upon the soil condition, under warm or cold condition. Ecological set up of the nature also plays significant role on this physiological process (Chachalis and Reddy, 2000). The factors significantly influencing the mechanism of seed germination includes the following:

9.2.1 *WATER*

Essential factors such as water are very much important for seed germination. Seeds under mature condition often tend to remain dry and optimum level of moisture is required to start the germination. Presence of water helps to initiate cell metabolism and growth processes. Seeds perform actively when water moistens them adequately but not soaking. Water imbibition promotes the breaking of seed dormancy and seed coat. Plants store reserve food material such as biological macromolecules essential for physiological progress towards maturity of seeds. Hydrolytic enzyme activation start with water imbibition by seeds and metabolic breakdown of the reserve food material occurs. Further useful chemical allows the embryonic cells to grow. Subsequently, the emergence of the seedlings initiates photosynthesis to produce glucose which would act as energy source for future growth.

9.2.2 *OXYGEN*

Oxygen is another important factor of metabolic activity in germinating seeds. Under the waterlogging condition of the soil as well, seed buried under soil may have deficiency of oxygen. Oxygen is mainly utilized under aerobic condition to promote growth in seedlings until leaf emerges from the seedlings which would make the seedling to photosynthesize to fulfill its energy requirement. Impermeability of seed coats prevents entry of oxygen the internal structure of seeds and thus making them dormant. This can be considered as physical dormancy which can be broken down through gas and water exchange propagules and its environment.

9.2.3 *TEMPERATURE*

Temperature influences the rate of cellular metabolism and growth. Different seeds have different temperature specificity such as some germinates at temperature just above the room temperature while some germinates above the freezing temperature and some responds differentially under varying temperature regimes. Some seeds have narrow temperature range above and below which they would not germinate. Some requires exposure to cold condition for breaking up of seed dormancy. Seeds under dormant condition would not germinate even the external condition is not favorable. Such type of temperature dependency of seeds is known as physiological dormancy.

9.2.4 LIGHT OR DARKNESS

This is also another factor of physiological dormancy, which triggers the seed germination process. However, this is not applicable for most seeds except species under forest canopy, which requires the triggering stimulus for seed germination and also for further growth.

9.2.5 DORMANCY

It is a physiological state of the seeds that the seeds go under a nascent stage and therefore would not germinate even after sufficient light or temperature. Factors regulating seed dormancy varies component wise of seed. Dormancy is usually broken through the mixed activity of different external and internal factors. Various environmental factors can contribute towards initiation of process of seed germination. Factors include ignition, animal uptake, photoperiod, and other associated factors that serve to end seed dormancy. Growth regulators in the form of hormones are very much critical from seed dormancy perspectives. Internal plant hormones (abscisic acid) and external hormones (gibberellins) may influence the process to a considerable extent.

Water is a big issue throughout the world. Last decade has put forward for us the question about decrease of usable water resource. The exponential growth rate of human population has increased the per capita water consumption. Both industrial application and irrigation in the agriculture sector has further aggravated the problem. Therefore, to combat the problem of water crisis reuse and recycling of water both at domestic, agriculture, and industrial sector appears to be suitable alternative. New potential source of water should be explored day by day to combat such issue of pressure imposed on water resource by the boosting human population. In country like India, rice a staple food crop and therefore, factors such as water is degenerated at a high level at domestic level after boiling of rice. Therefore, to assess the nutritive value and its irrigation potential the present investigation were carried out at laboratory scale.

9.3 APPLICATION OF INDUSTRIAL DISCHARGE FOR AGRICULTURAL CROP PRODUCTION

Arora and Chauhan (1996) have reported the influence of tannery discharge on germination physiology of *Hordeum vulgare*. Major parameters considered during this investigation include seedling growth, and other physiological parameters of *Hordeum vulgare* varieties. During the investigation tannery effluents were obtained from Mahajan Tannery situated in Agra which significantly reduced the germination physiological parameters.

Chidaunbalam et al., (1996) reported the negative impact of chemical industry effluent on germination physiology of *Vigna radiata* and *Vigna mungo*. In the experiment suitability were tested for the discharge emitted from chemical industry in terms of various germination physiology parameters of *Vigna radiata* and *Vigna mungo*. Results revealed that diluted effluent (10%v/v) was found to be effective in promoting germination, growth, chlorophyll, and protein content. The scientific exploration revealed the applicability of discharge of chemical industry for irrigation purpose at field level through suitable treatment.

Research report was found on the influence of crude oil on paddy cultivation along with changes in soil structure and microbial community. In-situ application design where paddy cultivation was undertaken in soil contaminated with crude oil from Borholla oil field. Results reveal positive influence of crude oil over soil organic carbon (SOC) and negative impact over water holding capacity and bulk density of soil after crop harvesting (Deka et al., 1997).

Germination physiology of rice seedlings were studied under the influence of paper mill effluent. Higher dose of paper mill effluent were found to be inhibitory from germination physiological point of view and the seed material losses their viability along with the phenomenon such as delayed germination in comparison to plants under controlled conditions (Dutta and Boissya, 1999).

Karpate and Choudhry (1997) reported the influence of fly ash and its associated effluent released from thermal power plant over wheat cultivar. The plants were grown under various dilution of fly ash water or grown in soil with fly ash amendments. Positive impact at lower level and negative impact at higher level were recorded in the concerned experiment. The study further generalized that fly ash water is toxic from germination physiology point of view.

Gram species were tested under the influence of discharge from daily industry through various germination physiology parameters. Results reveal that decrease in the growth attributes with gradual increment in the effluent concentration. 25% effluent concentration provided the best result in the investigation. Therefore, 25% dilution revealed positive influence to be applicable for irrigation at field scale (Prasanna Kumar et al., 1997).

An experimental study was done by using Talcher thermal power station fly ash on germination physiology of *Triticum aestivum*. Fifty percent of the amendment of soil through fly ash reflected better growth attributes, dry weight, girth, leaf area, leaf number, and spike length, etc. The study provided the general conclusion about the utilization potential of fly ash along with a ready solution for fly ash disposal problem (Anuradha and Sahu, 1997).

Pot experiments were carried out to evaluate metal mobilization and physiological mechanism of wheat and barley plants under fly-ash treatment. Fly ash reflecting pH 6.72 at low dose up to 30–20 g fly ash/kg reflected positive result in both of the crop and further revealed negative results with further increment in the pH level. It suggested that increase in macro and micronutrient level (K, Zn, Mg, and Fe) up to 30–20 g fly ash/kg soil treatment and further declining trend at higher level (Khan et al., 2000).

Chickpea and pea are the two *Rabi* crops which were tested in terms of their germination physiology under application of effluent from dye and printing industry and the results revealed negative impact over germination process under lower level of industrial effluent (Kumawat et al., 2001).

Pandey et al., (2009) reported the yield attributes of wheat under the problem of fly ash contamination. During this study, parameters considered includes growth attributes, yield attributes and various bio-chemical parameters. The findings of the observation were found to be negative from the concerned crop perspective.

Rampal and Dorjey (2000) made an assessment of impact about industrial effluent of foam industry on pulse crop (*Lens esculenta*). Growth attributes and yield attributes were measured during the study.

Ready and Borse (2001) reported about the influence of industrial effluent released from pulp and paper mill industry on germination physiology of methi (*Trigonella foenum graceum*). Upton 25% dilution of rice boiling water was highly promotive towards germination physiology and growth attributes. However, further increase in the concentration caused decline in the growth scenario of the crop. From the study, it revealed that dilution up to 25% level were recommended as to be utilized as liquid fertilizer. Further, a report of Sundari and Kankarani (2001) revealed the negative impacts of paper and pulp mill industry effluent on various environmental components and sustainability issues which even causes death of domestic cattlehood.

Javid et al., (2006) conducted a pot experiment in 2000 (summer season) in Aligarh to assess the impact of sewage and thermal power plant wastewater on physico morphology and yield of black gram (*Vigna mungo*) variety PU 19. Wastewater application promoted growth, enhanced level of enzyme activity, biochemical components, leaf nitrogen content and yield attributes than ground water treated plants. Physico-chemical properties of both the water samples (groundwater and wastewater) met the prescribed irrigation quality requirements and found to be within the national standard limits. Effluent released in the form of sewage and thermal power plant effluent could be effective for crop species (black gram) of short duration.

From the environmental perspective, reuse of wastewater to irrigate crops will minimize the pollution hazards, safer, and economically viable. Through research study it is known to all of us that sewage water and discharge from thermal power station contains harmful constituents like sodium and total solids in higher concentrations, which may pose problem during irrigation of long duration crops. Further, an experiment under controlled condition were done by using discharge from thermal power station to assess the effectivity of the discharge to be used for irrigation purpose. The crop species under consideration was linseed. In the experiment discharge of thermal power station or available groundwater were used for irrigation along with variable doses of nitrogen (N_0, N_{45}, N_{68}, N_{90}) keeping dose of phosphorus and potassium (each 30 kg/ha) constant. The thermal power plant wastewater contained elevated level of macro and micronutrient (nitrogen, potassium, sulfate, sodium, chloride, calcium, carbonates, and solids, etc.) compared with groundwater. Soil showed no alteration in water-soluble salts, cation exchange capacity, calcium, electrical conductivity, magnesium organic carbon, pH, potassium, sodium, etc. Thermal power plant wastewater proved advantageous over ground water reflecting higher growth, yield attributes of linseed. Thus, the study concludes that discharge of thermal power station is effective for linseed cultivation (Akhtar et al., 2006).

Apart from municipal sewage 13,468 MLD of discharge is generated from the industrial sector in India. Out of which about 60% of the discharge from industry level undergoes treatment. Medium and small-sized industry contributes significantly towards generating untreated effluent. From the context of toxicity municipal sewage is much more preferable than industrial wastewater. Such findings are further supported by several research works which reports about field trials related to sewage fed farming with less exploration of the industrial water use for farming. However, many reports are there about water use efficiency in industries and on site reuse. The emphasis has put focus over industries producing relatively less toxic effluent such as breweries sugar mills and fruit processing unit. Few industries already practice wastewater reuse for agriculture.

A proper monitoring set up for agriculture reuse of wastewater generated in the industrial sector needs to be developed. People from various sectors such as agriculture extension officers, local stakeholders, and industry people may work through joint collaboration to assess the characteristic and quality of discharge from the industrial sector (Derek Bewley et al., 2006).

9.4 GLOBAL CONTEXT OF INFLUENCE OF INDUSTRIAL DISCHARGE ON GROWTH AND GERMINATION OF SEEDLINGS

A study on the influence of dye industrial effluents on germination physiology and productivity of country bean (*Lablab niger*) were done. Seven stages of dye effluents were used to compare their effect on germination attributes and early growth stage of country bean. Neutralization stage (D5) promotes enhanced% germination substantially (100%) along with the treatment D1 (underground water). Other treatments of effluent from 2nd wash after bath (D4) and combined effluent released from ETP reflected hindrance in germination (73.33%). Further treatment also gave positive results. From the results, it has been concluded that neutralization stage of textile effluent might be used in the form of irrigation under farming system for sustainable production of country bean and leading to sustainable soil-environment (Hassan et al., 2013).

Effluent from different industries such as textile and paper industries are discharged into streams and land without any prior treatment. This therefore reveals significant negative impact on water quality of surface and groundwater affecting agricultural crop production. Mixed impact was observed among various crops. In the case of vegetables industrial effluents showed stimulatory impacts through improvement in seedling and root length. Higher concentration of effluent is imposing negative impacts by reduction in growth rate and attributes. Comparative studies between treated and untreated effluent of textile industry revealed positive and negative impacts. Untreated effluent reduced biomass content of shoot and root. Treated effluents promoted growth, increased level of biomolecules. Heavy metals present in industrial wastewater produces negative impacts by reduced growth and yield of vegetation (Ali et al., 2013).

Under laboratory condition an investigation was carried out by some workers to evaluate the impact of effluent released from pharmaceutical industry and wastewater generated from domestic sector on germination physiology of some selected plant species. A comparison was done between control and various treatments. Results revealed considerable impact of industrial wastewater and domestic waste on germination physiology of selected crop. *Nigella sativa* was highly affected under domestic and pharmaceutical discharge followed by *Coriandrum sativum* (Shahida et al., 2012).

Seed germination is the initiation of plant life on biosphere. Temperature, light, pH, and soil moisture are some of the key considerations influencing germination physiology (Chachalis and Reddy, 2000).

9.5 IMPACT OF INDUSTRIAL EFFLUENT ON SEEDLING GROWTH AND GERMINATION-INDIAN CONTEXT

From a global perspective, it was observed that pollution is mediated through population explosion, improvement in science and technology in the form of rapid industrialization and urbanization. A study was conducted to evaluate the suitability of discharge emitted from the paper mill industry to be used for irrigation purposes. Further, the experiment was accompanied under in-vitro condition through *Vigna mungo* seeds. Results reveal a positive impact on growth perspective at a 40% level of dilution. Higher level of dilution was inhibitory for crop growth (Paranthaman and Karthikeyan, 2015).

A study on the impact of sugar mill discharge situated in Haridwar district was assessed on various crop species in terms of germination physiology and growth attributes. Results revealed that 25% and 50% concentration were positive for both the crops (wheat and maize) studied (Saini and Pant, 2014).

Gradual development of science and technology has promoted industrial growth. Progress in industrial sector has put forward a challenge of eco-friendly disposal of various discharges from various industries. Further, an investigation was carried out on dairy effluent on germination physiology and biochemistry of *Phaseolus trilobus*. The findings revealed a remarkable effect in germination with treated dairy effluent on alterations in the morphology and biochemistry in *Phaseolus trilobus*. Twenty-five percent of dilution was proved to be stimulatory for germination and growth attributes. This suggests the utility of dairy effluent as liquid fertilizer at low level (Selvi and Sharavanan, 2013).

A study on impact of domestic wastewater (sewage) on germination physiology of rice and wheat revealed utility of sewage for irrigation because of its nutrient content and pollutant removal activity. 50% dilution of sewage effluent revealed positive influence on germination physiology and growth attributes. This, therefore, indicates the beneficial role of discharge of dairy industry with proper dilution for irrigation purpose as optimum water source (Dash, 2012).

Research on influence of distillery unit was assessed in relation to pea crop (*Pisum sativum*). Effluent characterization reveals alkaline nature and elevated level of TDS (Total Dissolved Solids) and chlorides. Low concentration of effluent of the distillery unit was inhibitive for germination and growth performance of the crop (Narain et al., 2012).

An experiment was conducted to evaluate the influence of textile discharge in relation to germination physiology and growth attributes of *Vigna mungo* (black gram). Low concentration promoted germination ratio and growth than control, with gradual reduction in germination, growth of black gram under higher concentration of textile mill effluent. 25% concentration revealed positive result than control. Further increase in the concentration inhibited growth attributes (Wins and Murugan, 2010).

9.6 MATERIALS AND METHOD

9.6.1 METHODOLOGY

Seeds of gram and wheat were obtained from the Directorate of Agriculture, Govt. of Chhattisgarh. Seeds were conserved further for experimental work.

9.6.2 PREPARATION OF DILUTED EFFLUENTS (TREATMENT SOLUTION)

Various concentrations (20%, 40%, 60%, 80%, and 100%) of rice boiling water were prepared by using the standard protocol along with control (only distilled water) treatment for comparison.

9.6.3 EXPERIMENTAL DESIGN

After subsequent washing of seeds through tap water and distilled water 10 seeds were then placed in Petri dish on discs having filter paper moistened with the various concentration of boiling rice water except control. The different treatment of boiling rice water under the present investigation includes T_1 control, T_2–20%, T_3–40%, T_4–60%, T_5–80%, and T_6–100%. Experimental setup was covered following standard protocol.

From the very beginning after sowing germination (%) and growth were recorded for 24 hours up to 7 days as the case might be. The Petri dish were kept moistened by regular adding required amount of (1 ml) treatment solution of boiling water of rice of respective treatments along with control every alternate day. The growth of root and shoot were recorded by using a centimeter-scale from 7 days old seedling and for fresh weight, 7 days old seedling were sampled, and they were freed from moisture to take the final weight. Dry root and shoot mass were taken by drying them in an oven at 80°C for 72 hrs.

The following attributes were considered during the present investigation:

1. **Total Seed Germination:** 7 days seedling was taken from Petri dishes after recorded for% germination (Datta et al., 2009).
2. **Root Length:** 7 days seedlings were taken from Petri dishes after removing cotton and blotting paper for measuring root length (Datta et al., 2009).
3. **Shoot Length:** 7 days seedlings were taken from Petri dishes after removing cotton and blotting paper for measuring shoot length (Datta et al., 2009).
4. **(%) Germination:** 100% germination was assessed for different dilutions of boiling rice water on two crops of two samplings, and (%) germination was calculated as follows:

$$\text{Germimation (\%)} = \frac{\text{Total number of seed germinated in particular treatment}}{\text{total number of seed treated in particular treatment}} \times 100$$

Seedlings of each treatment were measured for their radical and plumule length (cm). Data obtained were manually calculated for their Mean ± SE (Bazar et al., 2006).

5. **Seedling Length:** By using the following formula seedling length was measured:

Seedling length = mean shoot length + mean root length

6. **Seedling Vigor Index:** Formula given by Baki and Andersen (1972) was used to calculate vigor index:

Vigor index = (Average of root length + Average of Shoot length) ×% of seed germination (Haque et al., 2007)

7. **Germination Rate Index:** Rate of germination index (RGI) was determined by the formula given by Haque et al., (2007):

$$RGI = \frac{\text{Number of seedlings at 7 days}}{\text{Number of seedlings at 14 days}}$$

8. **Speed of Germination:** Krishnaswamy and Seshu, (1990) gave the formula of quantifying speed of germination (%) as below:

   Speed of germination (%) = Number of seed germinated at 72h/Number of seeds germinated at 168h.

9. **Root and Shoot Fresh Weight:** 7 days seedlings were taken out from the Petri dishes after removing cotton and blotting paper and the shoots and roots were separated and weighted separately for recording the said attributes.
10. **Root and Shoot Dry Weight:** 7 days old seedlings were taken out from the Petri dishes after removing cotton and blotting paper and the shoots and roots were separated and weighted separately for recording shoot and root fresh weight. By taking root and shoot fresh weight the separated plant part in the form of root and shoot were placed into incubator for 2 hours. Finally, the dry weight of different plant part (root and shoot) was taken after keeping in the desiccators for requisite times.
11. **Data Analysis:** Statistical analysis of data was done through ANOVA and significance of data was also tested.

9.7 RESULTS AND DISCUSSION

9.7.1 TOTAL SEED GERMINATION

In the plant life cycle, seed starts the new beginning of life and further matures to seedling stage. Further, it progresses and matures into an adult plant. Plant germination is a key for plant propagation (Idu, 1994). In other terms, germination leads to activation of the metabolic process within seeds, which produces plumule and radical convert into seedling (Khan, 1982). Water imbibition is the process, which initiates the process of seed germination and ends with growth and proliferation of axis of the embryo known as radical (Welbaum et al., 1998).

Higher respiration rate reveals that the physiological process such as germination is highly energy consuming process. Total seed germination in the present study shows variation among the various treatments. It varied from 1.666 (T_3) to 4.333 (T_2). Highest total seed germination were recorded for 20% dilution of pure boiling water of rice and least germination were recorded for 40% dilution of pure boiling water of rice compared to control (Figure 9.1). Most of the treatments (T_3–T_6) of diluted boiling water of rice reported lesser total seed germination compared to control.

The application of boiling water of rice have a positive influence on seed germination of gram from treatment T_3 to T_6 which is supported by Yadav et al., (2015) for barley under effluent from marble industry. Such result is obtained due to efficient water absorption by germinating gram seeds for their further progression towards maturity into seedling stage without which their growth and development would have hampered (Kelly et al., 1992; Debeaujan et al., 2000). In normal case energy producing molecules show good distribution within seed and along with unaltered level of biological macromolecules in the membrane might have promoted efficient water absorption by seeds (Yasmin et al., 2011).

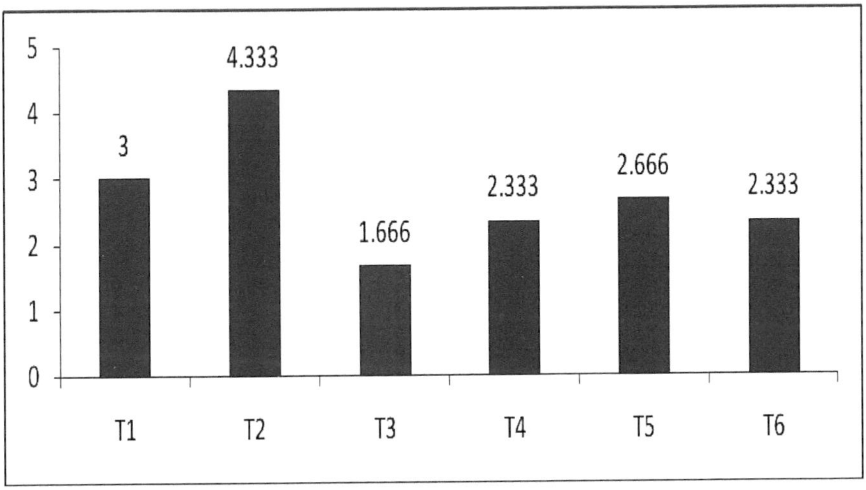

FIGURE 9.1 Total seed germination of gram under various treatments of boiling water of rice.

9.7.2 *ROOT LENGTH*

The gram root length varied significantly among treatments. The gram root length ranged between 4.833 (T4) to 12.333 (T_1). Highest root length was recorded for (T_1) distilled water (control) and least root length was recorded for 60% dilution of pure boiling water of rice (Figure 9.2). From the results it appears that in most of the treatments (T_2-T_6) lesser root length were recorded compared to control (Raju et al., 2015). The decrease in root length along with gradual increment of effluent concentration may have imposed negative influence. A regular decline of root length with gradual increment of concentrations of boiling water of rice was recorded (Yasmin et al., 2011). 20% dilution was promontory for root growth. Low level of rice boiling water may have optimum nitrogen and other mineral elements which contributed to root growth (Sahai et al., 1983b).

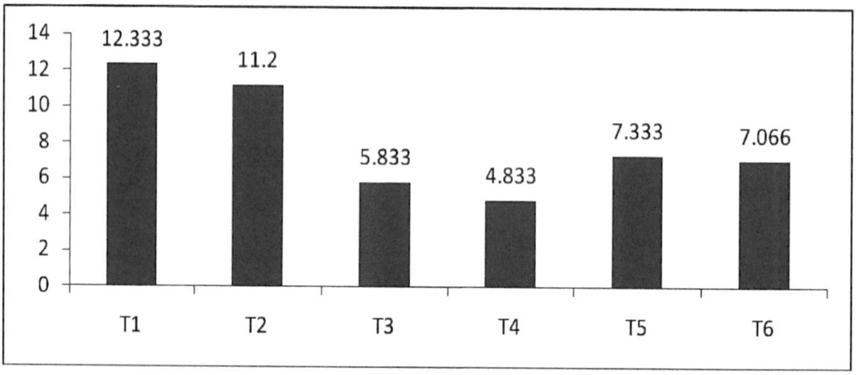

FIGURE 9.2 Root length of gram seedling under various treatments of boiling water of rice.

9.7.3 SHOOT LENGTH

During the present investigation significant level of variation were observed in case of shoot length of gram seeds among the various treatments. The shoot length of gram seeds ranged between 0.333cm (T_4) to 5 cm (T_1). Highest shoot length were recorded by T_1 treatment and least shoot length were recorded for 60% dilution of pure boiling water of rice (Figure 9.3). Treatments (T_2-T_6) diluted boiling water of rice reported lesser shoot length than control. High level of shoot growth were recorded during the study due to presence of adequate level of nitrogen and other mineral nutrients in the rice boiling water (Sahai et al., 1983b).

9.7.4 ROOT DRY WEIGHT

Data presented in Figure 9.4 showed gram root dry weight varied significantly. The gram root dry weight ranged between 0 gm (T_3, T_2) to 0.333 gm

(T_5). 80% dilution of pure boiling water of rice recorded highest gram root dry weight and lesser root dry weight were recorded for pure boiling water of rice (T_6) than control. 80% dilution of boiling water of rice was stimulatory through accumulation of biomass than control. 80% dilution of boiling water of rice reflected high level of dry weight of root because presence of adequate level of nutrient in the rice boiling water. This therefore promoted seedlings growth and development.

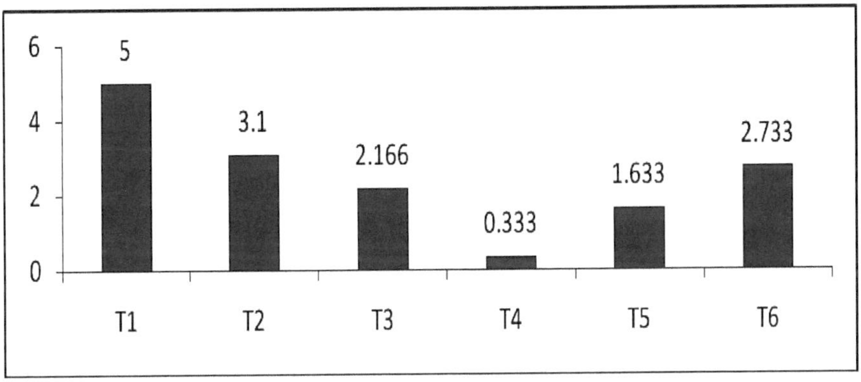

FIGURE 9.3 Shoot length of gram seedling under various treatments of boiling water of rice.

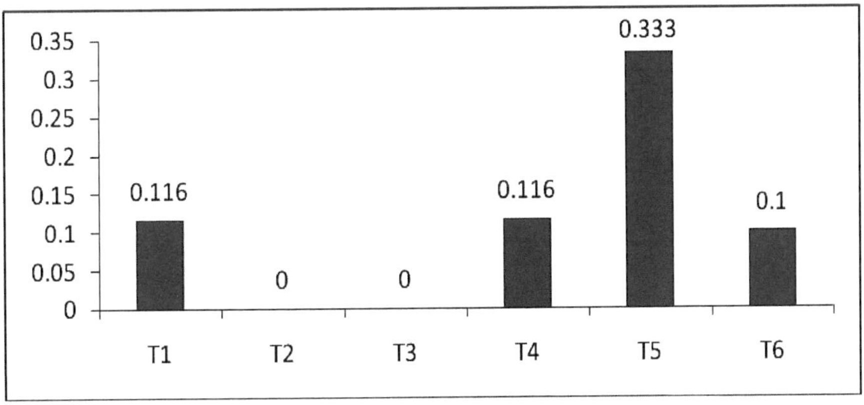

FIGURE 9.4 Gram root dry weight under various treatments of boiling water of rice.

9.7.5 *ROOT FRESH WEIGHT*

Gram root fresh weight reflects lesser variation among treatments which varied from 0.25 g (T_1, T_2, T_6, T_3) to 0.35 g (T_5). Highest root fresh weight were recorded for 80% dilution of pure boiling water of rice and least germination were recorded for 20%, 40%, 60%, 100% dilution of pure boiling water of rice compared to control (Figure 9.5). From the results it appears that most of the treatments of diluted boiling water of rice reported similar root fresh weight compared to control except T_5. The results suggests that boiling water of rice have promoted overall vegetative growth to gram seedlings irrespective of different treatments with little increase in T_5 treatment.

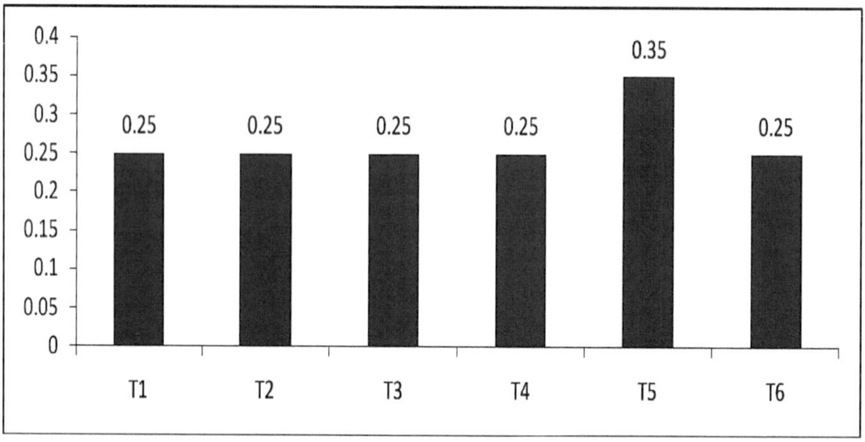

FIGURE 9.5 Root fresh weight of gram seedling under various treatments of boiling water of rice.

9.7.6 *SHOOT DRY WEIGHT*

Gram shoot dry weight reflects variability among treatments which varied from 0g (T_3, T_4, T_5) to 0.116g (T_1). T_1 (distilled water) recorded highest shoot dry weight. All the treatments reported lesser value than control (Figure 9.6). The shoot dry weight were significantly inhibited in various treatment of boiling water of rice which may be attributed towards osmotic inhibition of water absorption leading to inhibition of functionality of essential enzymes (Mayer and Poijakoff Mayber, 1982).

9.7.7 *SHOOT FRESH WEIGHT*

The gram shoot fresh weight reflected variability among treatments which ranged from 0 g (T_4) to 0.483 g (T_1). Highest shoot fresh weight were recorded for T_1 (distilled water) and least shoot fresh weight were recorded for 60% dilution of pure boiling water of rice (Figure 9.7). From the results it appears that most of the treatments of diluted boiling water of rice reported lesser shoot fresh weight compared to control. The shoot fresh weight were significantly inhibited in various treatment of boiling water of rice due to osmotic inhibition of water absorption inhibiting the essential enzymes functionality (Mayer and Poijakoff Mayber, 1982).

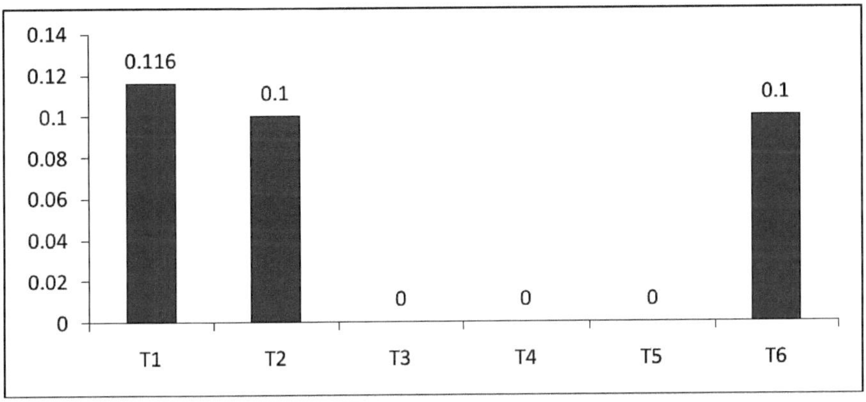

FIGURE 9.6 Shoot dry weight of gram seedling under various treatments of boiling water of rice.

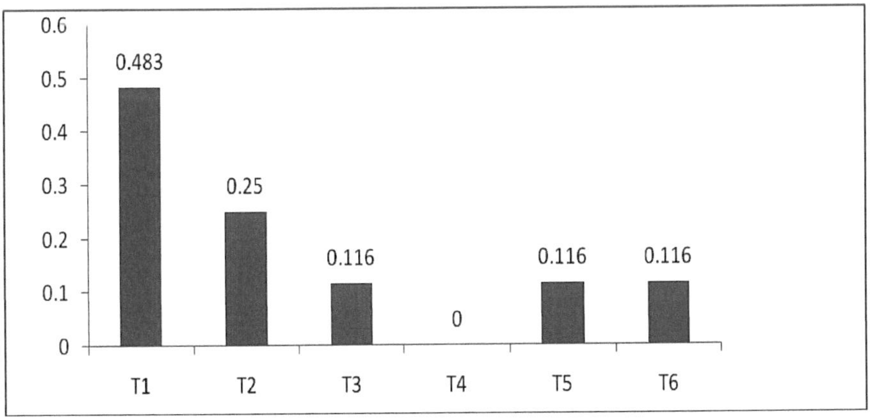

FIGURE 9.7 Shoot fresh weight of gram seedling under various treatments of boiling water of rice.

9.7.8 SEEDLING DRY WEIGHT

The seedling dry weight of gram seeds reflects variability among treatments. The seedling dry weight of gram seeds ranged between 0g (T_4) to 0.25g (T_2). Highest seedling dry weight were recorded for 20% dilution of pure boiling water of rice and least seedling dry weight were recorded for 60% dilution of pure boiling water of rice compared to control (Figure 9.8). From the results it appears that various treatments (T_3–T_6) of diluted boiling water of rice reported lesser seedling dry weight compared to control. Boiling rice water treated gram seedlings recorded lesser seedling dry weight except T_2 may be due to lesser synthetic activity of biological macromolecules for proper growth and differentiation as well as synthesis of new material as well as cell division associated with cell division (Sunderland, 1960).

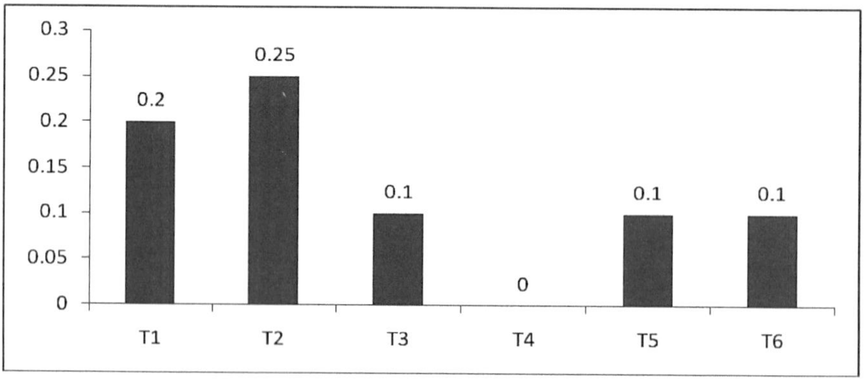

FIGURE 9.8 Seedling dry weight of gram seedling under various treatments of boiling water of rice.

9.7.9 (%) GERMINATION

The gram% germination reflects variability among treatments which varied from 16.666 (T_3) to 43.333 (T_2). Highest percentage germination were recorded for 20% dilution of pure boiling water of rice and least percentage germination were recorded for 40% dilution of pure boiling water of rice compared to control (Figure 9.9). It appears treatments (T_3-T_6) diluted boiling water of rice reported lesser germination percentage than control. During the present investigation a decline in% germination were observed than control due to lesser supply of oxygen to the germinating seeds. This reduces the supply of energy, which therefore retard the growth and developmental process (Hadas, 1976).

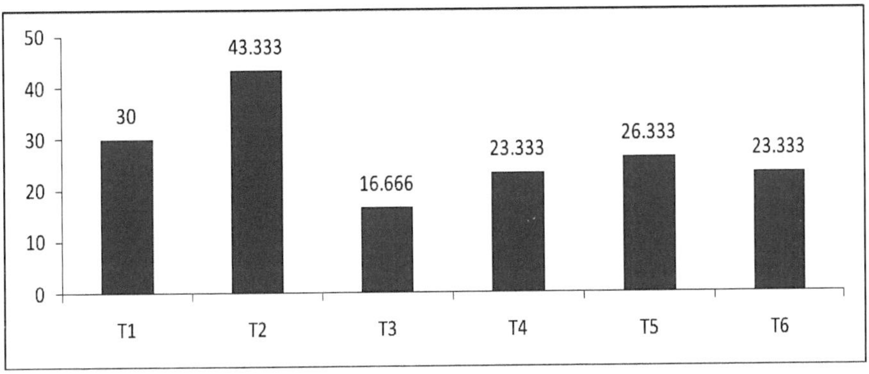

FIGURE 9.9 Germination (%) of gram seedling under various treatments of boiling water of rice.

9.7.10 SEEDLING LENGTH

The seedling length of gram seeds reflects significant level of variation among the various treatments. The seedling length of gram seeds ranged between 5.166 (T_4) to 17.333 (T_1). Highest seedling length were recorded for T_1 (distilled water) dilution of pure boiling water of rice and minimum seedling length were recorded for 60% dilution of pure boiling water of rice compared to control (Figure 9.10). From the results it appears that most of the treatments (T_3-T_4) diluted boiling water of rice reported lesser seedling length compared to control. There is a declining trend in seedling length of gram seedlings than control. Subramani et al., (1998) found a regular decline of growth attributes under the influence of higher concentration of fertilizer factory effluent. Mishra and Bera (1996) similarly reported that low effluent concentration of tannery reflects positive impact over growth and inhibitory effect at higher concentration due to less supply of oxygen rendering lesser supply of energy ultimately leading to retarded growth (Hadas, 1976).

9.7.11 SEEDLING VIGOR INDEX

Gram seedling vigor index reflected variability among treatments which varied from 133 (T_4) to 765.666 (T_1). Highest seedling vigor index were recorded for T_1 (distilled water) and least vigor index were recorded for 60% dilution of pure boiling water of rice compared to control (Figure 9.11). From the results, it appears that most of the treatments (T_3-T_4) diluted boiling water of rice reported lesser seedling vigor index compared to control. The vigor index of seedlings was significantly low in undiluted effluent than control. The vigor index value showed an increment along with decline in the effluent concentration because of optimum level of nutrient presence under low concentration. However, the excess minerals and nutrients inhibit the physiological process through alteration of metabolism during germination and growth (Verma and Verma, 1995).

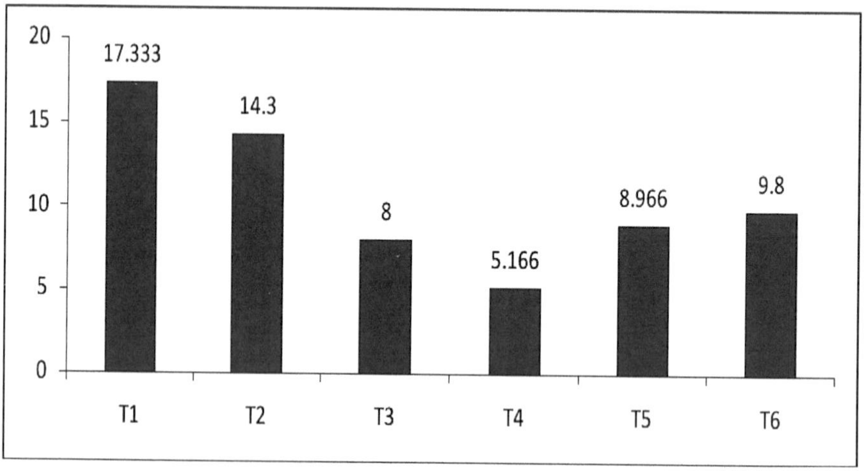

FIGURE 9.10 Gram seedling length under various treatments of boiling water of rice.

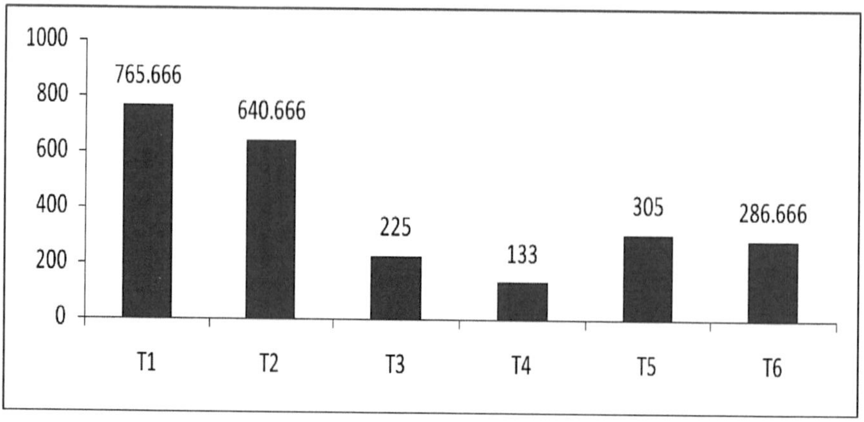

FIGURE 9.11 Gram seedling vigor index under various treatments of boiling water of rice.

9.7.12 GERMINATION SPEED

The germination speed of gram seeds reflected significant variation among treatments. The speed of germination of gram seeds ranged between 0.233 (T_3) to 0.616 (T_2). Highest speed of germination were recorded for 20% dilution of pure boiling water of rice and least speed of germination were recorded for 40% dilution of pure boiling water of rice compared to control (Figure 9.12). It appears treatments (T_3-T_6, T_4) diluted boiling water of rice recorded lesser germination speed than control. The highest speed of germination was observed at 20% dilution which reflects lesser influence of present of salt in the effluent.

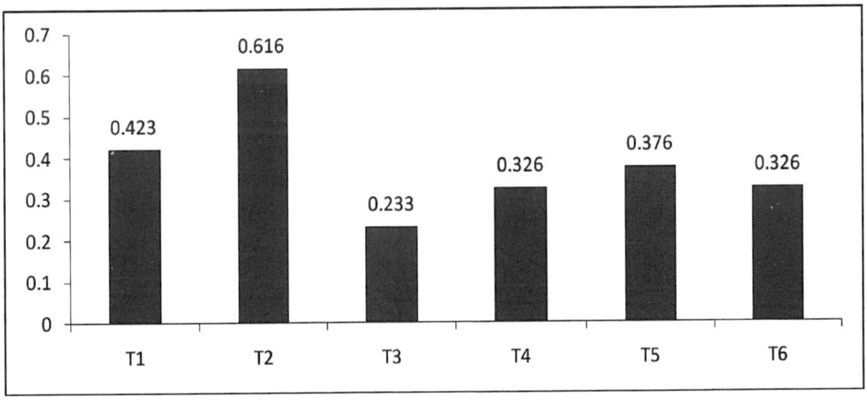

FIGURE 9.12 Speed of germination of gram seedlings under various treatments of boiling water of rice.

9.7.13 GERMINATION VALUE

The germination value of gram seeds reflects significant level of variation among the various treatments. The germination value of gram seeds varied from 6.189 (T_3) to 32.856 (T_2). Highest germination value were recorded for 20% dilution of pure boiling water of rice and least germination value were recorded for 40% dilution of pure boiling water of rice compared to control (Figure 9.13). It appears mostly (T_3, T_6, T_4) diluted boiling water of rice reported lesser germination value than control. A declining trend at high level of boiling water of rice was observed because of lesser supply of oxygen to the germinating seeds. Further, this has promoted lesser supply of energy to the plant leading to retarded growth and slow developmental process (Hadas, 1976).

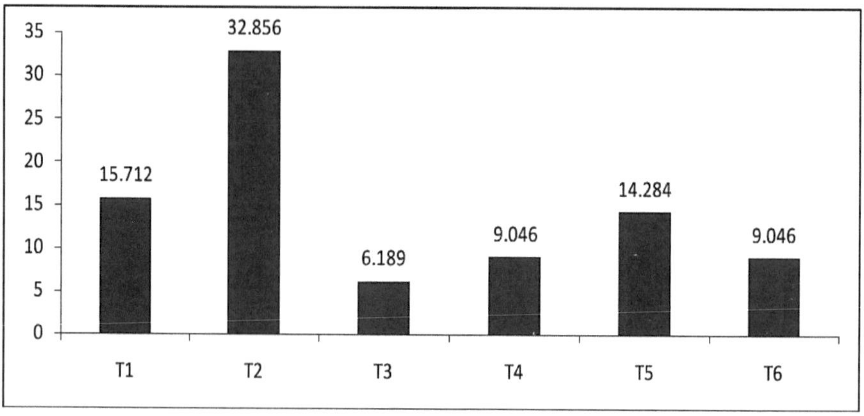

FIGURE 9.13 Germination values of gram seedlings under various treatments of boiling water of rice.

9.7.14 MEAN DAILY GERMINATION

The mean daily germination of gram seeds reflects significant level of variation among the various treatments. The mean daily gram seed germination varied from 2.38 (T_3) to 6.19 (T_2). Highest mean daily germination were recorded for 20% dilution of pure boiling water of rice and least mean daily germination were recorded for 40% dilution of pure boiling water of rice compared to control (Figure 9.14). From the results it appears that most of the treatments (T_3-T_6) diluted boiling water of rice reported lesser total seed germination than control. The mean daily germination value reduced in all the various treatments than control due to lesser supply of oxygen to the germinating seeds. This may be due to inhibition of energy supply which retards the process of development and growth (Hadas, 1976).

9.7.15 TOTAL SEED GERMINATION

The variation of total seed germination among various treatments of boiling water of rice on wheat seedlings were reflected in Figure 9.15. From the results, it appears that total seed germination value reduced among the different treatments than control. 80% dilution of boiling water of rice reflected least total seed germination. Khan et al., (2011) in their experiment on impact of effluent of textile industry on germination physiology reported that, in higher concentrations seed germination is affected while low level of effluent concentration supported 100% seed germination in kidney bean and millet, but osmotic pressure associated with high level of sugar factory effluent reduced the germination in kidney bean and millet (Ajmal and Khan, 1983).

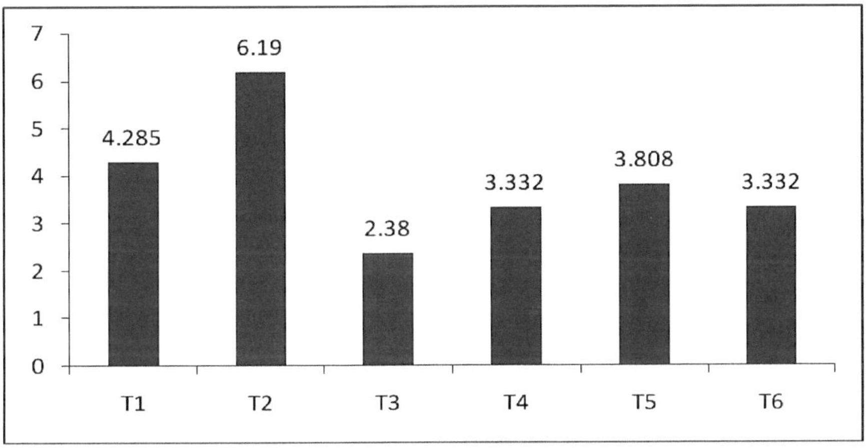

FIGURE 9.14 Mean daily germination of gram seedlings under various treatments of boiling water of rice.

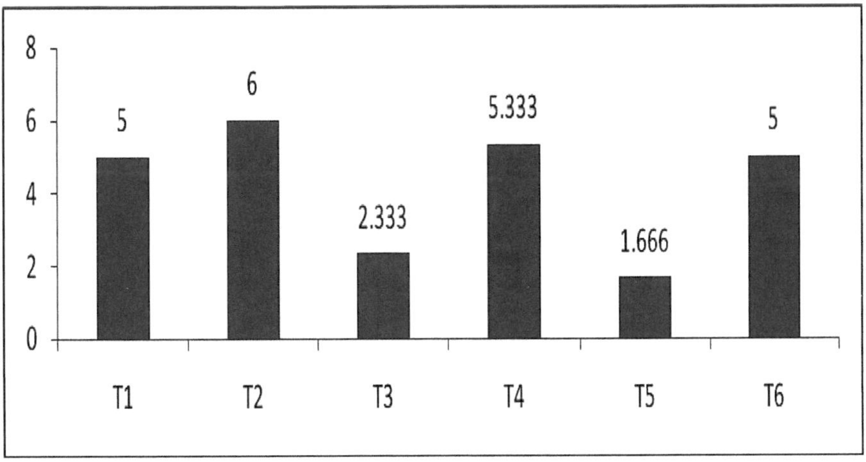

FIGURE 9.15 Total seed germination of wheat seedling under various treatments of boiling water of rice.

Total % germination increased in treatment T_6, T_4 and T_2 because of supply of moisture at optimum level and sufficient water absorption by germinating wheat seeds which is essential for proper seed germination without which seedling growth attributes and developmental process may be severely affected (Kelly et al., 1992; Debeaujan et al., 2000).

9.7.16 ROOT LENGTH (CM)

Data presented in Figure 9.16 represents the variability of root length among various treatments of boiling water of rice on wheat seedlings. From the results, it appears that root length value reduced in various treatments compared to control except T_2. 40% dilution of boiling water of rice reflected minimum level of root length. 20% dilution of boiling water of rice reflected highest root length. Increment and decline of root length on regular basis were recorded under gradual increase in effluent concentrations. This may be attributed towards negative influence of boiling water of rice on root (Yasmin et al., 2011). 20% dilution proved to be optimum for root growth which might be attributed towards low level of effluent concentration having stimulatory impact through presence of nitrogen and other necessary nutrient present in rice boiling water (Sahai et al., 1983b).

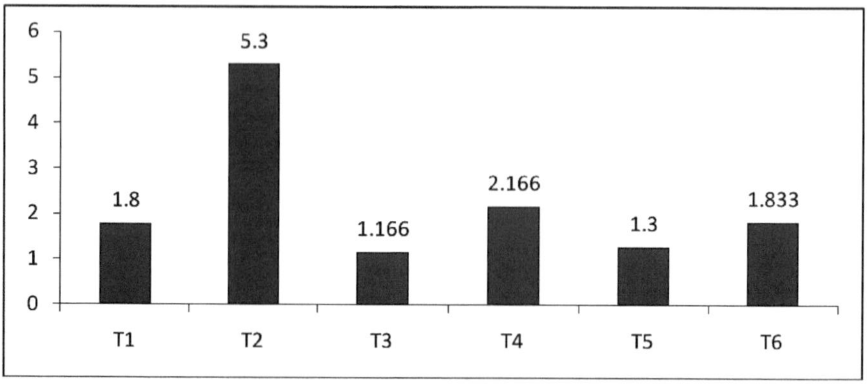

FIGURE 9.16 Root length (cm) of wheat seedling under various treatments of boiling water of rice.

9.7.17 SHOOT LENGTH (CM)

Data presented in Figure 9.17 represents the variability of total shoot length among various treatments of boiling water of rice on wheat seedlings.

From the results, it appears that total shoot length value reduced in various treatments compared to control. 80% dilution of boiling water of rice reflected minimum level of shoot length. Highest growth in shoot was found in 20% dilution of boiling water of rice. Nutrient present in the rice boiling water at 20% dilution may have promoted the growth of wheat to a considerable extent along with significant level of nitrogen (Sahai et al., 1983b).

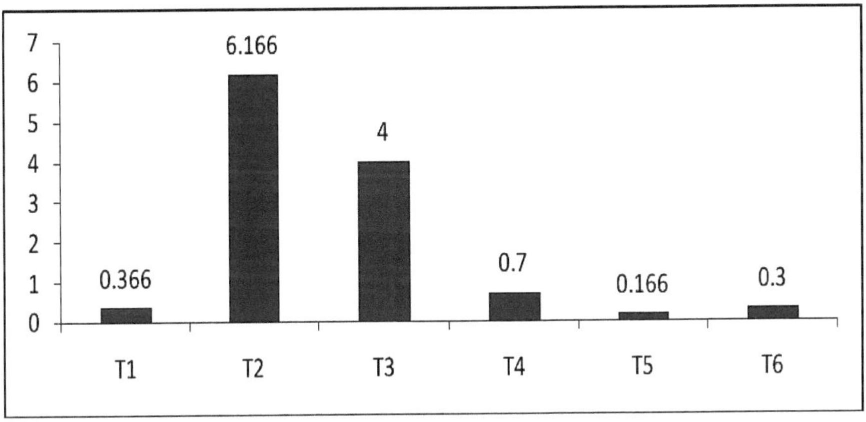

FIGURE 9.17 Shoot length (cm) of wheat seedling under various treatments of boiling water of rice.

9.7.18 (%) GERMINATION

Percentage germination of wheat shows substantial variation among various treatments of boiling water of rice were recorded (Figure 9.18). From the results, it appears that percentage germination value reduced in various treatments compared to control. 80% dilution of boiling water of rice recorded minimum level of percentage germination. The treatments T_1 recorded highest percentage germination for distilling water (control), and least (%) germination were recorded for 80% dilution of pure boiling water of rice in comparison to control. It appears that percentage germination value were highly statistically significant. During the present investigation a reduction in germination% were observed compared to control due to lesser supply of oxygen to the germinating seeds, which restricts the energy supply and retards the growth and development of seedling (Hadas, 1976).

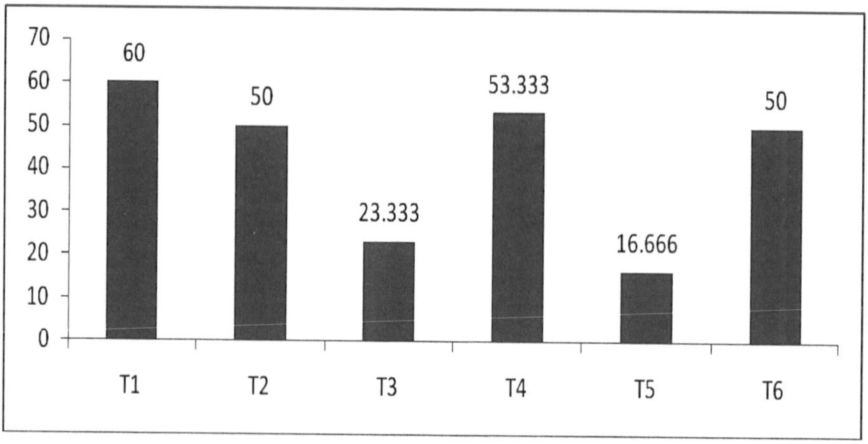

FIGURE 9.18 Germination (%) of wheat seedling under various treatments of boiling water of rice.

9.7.19 SEEDLING LENGTH

Figure 9.19 represents the variability of seedling length among various treatments of boiling water of rice on wheat seedlings. From the results, it appears that seedling length value reduced in various treatments compared to control. 80% dilution of boiling water of rice recorded a minimum level of total seedling length, and the highest seedling length was recorded for 20% dilution of boiling water of rice. There is a declining trend in seedling length of wheat seedlings from 20–60% dilution of boiling water of rice. Subramani et al., (1998) reported a progressive decrease in seedling growth with the increasing concentration of fertilizer factory effluent. Our findings were similar with the earlier findings of Mishra and Bera (1996). The work done by Mishra and Bera (1996) reflected inhibitory nature of higher concentration of tannery effluent and promontory effects on lesser concentration. This may be due to the fact that at higher concentration low oxygen level restricted the supply of energy and therefore inhibits the physiological process (Hadas, 1976).

9.7.20 SEEDLING VIGOR INDEX

Figure 9.20 represents the variability of total seedling vigor index among various treatments of boiling water of rice on wheat seedlings. From the results, it appears that seedling vigor index value reduced in various treatments compared to control. 40% dilution of boiling water of rice reflected minimum level of seedling vigor index. The vigor index of seedlings was significantly low in undiluted effluent in comparison to control. Gradual increment in the concerned parameters was observed with declined in treatment concentration which may be due to presence of optimum level of nutrients. However, the excess minerals and nutrients inhibit the germination and growth by interfering with metabolic activities during germination and growth (Verma and Verma, 1995).

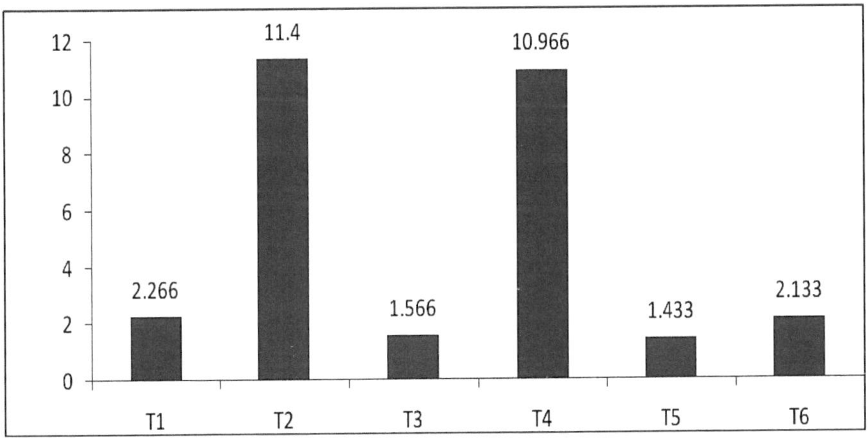

FIGURE 9.19 Seedling length of wheat seedling under various treatments of boiling water of rice.

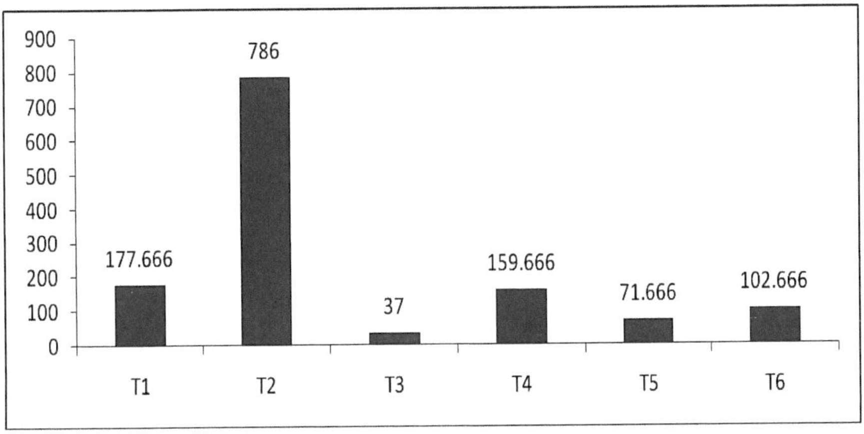

FIGURE 9.20 Seedling vigor index of wheat seedling under various treatments of boiling water of rice.

9.7.21 SPEED OF GERMINATION

Figure 9.21 represents the variability of speed of germination among various treatments of boiling water of rice on wheat seedlings. From the results it appears that speed of germination value reduced in various treatments compared to control. 80% dilution of boiling water of rice reflected minimum level of speed of germination (Figure 9.21). From the results it appears that most of the treatments (T_3-T_6, T_5) diluted boiling water of rice reported lesser speed of germination compared to control. The highest germination speed were observed at control compared to various treatments of boiling water of rice which reflects lesser influence of salt content of boiling water of rice.

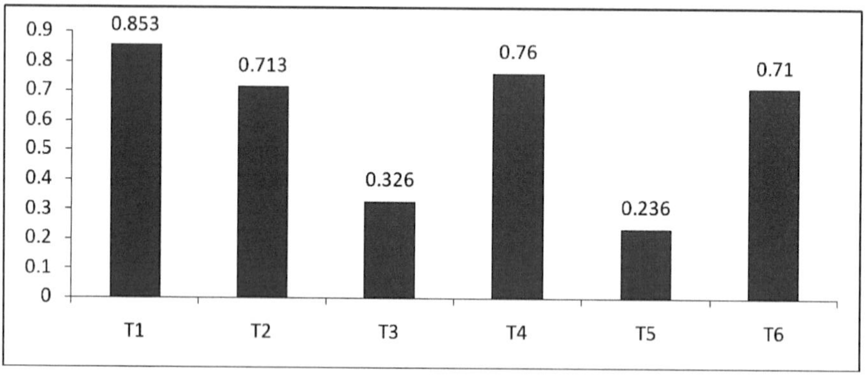

FIGURE 9.21 Speed of Germination of wheat seedlings under various treatments of boiling water of rice.

9.7.22 GERMINATION VALUE

Figure 9.22 represents the variability of germination value among various treatments of boiling water of rice on wheat seedlings. Forty percent of dilution of boiling water of rice reflected minimum level of germination value. Results revealed that various treatments (T_3-T_6) diluted boiling water of rice reported lesser germination in comparison to control and highest germination value for T_1 treatment (control). There was a declining trend at higher concentration of boiling water of rice which may be due to lesser supply of oxygen to the germinating seeds, which restricts the energy supply and retards the growth and development of seedling (Hadas, 1976).

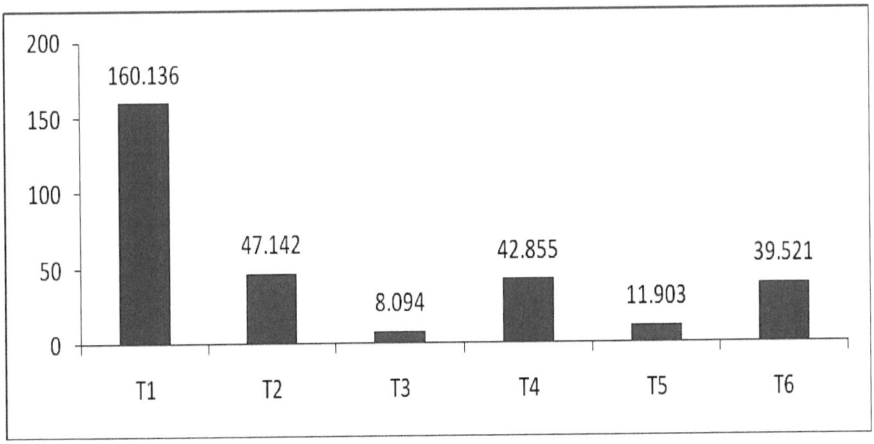

FIGURE 9.22 Germination value of wheat seedling under various treatments of boiling water of rice.

9.7.23 MEAN DAILY GERMINATION

Figure 9.23 represents the variability of mean daily germination among various treatments of boiling water of rice on wheat seedlings. From the results it appears that mean daily germination value reduced in various treatments compared to control. 80% dilution of boiling water of rice reflected minimum level of mean daily germination. The mean daily germination value reduced in all the various treatments than control due to lesser supply of oxygen to the germinating seeds inhibiting the supply of energy, growth, and development (Hadas, 1976).

9.8 CONCLUSION

The present chapter reflects the outcome of an experiment study to assess the influence of rice boiling water on germination physiology of gram and wheat seedlings under laboratory condition. The variations in terms of growth attributes and nutrient contents of boiling rice water effluent treatments on gram and wheat seedlings reflected significant variation. Higher dilutions of boiling water of rice reflected negative impact in some of the growth attributes (% germination and plant growth). Changes in morphological parameters reflected a gradual progression for both gram and wheat. Percentage germination was inhibited for both the crop at higher dilution of rice boiling water. Growth attributes such as length of root and shoot,% germination, germination, mean daily germination, germination value, vigor index value showed the maximum range at 20% dilution for both gram and wheat seedlings. As boiling water of rice contains sufficient nutrient content; therefore, at lower concentration, enhanced higher growth for rice and wheat. Under such circumstances, 20% of boiling water were found to be the most suitable dose to be applicable under field condition in terms of irrigation. This would serve the dual purpose of crop irrigation as well as conservation of water.

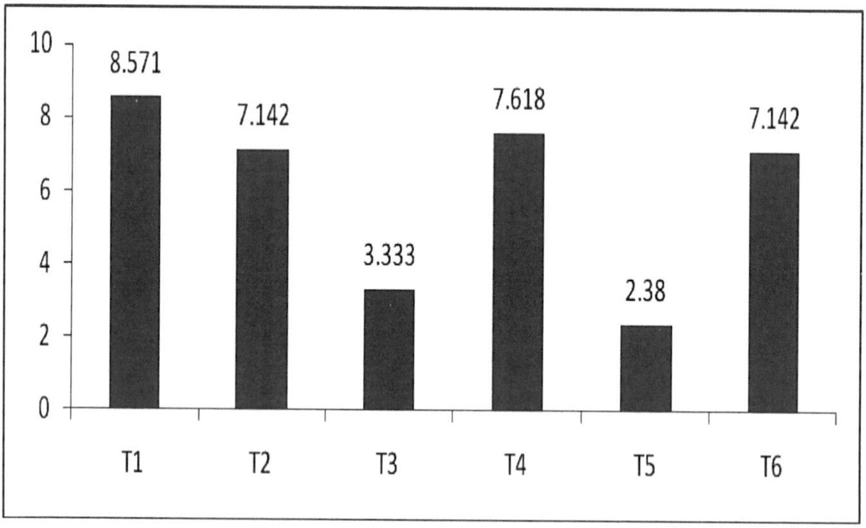

FIGURE 9.23 Mean daily germination wheat seedling under various treatments of boiling water of rice.

KEYWORDS

- **germination physiology**
- **germination**
- **growth traits**
- **paddy**
- **seedling vigor index**
- **shoot length**

REFERENCES

Ajmal, M., & Khan, A. U., (1983). Effects of sugar factory effluent on soil and crop plants. *Environmental Pollution, 30*, 135–141.

Akhtar, A., Saeed, S., Singh, S., Ahmad, I., Javid, S., & Inam, A., (2006). Effective use of thermal power plant wastewater as a source of irrigation and nutrients in crop productivity of linseed (*Linum usitatissimum* L.). *Asian J. Microbiol. Biotech. Env. Sci., 8*, 223–227.

Ali, H., Khan, E., & Sajad, M. A., (2013). Phytoremediation of heavy metals–concepts and applications. *Chemosphere, 91*, 869–881.

Baki, A. A., & Anderson, J. D., (1972). Physiological and biological deterioration of seeds. In: *Seed Biology* (Vol. II). Academic Press, New York.

Bazai, Z. A., Achakzai, A., & Kabir, K., (2006). Effect of wastewater from Quetta city on the germination and seedling growth of lettuce (*Lactuca sativa* L.). *Journal of Applied Sciences, 6*, 380–382.

Chachalis, D., & Reddy, K. N., (2000). Factors affecting *Campsis radicans* seed germination and seedling emergence. *Weed Science, 48*, 212–216.

Chidaunbalam, P. S., Pugazhendi, N., Lakshmanan, C., & Shanmagasundaram, R., (1996). Effect of chemical industry waste on germination, growth and some biochemical parameters of *Vigna rachata L. Wilseck* and *Vigna mungo L. Heppter. J. Environ. Pollut., 2–4*, 133–134.

Dash, A. K., (2012). Impact of domestic wastewater on seed germination and physiological parameter of rice and wheat, *IJRRAS, 12*(2), 280.

Datta, J. K., Ghanty, S., Banerjee, A., & Mondal, N. K., (2009). Impact of lead on germination physiology of certain wheat cultivar (*Triticum aestivum*), *J. Ecophysiol. Occup. Hlth., 9*, 145–151.

Debeaujan, I., Karen, M., & Koorneef, L. M., (2000). Influence of the testa on seed dormancy, germination and longevity in Arabidopsis. *Plant Physiol., 122*, 403–413.

Deka, S. Dev, I. A., Barthakur, H. P., & Kagti, L. C., (1997). Studies on the impact of crude oil pollution on physico-chemical properties, nature of micro-organisms and growth of rice plants in soil. *J. Environ Biol., 18*(2), 167–171.

Derek Bewley, J., Michael, B., & Peter, H., (2006). The encyclopedia of seeds: Science, technology and uses Cabi Series. *CABI*. p. 203.

Dutta, S. K., & Boissya, C. I., (1999). Effect of paper mill effluent on growth yield and N, P, K contents of transplanted rice (*Oryza sativa L var. Mahsuri*). *J. Environ. Biol., 20*, 29–32.

Hadas, A., (1976). Water uptake and seed germination in leguminous seeds under changing external water potential in charging osmotic solution. *J. Expt. Bot., 27*, 480–489.

Haque, A. H. M. M., Akon, M. A. H., Islam, M. A., Khalequzzaman, K. M., & Ali, M. A., (2007). Study of seed health, germination and seedling vigor of farmers produced rice seeds. *Int. J. Sustain. Crop Prod., 2*(4), 34–39.

Huma, Z., Naveed, S., Rashid, A., Ullah, A., & Khattak, A., (2012). Effects of domestic and industrial wastewater on germination and seedling growth of some plants. *Current Opinion in Agriculture, 1*(1), 27–30.

Idu, M., (1994). Seed germination in Bixa orellana L. *J. Trop. Agri., 32*, 17–21.

Javid, S., Singh, S., Ahmad, I., Saeed, S., Khan, N. A., & Inam, A., (2006). Utilization of sewage and thermal power plant discharged wastewater for the cultivation of a pulse crop. *Asian J. Microbiol. Biotech. Env. Sci., 8*, 217–222.

Karpate, R. R., & Choudhary, A. D., (1997). Effect of thermal power station's waste on wheat. *Journal of Environmental Biology*, *18*, 1–10.

Kelly, K. M., Staden, J. V., & Bell, W. E., (1992). Seed coat structure and dormancy. *Plant Growth Reg.*, *11*, 201–209.

Khan, A. A., (1982). *The Physiology and Biochemistry of Seed Development, Dormancy and Germination* (p. 547). Elsevier/North Holland Biomedical Press, Amsterdam, The Netherlands.

Khan, A. G., Kuek, C., Chaudhry, T. M., Khoo, C., & Hayes, W. J., (2000). Role of plants, mycorrhizae and phytochelators in heavy metal contaminated land remediation. *Chemosphere*, *41*, 197–207.

Khan, M. G., Danlel, G., Konjit, M., Thomas, A., Eyasu, S. S., & Awoke, G., (2011). Impact of textile wastewater on seed germination and some physiological parameters in pea (*Pisum sativum L.*), Lentil (*Lens esculentum L.*) and gram (*Cicer arietinum L.*). *Asian Journal of Plant Science*, *10*, 269–273.

Krishnaswamy, V., & Seshu, D. V., (1990). Germination after accelerate aging and associated characters in rice varieties. *Seed Science and Technology*, *18*, 147–150.

Kumawat, D. M., Tuli, K., Singh, P., & Gupta, V., (2001). Effect of dye industry effluent on germination and growth performance of two rabi crops. *J. Ecobiol.*, *13*, 89–95.

Mayer, A. M., & Poljakoff, M. A., (1982). *The Germination of Seeds*. Pergamon press, London.

Mishra, P., & Bera, A. K., (1996). Effect of tannery effluent on seed germination seedling growth in maize (*Zea mays L.cv. Vikram*). *Environ. Ecol.*, *14*(4), 752–754.

Narain, K., Bhat, M. M., & Yunus, M., (2012). Impact of distillery effluent on germination and seedling growth of *Pisum sativum* L., *Universal Journal of Environmental Research*, *2*(4), 269–272.

Pandey, V. C., Abhilash, P. C., & Singh, N., (2009). The Indian perspective of utilizing fly ash in phytoremediation, phytomanagement and biomass production. *Journal of Environmental Management*, *90*, 2943–2958.

Paranthaman, S. R., & Karthikeyan, B., (2015). Bioremediation of paper mill effluent on growth and development of seed germination (*Vigna mungo*), *CIBTech Journal of Biotechnology*, *4*(1), 22–26.

Prasannakumar, P. G., Pandit, P. R., & Meheshkumar, R., (1997). Effect of dairy effluent on seed germination, seedling growth and pigment of green gram (*Phaseolus aureus* L.) and black gram. (*Phaseolus mungo* L.). *Advances in Plant Sciences*, *2*, 146–149.

Rajini, A., & Chauhan, S. V. S., (1996). Effect of tannery effluent on seed germination and total biomass in some varieties of *Hordeum vulgare* L. *Acta Ecologica*, *18*(2), 12–15.

Raju, K., Vishnuvardhan, V., & Damodharam, T., (2015). Industrial effluents effect on seedling growth of rice and wheat (*Oryza sativa* L. *and Triticum vulgare* L.). *International Journal of Recent Scientific Research*, *6*(7), 4935–4939.

Rampal, R. K., & Dorjey, P., (2000). Effect of foam industry effluent on Lens esculenta Moench var. Malika. *Indian J. Environ. Protection*, *21*(1), 14–17.

Reddy, P. G., & Borse, R. D., (2001). Effect of pulp and paper mill effluent on seed germination and seedling growth of *Trigonella foenum-graceum* L. (Methi). *Journal of Industrial Pollution Control*, *17*, 165–169.

Sahai, R., Jabeen, S., & Saxena, P. K., (1983b). Effect of distillery effluent on seed germination and growth and pigment content of rice. *Ind. J. Ecol.*, *10*, 7–10.

Saini, S., & Pant, S., (2014). Physico-chemical analysis of sugar mill effluent and their impact on changes of growth of wheat (*Triticum aestivum)* and Maize (*Zea mays* L.). *Journal of Environmental Science*, *8*(4), 57–61.

Selvi, V., & Sharavanan, P. S., (2013). Effect of diary effluent on seed germination and pigment contents of *Phaseolus trilobus* (Ait). *IJCRD, 1*(1), 121–129.

Subramani, A., Sundaramoorthy, P., & Lakshmanachary, A. S., (1998). Impact of fertilizer factory effluent on the morphometrical and biochemical changes of cow pea (*Vigna ungiculta (Lino.)*. *Adv. Pl. Sci., 11*(1), 137–141.

Sundari, S., & Kankarani, P., (2001). The effect of pulp unit effluent on agriculture. *J. Ind. Pollu. Contl., 17*(1), 83–97.

Sunderland, N., (1960). Cell division and expansion in the growth of the leaf. *J. Exp. Bot., 11*, 68–80.

Tripathy, A., & Sahu, R. K., (1997). Effect of coal fly ash on growth and yield of wheat. *Journal of Environmental Biology, 18*(2), 131–135.

Verma, A., & Verma, A. P., (1995). Effect of tannery effluent on seed germination and chlorophyll content in *Phaseolus radiatus* L. *J. Industrial Pollution Control, 11*(1), 63–66.

Welbaum, G. E., Bradford, K. J., Yim, K. O., Booth, D. T., & Oluoch, M. O., (1998). Biophysical, physiological and biochemical processes regulating seed germination. *Seed Science Research, 8*, 161–172.

Wins, J. A., & Murugan, M., (2010). Effect of textile mill effluent on growth and germination of black gram - *Vigna mungo* (L.) Hepper. *International Journal of Pharma and Bio Sciences, 1*(1), 1–7.

Yadav, R. K., Saini, Y., & Sharma, S., (2015). Impact of marble slurry effluent on germination of seed growth of crops (Gram-RSG 888, Barley-RD 2035). *Bulletin of Environment, Pharmacology and Life Sciences, 5*(1), 68–72.

Yasmin, A., Nawaz, S., & Ali, S. M., (2011). Impact of industrial effluents on germination and seedling growth of *Lens esculentum* varieties. *Pak. J. Bot., 43*(6), 2759–2763.

CHAPTER 10

Importance of Forests and Agriculture in Global Climate Change

VISHNU K. SOLANKI,[1] VINITA PARTE,[2] and VINITA BISHT[3]

[1]*Department of Agroforestry, College of Agriculture, Ganjbasoda, Vidisha, Jawaharlal Nehru Krishi Vishwavidyalaya, Jabalpur, Madhya Pradesh, India, E-mail: rvishnu@hotmail.com*

[2]*College of Agriculture, Ganjbasoda, Vidisha, Jawaharlal Nehru Krishi Vishwavidyalaya, Jabalpur, Madhya Pradesh, India*

[3]*College of Forestry, BUAT, Banda (UP), India*

ABSTRACT

Forests not only playing a critical role to slow down or to stop the climate change (CC), it also decreasing the current as well as future effects on people. For example, forest goods and agriculture acting more climatic resilient whenever disaster occurs or occurrence of crop failure happens forests play a role like safety nets to protect the communities from losing all food resources and income also. Forests also help in regulation of waterways to protect soil, cool cities, and entire regions and more. Agriculture sector plays an inevitable role, for example, organic agriculture is the sustainable farming practices which play a positive role in term of storage and sequestration of carbon in soil and helps in maintaining healthy soil.

10.1 INTRODUCTION

Global warming and climate change (CC) are the big curse today's world, which can be minimized through the practices of the renewable biomass production. Renewable biomass is excellent source of energy as fuel and fire wood, plays multifarious role in productions of various timber and NTFPs such as production of grasses/pastures, fodder for livestock's, food production, conservation of biodiversity, biomass, and yields along with other ecosystem services for the welfare of society in the era of changing climate and global warming (Raj et al., 2018a, b). Sustainable agriculture practices comprises better agriculture system, agroforestry along with forestry and grassland systems are the good strategies for mitigating and minimizing global warming phenomenon (Singh and Jhariya, 2016; Jhariya et al., 2015, 2018).

For example, forestry sector has a tremendous potential to regulate and balance the atmospheric carbon (major source of GHGs) as they absorb and fix 2.6 billion tonnes of CO_2/yr, approximately from the fossil fuel burning about 1/3 CO_2 is released in the atmosphere. Similarly, forest absorbs a big source of atmospheric carbon through storage and sequestration process and their unsustainable harvesting and overexploitation proves removal of greenhouse gases (GHGs) to the atmosphere results global warming for example, alone deforestation activity adds approx. 20% of all the GHGs emissions which is more than the world's entire transport sector. Thus, declining forests result reduction in absorption of atmospheric carbon which is a major and potent source of GHGs and related global warming.

10.1.1 CLIMATE CHANGE (CC) IMPACTS

CC alters the forest and tree conditions in some areas through increasing growth rates while endangering the species survival and communities of forest in others. It was observed that along with variable geographic area, climatic conditions, species variety, and human interferences, there are some other limiting factors like temperature, availability of water, and seasonality changes.

Similarly, changes in climatic regimes such as uncertain rainfall, extreme temperature and related extreme weather conditions result emergence of insect pest and various infectious disease which destroy valuable crops and tree species (Raj et al., 2018a, 2018b). Changing temperature and rainfall patter leads to vegetation shifting phenomenon such as expansion of forest areas from temperate forest region to poleward region and in other cases like reduction of forest areas due to dieback disease under the condition of water unavailability and drought in the northeast region of Amazon. However, changing climate and extreme weather events affects forest goods and related ecosystem services in terms of reducing yield and productivity, affecting reproduction potential and health, which major concern today for the forest managers and policymakers. Similarly, unusual climate affects the forest productivity and its demands of forest-based renewable source of energy as a fuel and firewood which is the greater substitute of fossil fuels (cause of GHGs emission).

Fishlin et al., (2009) and Lucier et al., (2009) projected impacts and vulnerability of forests under CC in various continents. It varies as per forest type which is more vulnerable in comparison to others. Impacts cover the variable fire incidence, potential increase on the events of extreme weather severity, pests outbreaks, and natural calamities, e.g., droughts, wind, and rainstorms. Activities of a human-being which includes conservation of forest, protection, and management practices shows interaction of the management practices under changing climate and often proves to be difficult to differentiate between actual and projected changes. A vicious circle formed by the deforestation and fire incidence in Amazon in relation to changing climate (Nepstad et al., 2008; Aragão et al., 2008) which degrades more than half of the rain forest ecosystem of Amazon (Nepstad et al., 2008).

10.1.2 FOREST CONDITIONS AND AREA

Changing climatic regimes such as extreme temperature and uncertain precipitation affects both expansion and contraction of the forest covers. For example, expansion of forest area has been observed in temperate region whereas tropical, boreal, and mountainous forests showed the rate of contraction gradually. CC also affects and promotes the separation of forest area (Lucier et al., 2009). Natural regeneration and forest plantation activity have expanded the forest areas in the country like United States, China, Europe, Latin America, Caribbean, Cuba, Costa Rica, Chile, and Uruguay (FAO, 2010) whereas some anthropogenic (deforestation, livestock conversion, unsustainable land use practices) and natural factor like forest fires and others resulted declining forest areas in Asia, Africa, Latin America, Siberia's boreal forests and other tropical regions (FAO, 2009).

Expectation relies on migration of boreal forest towards northward side and it is also expected that temperate forests will improve its area in the same direction at a larger perspective from the boreal forests, therefore, reduces the boreal forests total area (Burton et al., 2010).

Land use conversion, unutilized land use practices and combined interaction of CC are expected in the future prospects. Similarly, availability of water resource affects both growth and survival of many forest tree species, although among various tree species and also among the different varieties of the same species (Lucier et al., 2009). Similarly, both intensity and frequency of fires affects by longer dry spells or lower precipitation in the boreal (Burton et al., 2010), sub-tropical, and Mediterranean forest (Fischlin et al., 2009) and the practices of land clearing in the Amazon forest area (Aragão et al., 2008; Nepstad et al., 2008).

10.1.3 HEALTH AND VITALITY

It is clear and remarkable impact on the health and vitality of the world's forest due to CC and global warming. Both favorable climate and CO_2 fertilization have impacted on the growth and vitality of forest crops. Increase in temperature has deleterious impact on the forest productivity and favors the population and growth of insect which damage the forest's health and productivity (Lucier et al., 2009). Some tree species are most susceptible to infectious disease and insect population which is control by particular temperature and moisture condition. For example, mountainous pine beetle (*Dendroctonus ponderosa*) has emerged due to absence of low temperatures and spread throughout the montane areas and entered into the boreal forests of cold region (Burton et al., 2010).

Similarly, certain conditions like longer harvesting periods, longer spore production season and increased storm damage favored the growth and populations of fungus (*Heterobasidion parviporum)* which increased the root and bud rots in the coniferous tree species in the Finland (Burton et al., 2010). Moreover, it is also reported that warming and higher temperature reduces many insect growth and populations (Lucier et al., 2009).

10.1.4 BIOLOGICAL DIVERSITY

The growth and survival of tree species depend on a large part of climate variables. Moreover, certain species survive in the specific range of climatic regimes and their growth varies as per varying environmental changes. As we know, all living organism and its diversity such as tree, animals, and other plants survive according to climatic conditions which resides together in any areas and maintains biodiversity. Diversity and composition of forest and its ecosystem services is strongly linked with prevailing climate which shows regional and global classification of the forest ecosystem. The Holdridge ecological life zones (Holdridge, 1967) are bounded by humidity, precipitation, and temperature. Various authors have been estimated the impact of CC and related shifting of the life zone boundaries in Central America (Mendoza et al., 2001; Jimenez et al., 2009).

It was observed that studies based upon projection under real condition is a great lacunae as because the influence of geographical shift due to CC do not occur on a particular species rather it occurs over a vegetation type. The background of this condition lies with the better adaptive capability in comparison to others reflecting overall changes in the composition of forest types than shifting of forest type geographically (Breshears et al., 2008).

As per Rosenzweig et al., (2007) and Breshears et al., (2008), several species have tendency to move up at the higher altitude and latitude due to their adaptability in the more favorable climatic conditions. More phonological changes like have been observed in various species in higher latitudes due to climate variability (Lucier et al., 2009). Flowering and bud breaking time are the common changes have been observed due to changing climate. Similarly, various attempts have been made on the studying phonological changes in number of tree species such as oak (Bauer et al., 2010), apple, and pears (Blanke and Kunz, 2009) and Mediterranean tree species (Gordo and Sanz, 2010). Moreover, these phonological changes affect the various life sustaining and supporting ecological processes viz., fruit setting, flowering time and pollination processes in number of forest tree species. Thus, natural disturbance has deleterious impact on forest ecosystem through damaging standing tree, health deterioration and its productivity (Chakraborty et al., 2008; Jepsen et al., 2008; Kurz et al., 2008; Nepstad et al., 2008).

10.1.5 CARBON STORAGE AND SEQUESTRATION

Indeed, forest work as potential carbon sink through the process of storage and sequestration of huge amount of atmospheric carbon which is stored and fix in vegetations and maintain carbon balance to secure ecosystem health (Jhariya, 2017; Jhariya and Yadav, 2018). Today, CC is becoming a major concern and well discussion topic in the world and many countries put their effort to resolve it through various national and international programme which is based on reducing an emission GHGs and maximizing forest covers which helps in enhancing carbon stocks and make global climate security.

An innovative financing mechanisms implementation for management of forest, planted forests and its conservation are recognized in Costa Rica during nineties century (Sánchez Chávez, 2009) which is a good effort towards maintenance and conservation of forest extent in both existing natural and planted forests. Increasing temperature and carbon dioxide concentration along with long duration of dry seasons favor low storage and sequestration of carbon (Nepstad et al., 2008; Ollinger et al., 2008; Saigusa et al., 2008; Clark et al., 2003). Similarly, short-term increase in temperature reduces the carbon storage potential and this may very which depends on the temperate regions season.

10.1.6 SOIL AND WATER PROTECTION

Forest as a good contributor to soil and water protection and its conservation in several countries and provides ecosystem system services for the wellbeing of biodiversity (Postel and Thompson, 2005). The soil protection and water conservation play a key role in growth and productivity of tree species and many other living organisms which becomes more important under changing climatic conditions (Jhariya, 2014).

Total annual rainfall reduction along with evaporation causes problems for water absorption by forests. This therefore reflects a decline in the flow of water during rainy season along with increase in the evaporation rate during summer. Areas under cloudy and foggy condition influence the water absorption rate of the forest ecosystem depending upon the total rainfall of the region (Stadtmüller, 1994). Palaeoecological study on vegetation of Amazonreflects an interesting feature in terms of warming of forest area. Mayle and Power (2008) reported that forest under cloud cover and under intense fog condition, warming of the forest takes place as the cloud remains above the tree. Further, this would create reduction in the horizontal precipitation pattern of the forest area.

10.1.7 MULTIPLE SOCIOECONOMIC BENEFITS

CC increases the growth of tree species in some areas while decrease the growth potential in other areas. Increasing wood production favors lower prices and which is advantageous to consumers. Therefore, combination of lower prices and regionally differentiated effects on productivity will lead to differentiated effects on harvest of timber relevant income and employment (Osman-Elasha et al., 2009).

NWFP (non-wood forest products) has three important functions to perform due to which its harvesting has a huge importance. It includes full-filling daily needs of people who is dependent upon forests, acts as a source in the form of off-farm income along with a cushion for people under adverse situation when agricultural production declines and survivality of people become questionable.

Moreover, Osman Elasha et al., (2009) has concluded that CC has deleterious impact on both productivity of NWFPs which affects users (tribal people and farmers) of forest fringe areas in term of declining incomes and poor socioeconomic status of people. Increased frequency and extreme CC along with some disturbances like emergence of insect pest related disease and fires will affect poverty areas where people are highly dependent upon NWFP for their socioeconomic upliftments and livelihood generation. CC impact on these products provision and the subsequent socioeconomic consequences however, requires core studies (Painkra et al., 2016).

10.1.8 CONSEQUENCES OF GLOBAL WARMING

World climate has been continuously changing at rates which are unprecedented projected. Ecosystem disruption, glacial ice melting and inundation of coastal areas are observed due to global warming impact through changing climate due to rising sea level and uncertain precipitation along with flood and severe and frequent storms. Therefore, human society, ecology, and national economy are suffered consequently. As per the IPCC report world average surface temperature can rise by 1.4°C to 5.8°C along with rising seas level under changing climate during 21st century (Kauppi et al., 2001). Calamities such as flood, drought, and storms are happening with a great frequency and higher intensity due to extreme weather condition. Coastal regions are more prone to rising sea level which results severe coastal erosion and inundation, declining the quality of ground and surface water, losses of coastal habitats, health, and property, losses of nonmonetary cultural resources and values, impacts on agriculture land and its production potential, increases risks of flood, declining soil quality along with loss of tourism, transportation, and recreation systems. As per IPCC 4th report world sea level will rise from 0.18 to 0.59 m at the end of this century which will result loss of humans, ecosystem degradation, problematic for urban cities and rural areas and coastal areas of wetlands.

Stern (2009) emphasized the fast action importance otherwise in near future many low-lying areas around the world will be underneath of water. Impact of global warming in terms of elevate the costs all around is becoming evident. Most of the time poor people suffer (Simms et al., 2004).

CC impact may vary nation-to-nation or even in the same nation across the people, place, and times. As per Scott et al., (2003), the death of approximate

30,000 premature were linked with summer the heatwave in the European country, which has been linked with 90% probable global warming and CC situations. As we know, extreme climate has severe impacts on both health and property of human and other organisms.

It was observed that disease vectors and their geographic distribution are very much important under the changing climate scenario which helps to spread the diseases in a dynamic way. Alteration in the life cycle of vectors due to the impact of CC leads to aggregation of microbes and other disease causing organism which lead to potential increase in the vector borne diseases. Further, along with vector borne disease, the non-vector borne disease also emerges under tropical and sub-tropical climatic conditions due to climatic perturbations over water availability, regime of temperature as well as microbe (Shaffer et al., 2009).

Agriculture sector is more prone to changing climate and extreme weather. The multiple impacts on agricultural productions are:

i. Rising temperature, emergence of pest and water scarcity due to CC affect health and productivity of plant.
ii. Uncertain rainfall pattern and moving rainy season affect both irrigated and rainfed agriculture due to unavailability of water.
iii. Enhanced weather extremes frequency and increase variability supply.
iv. In some cases, CO_2 concentration increases in the atmosphere which may improve the crop yield and crop productivity.
v. Loss of farmer's life or even world production system is due to rising sea level and more frequent flooding.

The climatic impact on agriculture production system are varies region to region. For example, the cases of increased agricultural productivity have been seen in the temperate zones of North America, Asia, and Northern Europe whereas loss has been seen in the agriculture practices of Mediterranean and tropical zones. The yield and productivity of agriculture system in the semi-arid, tropical zones and arid climate are decline and it makes tougher to satisfying and fulfilling the growing food demands (Scialabba Nadia El Hage, 2007).

Under the changing climate scenario it was observed there is a shifting of production possibilities in the agriculture sector across various agro-ecological zones and as a consequence the global market and trade dwindles around dynamically. In developing countries the scenario even worse due to decline in the agricultural production, drastic alteration of food demand which would promote the increase in the financial market export phenomenon.

Changing climate affect agricultural productions and which have negative impacts on food security. Thus, CC favors biodiversity loss and extinction of several valuable forest and plant resources and also increase the extinction risk for many plants and animal species particularly those that are already at risk and adversely impact on several valuable ecosystem services which is vital for sustainable development (Rathore et al., 2014).

10.2 AGROFORESTRY

The most promising and best option for carbon sequestration (CS) on agricultural land is agroforestry because it could help for the significant amount of CS while leaving the maximum of the land for agricultural production. In removal of atmospheric accumulation of GHGs, the important role-plays by agroforestry (IPCC, 2001). Complex agroforestry, boundary planting, windbreak, hedgerow intercropping and home gardens are improved CS rates (Calfapietra et al., 2010). Nevertheless, highest storage of carbon are reported in multi-storey/complex agroforestry systems (AFs) (Sajwaj et al., 2008) that have numerous diverse species using ecological niches from the highest canopy to bottom story shade of tolerant crops. Shade grown coffee and cocoa plantations are some examples where cash crops are grown under tree's canopy that helps in CS and also provide shelter for wildlife. Where tree competition is minimum some simple intercrops are used. As per soil and climate differences and differential value of tree crop species, there is need to develop agroforestry (Jhariya et al., 2015, 2018). The agroforestry potential seems to be substantial but it remains known as an option for GHGs mitigation for agriculture in the world (Sileshi et al., 2007).

Agroforestry is a sustainable land use farming practices and a better option to enhancing carbon density through absorption and sequestration of atmospheric carbon dioxide into the both vegetation and soils (Dixon, 1995). Agroforestry play an important role in soil conservation, water regulation, and ecological services to livestock's/animals and reduce the pressure on natural forest by providing multifarious products and potential ecosystem services. Adoptability of improved AFs would help in reducing small-scale farmer's vulnerability and farmers are able to adapt adverse and altering condition to meet need of fuelwood, fodder (for livestock's), timber, and NTFPs. To create synergies between mitigation and adaptation and to fulfill the requirements, CDM projects endorse these systems and CDM projects create social as well as environmental benefits both.

10.3 FORESTS AND CLIMATE CHANGE (CC)

The occurrence of world CC is due to the equilibrium of the world's carbon cycle has been disturbed. Human activities resulted continuous emission of carbon dioxide along with other GHGs amount is pick up and surpass the earth's potential to stock up carbon in oceans, forests, and biomass. Most climate policy mainly focused on dropping industrial emissions which account 70% of total emission of GHGs in the atmosphere whereas the rest 30% emission is contributed by combination of non industrial areas land use practices like forestry and agriculture sectors (Table 10.1). Both agriculture and forestry sectors have remarkable potential for storage and sequestration of the atmospheric carbon and other harmful GHGs, which helps in mitigating CC issues but adoption of good policy is also a better strategies to resolve this problem in national and international level. In this context, REDD (reduction of emissions from deforestation and degradation) and LULUCF (Land Use, Land Use Change and Forestry) are working on this issue of different aspects of land use practices along with forestry and agriculture practices.

Globally, forests stored approximate >650 billion tons of carbon of which biomass, dead wood and litter stored 44 and 11% of carbon and rest (45%) is contributed by soil. IPCC (2007) has reported that 17–18% of GHGs emissions due to the activity of deforestation whereas 26% of the emission contributed by power stations and fossil fuel combustion.

TABLE 10.1 Percent of Total GHGs Emissions Through Forestry and Agriculture

Forestry and Agriculture	Total GHGs Emissions (%)
Through power supply	21
Through industry	19
Through forestry	17
Through agriculture	14
Through transport	13
Through building	8
Through fossil fuel burning	5
Through waste	3
Total	100

Source: UNFCCC, 2007.

Both deforestation and timber harvesting practices contribute 20.8% of greenhouse emissions which is compensated by 2% through the practices of afforestation and reforestation activity. About 13.5% of the total global GHGs emissions are contributed by deforestation (release CO_2) and agriculture (release CH_4 and NO_2). Table 10.1 point-out that the decrease of emissions from deforestation and agriculture can be a significant part of worldwide efforts to battle CC and reduce GHG emissions. Therefore, reduction in carbon emissions through the afforestation and other sustainable forest management is burning topic today and noteworthy feature of worldwide discussions on responses to CC. Increasingly, forest being managed and conserved for multiple values and use often in combination. Approx. 949 m ha forests (24% of the total) are designated for multiple uses such as provision services of good supply, soil, and water conservation, biodiversity conservation and other social services (Table 10.2).

Agricultural land is primarily used for food and other products and can also provide other services like soil and water protection and storage of carbon, which is based on type of practices.

TABLE 10.2 Designated Functions of Forests 2010

S. No.	Functions of Forests	Percentage
1	Production	29%
2	Soil and water conservation	8%
3	Biodiversity conservation	12%
4	Provide the social services	4%
5	Multiple use of forest	24%
6	Other function	7%
7	Unknown function	16%

Source: FAO, 2010.

10.3.1 SIGNIFICANCE OF FORESTS IN THE CARBON CYCLE

Consider a closer connection between CC and forestry and clear some relevant definitions in order to enhance the knowledge of the role of forests in CC combating. In two different ways, forests can affect the world carbon cycle:

- **Forests as Stock of Carbon:** Forest has potential to absorb carbon (CO_2) from the atmosphere and dissociate into carbon and oxygen.

Moreover, this carbon storage in the bole, branch, leaf, and twig whereas some parts of carbon stored by soils living and non-living components. Similarly, tree dry biomass of two tons can hold around one ton of carbon. As per one figure, above-ground biomass in tropical wet forests can hold up to 430 tC/ha (CIFOR, 2009).

- **Forests as Carbon Fluxes:** Forests create carbon flux and influence the process of carbon cycle. Through the process of photosynthesis and presence of sunlight, from atmosphere, leaves has potential to absorb CO_2 inbound flux. Moreover, stored carbon in the plant is transfer and move into the soil through its fall and decaying which release free carbon into atmosphere and also some part of C release into the atmosphere through the process of plant respiration and soil mineralization outbound flux. Therefore, inbound, and outbound differences results net absorption flux (CIFOR, 2009).

The separation of CO_2 from the atmosphere is represented by net flux which is inbound flux during the growth of forest. This process is called fixation, absorption or removal and ecosystem is termed as carbon sink. Further, it was observed that the stock declines due to forest fire which would lead to further addition of GHG into the atmosphere leading to CC (Kittur et al., 2014a, b). This event is popularly known a carbon emission, and therefore the ecosystem is considered as the source (CIFOR, 2009).

10.3.2 SOCIAL AND ECOLOGICAL FUNCTIONS OF FORESTS

Besides the function of carbon storage and sequestration, forest provides various ecosystem services for the welfare of social and ecology. As per one estimate, forest supports 90% of the world terrestrial biodiversity along with livelihood supportive function of 1.2 billion people living in extreme poverty areas where they utilize forest tangible products for their livelihood upliftments and socioeconomic development. Similarly, the forest also provides intangible services like rainfall promotion, watershed management, soil fertility enhancement, efficient nutrient cycling, disease management, and water regulation (Parker et al., 2008; Jhariya and Raj, 2014).

Several economists have attempted for total economic valuation of forest ecosystem. During their estimation, it was found that the entire benefit of a forest ecosystem can be monetized by applying specific methodological approaches dealing with ethical concern which would help the forest people to generate specific policies considering this information. In this connection, a suitable comparison can be made between the productivity value of the forest and the preservative value of the forest. Various studies have been suggested to find out the suitability of extractive use in comparison to preservative use. For these non-marketable goods such as storage of carbon, watershed benefits, and other recreational benefits are being combined with the productive values of forests in terms of food, fodder, and timber production in Mediterranean forests. Research reports revealed that around 5 trillion dollar price value lies for the human prosperity of which up to 20% of benefits were obtained from forest ecosystem across the globe. Apart from economic benefits forest also serves in the form of ecological mode by controlling erosion, cycling of nutrients, regulating climate over an area and above all through waste treatment (Costanza et al., 1997).

10.3.3 SCALE OF WORLD FOREST COVERAGE AND FOREST LOSS

Forest cover contributes 31% of land areas of the globe where Brazil and the Russian Federation contributed the largest forest areas (Table 10.3). Also, the region of the Eastern U.S., Canada, South-East Asia, and Central Africa has potentially large forest areas (Table 10.4). Various authors have reported the reasons behind shrinkage of the forest cover, which are presented in Table 10.5.

TABLE 10.3 Top Countries With the Largest Forest Area

Countries	Forest Area (Million ha)
Russian federation 1st position	809
Brazil 2nd position	520
Canada 3rd position	310
United States of America 4th position	304
China 5th position	207
Democratic Republic of the Congo 6th position	154
Australia 7th position	149
Indonesia 8th position	94
Sudan 9th position	70
India 10 position	68
Others countries	1347 million ha

Source: FAO, 2010

TABLE 10.4 The World's Forest Coverage

S. No.	Region	Total Forest Area (Million ha)
1	Africa	674
2	Asia	593
3	Europe	1005
4	North and South America	705
5	Oceania	191
6	South America	864
World		4033

FAO, 2010.

TABLE 10.5 Causes of Forest Decline

S. No.	Causes
A	**Direct**
I	Natural causes
1	Hurricanes'
2	Natural fires
3	Pests
4	Flood
II	Human (anthropogenic cause)
1	Agricultural expansion
2	Cattle ranching
3	Logging
4	Coal mining and oil extraction
5	Dams constructions
6	Roads
B	**Underlying**
I	Market failure
1	Un-priced forest goods
2	Monopolies and monopsonistic forces
II	Mistaken policy Interventions
1	Wrong Incentives
2	Regulatory mechanisms
3	Government Investment

TABLE 10.5 *(Continued)*

S. No.	Causes
III	Governance weaknesses
1	Land ownership concentration
2	Arrangement of land tenure system and fragile or non-existing land ownership
3	Corruption and related illegal activities
IV	Widersocio-economic and unhealthy political causes
1	Population growth and density
2	Economic growth
3	Unhealthy economic distribution and political power
4	Excessive consumption
5	Toxification
6	Global warming
7	War
V	Agents forest decline
1	Slash and burn farmers
2	Agribusiness
3	Cattle ranchers
4	Miners
5	Oil corporation
6	Loggers
7	Non-timber commercial corporations

Source: Contreras Hermosilla, 2000.

10.4 AGRICULTURE AND CLIMATE CHANGE (CC)

Agriculture itself a cause of CCs and contributed 13–15% GHGs (Table 10.6).

The use of chemical fertilizer, livestock management, and other energy sources are the reasons for emissions. Rice cultivation in agriculture is the most potent cause of methane emission. Animal manure production and continuous application of nitrogen fertilizer are the major sources of N_2O emissions, which is projected to increase by 35–60% up to 2030 (FAO, 2003). Diversion of forestland to cultivated agriculture land promotes the activity of deforestation, which results in loss of biodiversity and valuable ecosystem services. Application of biochar to the soil has potential to enhance productivity and helps to improve the soil health (Scher and Sthapit, 2009).

TABLE 10.6 From Agriculture Global Greenhouse Gas Emissions

S. No.	Global GHG Emissions	Percentage
1	Agriculture	15%
1a	Subsector like	
	• Soil (Nitrous oxide)	40%
	• Enteric fermentation (Methane)	27%
	• Rice (Methane)	10%
	• Energy-related (Carbon dioxide)	9%
	• Manure management (Methane)	7%
	• Other activities (Methane and Nitrous oxide)	6%
1b	Gas	
	• Nitrous oxide	46%
	• Methane	45%
	• Carbon dioxide	9%
2	Rest of world greenhouse gases	85%

Source: WRI, 2011.

10.4.1 POTENTIAL FOR GREENHOUSE GAS (GHG) MITIGATION IN AGRICULTURE

As per the 4th assessment report of IPCC's (Metz et al., 2007), the mitigation opportunities of GHGs in the agriculture sectors (Table 10.7).

- By managing the flowing ratio of carbon and nitrogen in agricultural ecosystems which helps in minimizing and controlling the GHGs emission in agriculture.
- Through promoting and enhancing storage and sequestration of carbon in the both plant and soils.
- Replacing fossil fuels-Good strategy to mitigate CC and global warming through reducing GHGs emissions by using agricultural and forest crops as fuel other than fossil fuel. However, this bio-energy also contributes to carbon dioxide emission through combustion (Metz et al., 2007).

10.4.2 THE IMPACT OF BIOFUELS

As we know, biofuel is deriving from organic waste, plants, animals, and various microorganisms and a good substitute for the fossil fuels. This will help in excessive carbon reduction, which is emitted through the burning and combustion of fossil fuels. Biofuels have some inherent drawbacks. It was observed that energy input from biofuel has a positive influence through reduction in net carbon emission which appears to be sometime negative. Under the circumstance of land and water scarcity it was observed that biofuel can be utilized to provide additional benefits either as energy or proper utilization of land area which has been converted into wasteland. It is, therefore, necessary for the government to provide subsidy to promote biofuel production for community development. Biofuels of different types are as follows:

- Various form of carbohydrates (sugar, starch), vegetable oil or animal fats are utilized for first-generation biofuel production through the conventional process.
- A variety of non-food crops including waste biomass becomes the source of generation of second-generation biofuels.
- Third generation biofuels are generally produced from algae.

TABLE 10.7 Global GHG Mitigation from Different Source of Agriculture

S. No.	Different Sources of Agriculture
1	GHG mitigation from cropland management
2	GHG mitigation from grazing land management
3	GHG mitigation from restore cultivated organic soils
4	GHG mitigation from restore degraded lands
5	GHG mitigation from rice management
6	GHG mitigation from livestock
7	GHG mitigation from bio-energy (soils component)
8	GHG mitigation from water management
9	GHG mitigation from set-aside, LUC, and agroforestry
10	GHG mitigation from manure management

Source: Metz et al., 2007; Smith et al., 2008.

10.4.3 BIOFUELS, FOOD SUPPLY, AND FORESTS

Studies reveal that some food crops are used as biofuels. Increasing demand for biofuels often creates competition with food crops as cultivable land area increases the price of the food. As per UNEP (2009), 2% of cropping land is being utilized for biofuel production. There would be a gradual build-up of conflict for increasing demand of biofuels along with expanding global food requirements. It was observed that there is a trend of conversion of agricultural production sector towards forested areas which has indirect effects. Further biofuels have different impacts. The method of production of biofuel influences the effectivity and their potential role towards carbon emission reduction. However, expansion of biofuel production may not always be good but maybe harmful towards the environment.

In the upcoming years, biodiversity would be severely affected due to ecological invasion, habitat alteration, and gradual build-up nutrient load (Yadav et al., 2017; Jhariya and Yadav, 2017). As, for instance, expansion of farming area often leads to degeneration of forest area in the form of habitat loss. It was observed that grass species and other allied species used as feedstock for biofuel production may become potential invaders. Cropping of biofuel may lead to higher nutrient load in water and air, and subsequently would impact upon species on ecological systems both in land and water.

10.4.4 RELATION OF WEATHER WITH AGRICULTURE

1. **Solar Radiation:** It includes light intensity, light quality, and duration of sunlight. Out of received total radiation, only 50% in photosynthetically active radiation (PAR), which lies in between the range of 400–700nm, remaining is UV or IR. Now there is an exponential relation between amount of light intercepted by canopy and leaf area index (leaf area/ground area). The sum of these values for individual days is directly proportional to crop yield.
2. **Temperature:** In crop from flowering to harvest date, a term growing degree-days is used. Every crop has a temperature range below or above which GDD for that particular day is zero. GDD is the sum of difference of daily temperature and base temperature. So that prediction of harvest date can be done which depends on the time period when the GDD is achieved.
3. **Precipitation, Evaporation, and Transpiration:** For calculation of period of growth a parameters named evapotranspiration was used. For this precipitation amount and water requirement of crop needs to be incorporated. Further, physiological process such as transpiration helps to regulate the temperature which in turn prevents crop damage. There are some factors that regulate the development and growth of crops, which, if varies would significantly affect the crop physiology. Such event may frequently occur under changing climate.

10.4.5 AGRICULTURE AND WEATHER RELATION

Yield of crops are highly influenced by elevated CO_2 level. Alteration in the temperature, stress in the form of water, nutrient, and ozone concentration may influence the potential yield of crops. During research it was observed that if optimum level of water and nutrient concentration were not found and if the temperature rises beyond the crop acceptability level the yield output declines. For crops like soybean and alfa-alfa it was observed that reduced nitrogen and protein content accompanied with CO_2 level results into loss of quality in the product. Reduction in the rangeland and pastureland for feeding of livestock may reduce the quality of forage and grains. Further crop growth is regulated under extreme climatic condition such as precipitation pattern and temperature regime in the form of drought and flood.

Conditions such as elevated temperature during summer season may cause drying up of soils leading to drought condition which is a very tough challenge to combat especially in tropics. However, this problem can be managed through higher irrigation along with watershed management, soil water conservation and many more. It was found that infestation of weed, pest, and fungi takes place under elevated level of CO_2, moist climatic condition and high temperature. It was found rising temperature may impact the crop through nutrient reduction and stimulation of plant growth. For crops like soybean, wheat, and paddy it was found that nutrient content declines with rising CO_2 level. This may prove to be a potential threat in human health in terms of low nutrition level accompanied with rising CO_2. The problem was further aggravated through pest outbreaks and inefficiency of pesticides leading to higher pesticide use.

10.4.6 MODIFICATION OF CROP MICROCLIMATE

In the unfavorable condition of environment, many vegetable crops the performance of the crop is affected. Producers can, however, modify the environment a small scale, creating microclimate more suitable for growing high value, warm-season crops.

Artificial control of field environment is required for better production and growth of crop plants. A complete knowledge of plant physiology and physical environment is needed towards eco-friendly practices which are as follows:

1. Controlling wind velocity.
2. Controlling heat load.
3. Controlling water balance.

10.4.7 CLIMATIC NORMAL'S FOR CROP AND LIVESTOCK PRODUCTION

10.4.7.1 CLIMATIC NORMAL'S FOR CROP PRODUCTION

1. **Rice:** Research study reveals that yield of rice is linked with the physiology of grain production as well as pest and disease outbreak due to elevated temperature and radiation. It was observed that differencing temperature regime and solar insolation influences yield at the growing season of rice. For optimum growth and development rice requires elevated temperature, adequate supply of water with high humidity level. It was reported that if water is not scarce then rice may adjust up to 40°C temperature. However, an average temperature of 22°C is required during the growing phase of rice. It was reported that if temperature drops below 15°C it produces negative impact by reducing rice yield through sterile spikelet formation. Low temperature often becomes very critical issue before 10–14 days of the active growth phase. At vegetative growth stage low solar insolation and photo-period has minimum influence over yield. On the same time in the reproductive phase under similar condition a drastic reduction in yield takes place due to lesser spikelet formation. Research report revealed 5 t/ha yield of rice requires 300cal cm2/day of solar insolation. Very interestingly it was observed that a combination of average daily temperature with high insolation promotes better yield and productivity.

   Rainfall-Rice happens to be hydrophytic in nature due to its higher moisture requirement. During the growth phase rice always requires a submerge condition (125 cm water). For better growth and production of rice in low land and upland the land-use requires 200 mm and 100 mm rainfall, respectively.

2. **Wheat:** For sowing, it was observed that up to 20°C temperature is optimum for wheat cultivation. During the maturity of crop a higher temperature of 25°C is required. During harvesting a temperature range of 30–35°C and a photoperiod of 9–10 hours are required. As per an estimate one hectare of wheat can consumes up to 2500–3000 tons of water. Heading stage of wheat may be critical if there is water scarcity leading to low yield.

3. **Maize:** It requires an intermediate climatic condition for best performance. As per its life cycle the average temperature should be round about 24°C and during nighttime it should be 15°C. The average

temperature in summer should be less than 19°C and during night-time it should be always above 21°C and therefore if it declines then maize production would not be possible. Further researches revealed that yield of maize declines due to high temperature at nighttime. Maize as a crop is adapted to humid condition. In its growth time it requires 75 cm of rainfall. Maximum utilization of water by maize takes place when the rainfall is between 41 and 64 cm. It was also reported that maize requires little amount of water from its germination stage to earning stage. As the crop progresses towards maturity it requires the higher amount of water.

10.4.7.2 CLIMATE NORMAL'S FOR LIVESTOCK PRODUCTION

1. **Direct Impacts of CC on Livestock:** Heat stress is the key factor affecting livestock production as a consequence of changing climate. The financial cost increases in terms of processing and production of milk, decrease in milk and meat production, health, and reproductive ability of animals. Therefore, it is a very serious issue that CC drastically alters the performance of the animals in animal husbandry.
2. **Indirect Impacts of CC on Livestock:** Considering the indirect impacts water and food scarcity significantly hampers the production process in livestock. CC has been found to influence various economic activities such as alteration in the water demand of forage crop during cultivation, alteration in the forage quality, changes in the species composition of rangeland vegetation and many more. Crops and forage plant would suffer due to CC because the changing climate would result into high CO_2 level, elevated temperature regime as well as alteration in the precipitation pattern leading to lesser water availability. Further the impact of CC would also show negative consequences on grassland ecology. Across the globe changing precipitation pattern would severely affect the forage production. Therefore, livestock production would be hampered severely through CC event. Alteration in the geophysical setup of earth would significantly alter the climate and weather conditions. Places with high intensity rainfall, caused due to CC produce more runoff and lesser recharge of groundwater. Further lesser recharge of groundwater would inevitably alter water resource availability, problem in agricultural irrigation facility as well as drinking water supply during summer months. This would in turn cause loss of homeostasis in the

living organism in the form of low body weight, lesser reproductive potential and highly susceptible towards disease. Considering this, future research and developmental activities need to be done in the area of water resource vulnerability in the context of CC.

10.5 CONCLUSION

Imbalance of global cycle of carbon has led the problem of changing climate globally which occurs due to anthropogenic influence and GHGs emission to a considerable amount which surpasses the carrying capacity of the earth to accommodate the emissions through various ecosystem components. However, there is a huge prospect for carbon assimilation and storage in the sectors of agriculture and forestry. Various schemes such as zero tillage, intercropping, agroforestry, row planting, grazing management, and other methodologies help in the reduction of storage of carbons in soils and also reduce the carbon emissions. Reduction in fertilizer uses, manure management along with utilization of biogas, etc. can minimize the emission of N_2O and CH_4 at significant level.

KEYWORDS

- **agriculture sector**
- **climate change**
- **greenhouse gasses**
- **intergovernmental panel on climate change**
- **nitrous oxide**

REFERENCES

Aragão, L. E. O. C., Malhi, Y., Barbier, N., Lima, A., Shimabukuro, Y., Anderson, L., & Saatchi, S., (2008). Interactions between rainfall, deforestation and fires during recent years in the Brazilian Amazonia. Philosophical transactions of the royal society. *Biological Sciences, 363*(1498), 1779–1785.

Bauer, Z., Trnka, M., Bauerova, J., Mozny, M., Stepanek, P., Bartosova, L., & Zalud, Z., (2010). Changing climate and the phenological response of great tit and collared flycatcher populations in floodplain forest ecosystems in Central Europe. *International Journal of Biometeorology, 54*(1), 99–111.

Blanke, M., & Kunz, A., (2009). Einflussrezenter Klimaveränderungen auf die Phänologiebei Kernobst am Standort Klein-Altendorf–anhand 50-jähriger Aufzeichnungen. Erwerbs-*Obstbau, 51*(3), 101–114.

Breshears, D. D., Huxman, T. E., Adams, H. D., Zou, C. B., & Davison, J. E., (2008). Vegetation synchronously leans upslope as climate warms. *Proceedings of the National Academy of Sciences, 105*(33), 11591–11592.

Burton, P. J., Bergeron, Y., Bogdansky, B. E. C., Juday, G. P., Kuuluvainen, T., McAfee, B. J., Ogden, A., Teplyakov, V. K., Alfaro, R. I., Francis, D. A., Gauthier, S., & Hantula, J., (2010). Sustainability of boreal forests and forestry in a changing environment. In: Mery, G., Katila, P., Galloway, G., Alfaro, R., Kanninen, M., Lobovikov, M., & Varjo, J., (eds.), *Forests and Society–Responding to Global Drivers of Change* (Vol. 25, pp. 249–282). IUFRO World Series.

Calfapietra, C., Gielen, B., Karnosky, D., Ceulemans, R., & Mugnozza, G. S., (2010). Response and potential of agroforestry crops under global change. *Environ. Pollut., 158*, 1095–1004.

Chakraborty, S., Luck, J., Hollaway, G., Freeman, A., Norton, R., Garrett, K. A., Percy, K., Hopkins, A., Davis, C., & Karnosky, D. F., (2008). Impacts of global change on diseases of agricultural crops and forest trees. CAB Reviews: *Perspectives in Agriculture, Veterinary Science, Nutrition and Natural Resources, 3*(54), 1–15.

CIFOR, World agroforestry centre and USAID, (2009) forest and climate change toolbox, [power point presentation]. In: *Forest and Climate Change Toolbox: Topic 2 Section B [2009–2011]*. Available from: http://www.cifor.cgiar.org/fctoolbox/download/Topic-2-Section-B.pdf (Accessed on 10 October 2019).

Clark, D. A., Piper, S. C., Keeling, C. D., & Clark, D. B., (2003). Tropical rain forest tree growth and atmospheric carbon dynamics linked to inter annual temperature variation during 1984–2000. *PNAS, 100*(10), 5852–5857.

Contreras, H. A., (2000). *The Underlying Causes of Forest Decline Cite Seer.*

Costanza, R., Arge, R., Groot, R., De. Farber, S., Grasso, M., Hannon, B., Limburg, K., Naeem, S., ONeill, R. V., & Paruelo, J., (1997). The value of the world's ecosystem services and natural capital. *Nature, 387*(6630), 253–260.

Dixon, R. K., (1995). Agroforestry systems: Sources or sinks of greenhouse gases? *Agroforestry System, 31*, 99–116.

FAO, (2003). *World Agriculture: Towards 2015/2030: An FAO Perspective*. London: Earthscan Publication Ltd. ftp://ftp.fao.org/docrep/fao/005/y4252e/y4252e.pdf (Accessed on 10 October 2019).

FAO, (2009). *Situación de los Bosquesdelmundo*. FAO, Rome.

FAO, (2010). *Global Forest Resources Assessment Full Report*. FAO Forestry Paper 163. Rome.

Fischlin, A., Ayres, M., Karnosky, D., Kellomäki, S., Louman, B., Ong, C., Plattner, C., Santoso, H., Thompson, I., Booth, T., Marcar, N., Scholes, B., Swanston, C., & Zamolodchikov, D., (2009). Future environmental impacts and vulnerabilities. In: Seppala, R., Buck, A., & Katila, P., (eds.), *Adaptation of Forests and People to Climate Change*. IUFRO World Series 22.

Gordo, O., & Sanz, J. J., (2010). Impact of climate change on plant phenology in Mediterranean ecosystems. *Global Change Biology, 16*(3), 1082–1106.

Holdridge, L. A., (1967). *Life Zone Ecology*. Tropical Science Center, San José, Costa Rica.

IPCC, (2001). *Climate Change: Impacts, Adaptation and Vulnerability* (p. 967). Report of the working group II. UK: Cambridge University Press.

IPCC, (2007). *Climate Change: Synthesis Report*. Geneva, Switzerland: Intergovernmental Panel on Climate Change (IPCC) Secretariat.
Jepsen, J. U., Hagen, S. B., Ims, R. A., & Yoccoz, N. G., (2008). Climate change and outbreaks of the geometrids Operophterabrumata and Epirritaautumnatain subarctic birch forest: Evidence of a recent outbreak range expansion. *Journal of Animal Ecology, 77*(2), 257–264.
Jhariya, M. K., & Raj, A., (2014). Human welfare from biodiversity. *Agrobios Newsletter, XIII*(9), 89–91.
Jhariya, M. K., & Yadav, D. K., (2017). Invasive alien species: Challenges, threats and management. In: Sudhir, K. R., & Sarju, N., (eds.), *Agriculture Technology for Sustaining Rural Growth* (pp. 263–285). ISBN: 978–81–7622–381–2. Biotech Books, New Delhi, India.
Jhariya, M. K., & Yadav, D. K., (2018). Biomass and carbon storage pattern in natural and plantation forest ecosystem of Chhattisgarh, India. *Journal of Forest and Environmental Science, 34*(1), 1–11. doi: 10.7747/JFES.2018.34.1.1.
Jhariya, M. K., (2014). Effect of forest fire on microbial biomass, storage and sequestration of carbon in a tropical deciduous forest of Chhattisgarh. *PhD Thesis* (p. 259). I. G. K. V., Raipur (C. G.).
Jhariya, M. K., (2017). Vegetation ecology and carbon sequestration potential of shrubs in tropics of Chhattisgarh, India. *Environmental Monitoring and Assessment, 189*(10), 518. https://doi.org/10.1007/s10661–017–6246–2 (Accessed on 10 October 2019).
Jhariya, M. K., Banerjee, A., Yadav, D. K., & Raj, A., (2018). Leguminous trees an innovative tool for soil sustainability. In: Meena, R. S., Das, A., Yadav, G. S., & Lal, R., (eds.), *Legumes for Soil Health and Sustainable Management* (pp. 315–345). Springer, ISBN 978–981–13–0253–4 (eBook), ISBN: 978–981–13–0252–7 (Hardcover). https://doi.org/10.1007/978–981–13–0253–4_10 (Accessed on 10 October 2019).
Jhariya, M. K., Bargali, S. S., & Raj, A., (2015). In: Miodrag, Z., (ed.), *Possibilities and Perspectives of Agroforestry in Chhattisgarh, Precious Forests - Precious Earth* (pp. 237–257). ISBN: 978–953–51–2175–6, In-Tech, doi: 10.5772/60841.
Jimenez, M., Finegan, B., Herrera, B., Imbach, P., & Delgado, D., (2009). *Resiliencia de laszonas de vida de Costa Rica al Cambioclimático*. Presentación XIII World Forestry Congress, Argentina.
Kauppi, P., Sedjo, R. J., Apps, M., Cerri, C., & Fujimori, T., (2001). Technical and economic potential of options to enhance, maintain and manage biological carbon reservoirs and geo-engineering. In: Metz et al., (ed.), *Mitigation: The IPCC Third Assessment Report*. Cambridge University Press.
Kittur, B., Jhariya, M. K., & Lal, C., (2014b). Is the forest fire can affect the regeneration and species diversity. *Ecology, Environment and Conservation, 20*(3), 989–994.
Kittur, B., Swamy, S. L., Bargali, S. S., & Jhariya, M. K., (2014a). Wildland fires and moist deciduous forests of Chhattisgarh, India: Divergent component assessment. *Journal of Forestry Research, 25*(4), 857–866, doi: 10.1007/s11676–014–0471–0.
Kurz, W. A., Stinson, G., Rampley, G. J., Dymond, C. C., & Neilson, E. T., (2008). Risk of natural disturbances makes future contribution of Canada's forests to the global carbon cycle highly uncertain. *Proceedings of the National Academy of Sciences, 105*(5), 1551–1555.
Lucier, A., Ayres, M., Karnosky, D., Thompson, I., Loehle, C., Percy, K., & Sohngen, B., (2009). Forest responses and vulnerabilities to recent climate change. In: Seppala, R., Buck, A., & Katila, P., (eds.), *Adaptation of Forests and People to Climate Change*. IUFRO World Series 22.

Mayle, F. E., & Power, M. J., (2008). Impact of a drier early–mid Holocene climate upon Amazonian forests. Philosophical transactions of the royal society B. *Biological Sciences, 363*(1498), 1829–1838.

Mendoza, F., Chévez, M., & González, B., (2001). Sensibilidad de laszonas de vida de Holdridge en Nicaragua en funcióndelcambioclimático. *Revista Forestall Centro Americana, 33*, 17–22.

Metz, B. O., Davidson, P., Bosch, R. D., & Meyer, L., (2007). *Climate Change 2007-Mitigation of Climate Change.*

Nepstad, D. C., Stickler, C. M., Soares-Filho, B., & Merry, F., (2008). Interactions among Amazon land use, forests and climate: prospects for a near-term forest tipping point. *Philosophical Transactions of the Royal Society B. Biological Sciences, 363*(1498), 1737–1746.

Ollinger, S., Goodale, C., Hayhoe, K., & Jenkins, J. P., (2008). Potential effects of climate change and rising CO2 on ecosystem processes in northeastern U. S. forests. *Mitigation and Adaptation Strategies for Global Change, 13*(5), 467–485.

Osman-Elasha, B., Parrotta, J., Adger, N., Brockhaus, M., Pierce Colfer, C. J., Sohngen, B., Dafalla, T., Joyce, L. A., Nkem, J., & Robledo, C., (2009). Future socioeconomic impacts and vulnerabilities. In: Seppala, R., Buck, A., & Katila, P., (eds.), *Adaptation of Forests and People to Climate Change*. IUFRO World Series 22.

Painkra, G. P., Bhagat, P. K., Jhariya, M. K., & Yadav, D. K., (2016). Beekeeping for poverty alleviation and livelihood security in Chhattisgarh, India. In: Sarju, N., & Sudhir, K. R., (ed.), *Innovative Technology for Sustainable Agriculture Development* (pp. 429–453). ISBN: 978–81–7622–375–1. Biotech Books, New Delhi, India.

Parker, C., Mitchell, A., Trivedi, M., & Mardas, N., (2008). *A Guide to Governmental and Non-Governmental Proposals for Reducing Emissions from Deforestation and Degradation.* The Little REDD Book: A Guide to Governmental and Non-Governmental Proposals for Reducing Emissions from Deforestation and Degradation.

Postel, S. L., & Thompson, B. H., (2005). Watershed protection: Capturing the benefits of nature's water supply services. *Natural Resources Forum, 29*(2), 98–108.

Raj, A., Jhariya, M. K., & Bargali, S. S., (2018a). Climate smart agriculture and carbon sequestration. In: Pandey, C. B., Mahesh, K. G., & Goyal, R. K., (eds.), *Climate Change and Agroforestry: Adaptation Mitigation and Livelihood Security* (pp. 1–19). ISBN: 9789–386546067. New India Publishing Agency (NIPA), New Delhi, India.

Raj, A., Jhariya, M. K., & Harne, S. S., (2018). Threats to biodiversity and conservation strategies. In: Sood, K. K., & Mahajan, V., (eds.), *Forests, Climate Change and Biodiversity* (pp. 304–320, 381). Kalyani Publisher, India.

Rathore, A. K., Jhariya, M. K., Jain, R., & Kumar, S., (2014). Agriculture: Cause, victim as well as mitigator of climate change. *Ecology, Environment and Conservation, 20*(3), 995–1000.

Rosenzweig, C., Casassa, G., Karoly, D. J., Imeson, A., Liu, C., Menzel, A., Rawlins, S., Root, T. L., Seguin, B., & Tryjanowski, P., (2007). Assessment of observed changes and responses in natural and managed systems. In: Parry, M. L., Canziani, O. F., Palutikof, J. P., Van Der Linden, P. J., & Hanson, C. E., (eds.), *Climate Change 2007: Impacts, Adaptation and Vulnerability, Contribution of Working Group II to the Fourth Assessment Report of the Intergovernmental Panel on Climate Change*. Cambridge University Press, Cambridge, UK.

Saigusa, N., Yamamoto, S., Hirata, R., Ohtani, Y., Ide, R., Asanuma, J., Gamo, M., Hirano, T., Kondo, H., Kosugi, Y., Li, S. G., Nakai, Y., Takagi, K., Tani, M., & Wang, H., (2008). Temporal and spatial variations in the seasonal patterns of CO2 flux in boreal, temperate, and tropical forests in East Asia. *Agricultural and Forest Meteorology, 148*(5), 700–713.

Sajwaj, T., Harley, M., & Parker, C., (2008). Report to the office of climate change. EAT/ENV/R/2623/Issue. *Eliasch Review: Forest Management Impacts on Ecosystem Services, 1*, 1–31.
Sánchez, C. O., (2009). El pagoporserviciosambientalesdel Fondo Nacional de Financiamiento Forestal (FONAFIFO), unmecanismoparalograr la adaptación al cambioclimático en Costa Rica. In: Sepúlveda, C., & Ibrahim, M., (eds.), *Políticas y Sistemas de Incentivospara el fomento y adopción de Buenasprácticasagrícolas, Comounamedida de Adaptación al Cambioclimático en América Central*. Serietécnica No. 37, CATIE, Turrialba, Costa Rica.
Scher and Sthapit, (2009). *Mitigating Climate Change Through Food and Land Use, Word Watch Institute, Report 179*. http://www.wocan.org/files/all/mitigating_cc_through _food_and_land_use.pdf (Accessed on 10 October 2019).
Scialabba, N. E. H., (2007). Organic agriculture and food security. In: *Food and Agriculture Organization Paper* (p. 25). Rome: FAO.
Scott, P. A., Stone, D. A., & Allen, M. R., (2003). Human contribution to the European heat wave of 2003. *Nature, 432*, 610.
Shaffer, G., Olsen, S. M., & Pederson, G. O. P., (2009). Long-term ocean oxygen depletion in response to carbon dioxide emissions from fossil fuels. *Nat. Geosci., 2*, 105–109.
Sileshi, G., Akinnifesi, F. K., Ajayi, O. C., Chakeredza, S., Kaonga, M., & Matakala, P., (2007). Contributions of agroforestry to ecosystem services in the Miombo eco-region of Eastern and Southern Africa. *Afr. J. Environ. Sci. Tech., 1*, 68–80.
Simms, A., John, M., & Hannah, R., (2004). *Up in Smoke? New Economics Foundation*. http://www.neweconomics.org (Accessed on 10 October 2019).
Singh, N. R., & Jhariya, M. K., (2016). Agroforestry and agrihorticulture for higher income and resource conservation. In: Sarju, N., & Sudhir, K. R., (ed.), *Innovative Technology for Sustainable Agriculture Development* (pp. 125–145). ISBN: 978–81–7622–375–1. Biotech Books, New Delhi, India.
Smith. P., Martino, D., Cai, Z., Gwary, D., Janzen, H., Kumar, P., McCarl, B., Ogle, S., O'Mara, F., & Rice, C., (2008). Greenhouse gas mitigation in agriculture. *Philosophical Transactions of the Royal Society B: Biological Sciences, 363*(1492).
Stadtmüller, T., (1994). El impactohidrológicodelmanejoforestal de bosquesnaturalstropicales: medidasparamitigarlo. Unarevisiónbibliográfica. Serietécnica, informetécnico No. 240, CATIE, Turrialba, Costa Rica.
Stern, A., (2009). *Blueprint for a Safer Planet*. London: The Bodley Head.
UNEP, (2009). *Towards Sustainable Production and Use of Resources: Assessing Biofuels*. Towards sustainable production and use of resources: Assessing biofuels.
UNFCCC, (2007). *Investment and Financial Flows to Address Climate Change*. UNFCCC, Bonn, Germany.
Yadav, D. K., Ghosh, L., & Jhariya, M. K., (2017). *Forest Fragmentation and Stand Structure in Tropics: Stand Structure, Diversity and Biomass* (p. 116). Lap Lambert Academic Publishing. Heinrich-Bocking-Str. 6–8, 66121, Saarbrucken, Germany. ISBN: 978–3–330–05287–1.
WRI, World Resources Report 2010-2011. World Resources Institute. ISBN: 978-1-56973-774-3 (2011). http://www.wri.org (Accessed on 10 October 2019).

CHAPTER 11

Solid Waste Management Scenario in Ambikapur, Surguja, Chhattisgarh: A Sustainable Approach

KEERTI MISHRA,[1] ARNAB BANERJEE,[2] MADHUR MOHAN RANGA,[3] MANOJ KUMAR JHARIYA,[4] DHIRAJ KUMAR YADAV,[4] and ABHISHEK RAJ[5]

[1]*Post Graduate Student, University Teaching Department, Department of Environmental Science, Sarguja Vishwavidyalaya, Ambikapur, Chhattisgarh–497001, India*

[2]*Assistant Professor, University Teaching Department, Department of Environmental Science, Sarguja Vishwavidyalaya, Ambikapur, Chhattisgarh–497001, India, Mobile: +919926470656, E-mail: arnabenvsc@yahoo.co.in*

[3]*Professor, University Teaching Department, Department of Environmental Science, Sarguja Vishwavidyalaya, Ambikapur, Chhattisgarh–497001, India*

[4]*University Teaching Department, Department of Farm Forestry, Sarguja Vishwavidyalaya, Ambikapur, Chhattisgarh–497001, India, E-mail: manu9589@gmail.com (M. K. Jhariya)*

[5]*PhD Scholar, Indira Gandhi Krishi Vishwavidyalaya, Raipur, Chhattisgarh, India*

ABSTRACT

Waste management is a major issue and challenge from an environmental perspective across the world. It is a systematic process in which waste items from different sources were collected, segregated, and managed in a systematic way. Solid waste comprises of a variety of items under the category of dry and wet waste. Waste items originate from various sectors such as medical, healthcare facilities, market place, agriculture, industry, domestic, and commercial sectors. Accumulation and mismanagement of waste often leads to groundwater contamination as well as spread of diseases. Therefore, proper management of waste is the need of the hour. Processes such as incineration, open burning, dumping of waste, vermin-composting are being widely practiced across the world for effective management of waste. Management policy involves reduce, reuse, refuse, and recycling practices for effective management of waste items. The present investigation reports the different waste items and their subsequent management in Ambikapur Township, Surguja through the establishment of various SLRM centers. In the study area, a diverse solid waste management system is in practice. Open dumps are responsible for so many negative environmental impacts in the study area. Due to lack of funding and proper management, the existing solid waste management system is not working successfully in Ambikapur city. Due to the shortage of storage bins, collection efficiency is very low in the study area. Special wastes like hospital waste and other hazardous materials are being disposed of along with the municipal solid waste. Considering the overall negative impacts associated with open dumping and open burning, these practices must be strongly discharged.

11.1 INTRODUCTION

Management of waste items helps towards maintaining sustainability of environment, economy, and human health. The major problem of managing solid waste includes its proper collection, transportation to respective site, and their final disposal in third world nations. As per research reports it was observed that solid waste is becoming a biggest issue for Pakistan due to improper planning and funding (Ejaz and Janjua, 2012).

Throughout the world, it has observed that human population growth, increased rate of urbanization, changing land use and food habits are the significant factors contributing significantly towards increase in the volume of solid waste. There is a rapid growth of commercial appliances such as shopping malls, cafeteria, and theatre. They are also contributing significantly towards solid waste accumulation (Saxena and Oraon, 2016).

Solid waste includes waste material of solid and semi-solid nature excluding the night soil. Solid waste comprises of organic as well as inorganic material. It is an integrated process for waste collection, treatment in eco-friendly way and ultimately their final safe disposal so that it does not poses any harm to human civilization and environment. The entire management process includes waste management during generation, proper storage, collection at primary level, temporary storage at respective sites and their subsequent movement towards disposal site or treatment site for the waste (Kumar and Pandit, 2013).

Typically the term solid waste is defined as items without future use. Reuse of solid waste may not be possible always as some may pose hazard towards well-being of human beings. Waste items such as vegetable part, cooked item, fruits may leads in contamination of pathogenic microorganisms (Rajput et al., 2009).

There are diverse sources of solid waste, which includes waste generated from residence, commercial hubs, factories, construction, and demolition activities, farming activities, treatment plants and other sources. Solid waste collected by Municipal Corporation is different in their nature as well as from the source from where they had been collected (Yadav and Linthoingambi, 2009).

11.1.1 GENERAL DESCRIPTION

Substances with no future use possibility are considered as solid waste. It can be considered as a direct pollutant in terms of its impact as well as it also exhibits some indirect or secondary impacts also. The solid waste has a diverse source of origin coming from commercial units such as market places, shops as well as industrial waste material, waste items generated from day to day use from residential complexes, etc. which ultimately accumulates beside roads and other public places. Solid waste is being generated from ancient times, but the nature to cause harm to human civilization has changed completely. The main reason behind this is that solid waste of the ancient time period was biodegradable in nature. With the gradual development of science and technology, the biodegradable nature has changed into non-biodegradable, highly dangerous, toxic, and hazardous in nature. This has promoted to explore the world of waste processing and its safer disposal (Dwivedi et al., 2014).

11.1.2 WHAT IS WASTE?

Any unwanted discarded material which is of no future use is considered as waste.

1. **What is Solid Waste:** Any unwanted, discarded, useless material in solid-state. It also includes waste items as food materials as well as sludges originating in the municipality (Singh et al., 2014).

2. **Characteristics of Solid Waste:** Proper information in terms of physic-chemical characterization of solid waste is very much essential in order to design the treatment facility and proper management of solid waste along with recovery options. The solid waste components vary depending upon the location as well as the time of collection. It may include materials such as paper, glass material, food items, cloth material, etc. (DCC, 1999). Sources may be of different types, such as industrial, institutional, domestic, and commercial sectors. Some of the waste items capable of showing significant deleterious impact are considered to be hazardous waste.

11.2 TYPES OF WASTE

11.2.1 SOLID WASTE

These are the type of items which is of no use and therefore discarded. This waste includes items generated from factories, agriculture, urban sector, wastes generated from medical health facilities as well as nuclear waste. Alternately solid waste is also known as refuse.

11.2.2 LIQUID WASTE

These are the waste items usually generated from flushing, cleaning, and processing in industries. They are also known as sewage. The major problem associated with them includes disposal of liquid waste items to nearby surface water bodies such as a river, water channels, etc., without and treatment. This would, therefore, pose a significant level of damage to the particular ecosystem.

11.2.3 GASEOUS WASTE

The waste items which are generated in gaseous form through vehicular emission, emission from industrial chimneys as exhaust material, fossil fuel burning, and therefore, gets easily mixed with the atmosphere causing significant level of air pollution. Gaseous emissions include SO_2, carbon dioxide, ozone, NO_2, and greenhouse gas, such as CH_4.

On the basis of the nature of the waste, i.e., dry, and wet are given below in Table 11.1.

TABLE 11.1 Composition of Dry and Wet Waste

Dry Waste	Wet Waste
1. Cardboards	1. Medical waste
2. Metals items	2. Food packets
3. Plastic items	3. Cattle enable items
4. Rubber and Leather items	4. Cattle non-enable items
5. Tablet covers	5. Food waste
6. Non Recyclable items	6. Wet paper/cardboards
7. Carbon papers	7. Plastic/washing items
8. Papers	8. Egg Shells
9. Plastic covers	9. Bone
10. Plastic cops	10. Non-recyclable items
11. Plastic/Alu items	11. Non-veg items
12. Hair	12. Citric fruits and peel
13. Tharmacoal	13. Organic matter
14. Organic waste	14. Twigs
15. Medical waste	15. Garden waste
16. Wet paper/Cardboards	
17. Waste Bottles glasses	

11.3 SOURCES OF WASTE

11.3.1 MUNICIPAL WASTE

As defined by the Municipal solid wastes (Management and Handling) Rules 2000 under Environment Protection Act, 1986 enacted by Government of India (GoI) municipal waste comprises of waste items generated from various commercial sectors and domestic sectors including bio-medical waste that has undergone treatment and excluding dangerous industrial waste posing significant damage to human civilization in solid or semi-solid form. These waste items are biodegradable and decomposable in nature and mostly comprises of various food items, vegetable waste along with waste generated in the hotels, shops, offices, and other commercial sectors.

11.3.2 MEDICAL WASTE

These include waste items generated from medical healthcare facilities such as nursing homes, polyclinic, medical testing laboratories, and hospitals. Other items include human anatomical, pathological, infectious materials. These waste items are the consequences of treatment, medical diagnosis of diseases as well as immunization programme, research activities. Wide variety of waste items such as outdated and expired medicines, cottons, bed linens, scalpel, blades, and disposable needles, bandages, human excretory material, human body parts and body fluids. Such waste items pose high danger for human health if they are not managed scientifically and therefore can spread the contamination easily.

11.3.3 AGRICULTURAL WASTE

This includes waste items originating from various sectors of agriculture and animal husbandry houses. It comprises of waste items organic in origin such excretory products of animals named as slurries, FYM (Farmyard manure), compost made up from spent mushroom along with waste items such as medicines used in veterinary, scrap machinery pesticides and other items. The potential hazards associated with agricultural waste items include eutrophication of surface water body due to nutrient run-off from agricultural waste material. Modern days chemical-based farming systems lead to emission of ammonia contributing to soil acidification and methane-a potential greenhouse gas.

11.3.4 INDUSTRIAL WASTE

These are the waste items generated during the manufacturing and processing steps within a factory, industry, mills, and from mines. Examples of this category of waste include various toxic and hazardous waste, organic, and inorganic chemical solvents, paints, paper products, waste items of radioactive nature, and other products. To treat industrial waste material conventional sewage treatment plants are being build which are capable only to disintegrate BOD (Biochemical Oxygen Demand). Rest of toxic and hazardous items were treated under specially designed treatment items.

11.3.5 DOMESTICS WASTES

These are waste items that are non-putrescible in nature and are therefore easily combustible. This includes cardboard, paper, yard waste< wood material, and other allied products.

11.3.6 COMMERCIAL WASTE

With gradual development of science and technology, rapid urbanization, improved vehicular transport system, huge amount of waste items are generated per day basis. These include commercial wastes, construction, and demolition waste, biomedical waste items from health care clinic which are hazardous, toxic in nature, disposable, and chemical compounds, etc. such wastes are often disposed in areas that may cause death and damage to human civilization to a considerable extent. Apart from these waste items waste generated due to mining and nuclear power plant may pose significant level of danger for the entire society and human civilization (Rajput et al., 2009).

11.4 CLASSIFICATIONS OF WASTES

On the basis of physical, chemical and biological characteristics and their nature waste items are categorized in the following types.

11.4.1 BIODEGRADABLE WASTES

Biodegradable wastes are the waste items that are organic in nature and are readily degraded and decomposed by micro-organisms. Waste generated in the meat shops, slaughterhouses, sugar mill industry, rice mill, cotton mill, paper mill industry are considered to be the biodegradable waste. Waste items generated from these industries are frequently reused. However, such waste items in excess can act as potential pollutant and they become less biodegradable.

11.4.2 NON-BIODEGRADABLE WASTES

These are mostly inorganic in nature and therefore not degraded by microorganisms. However, they can undergo physic-chemical disintegration through oxidation and auto-dissociation. Waste items generated from petroleum oil refineries, colliery operations, oil residue in sludges, dry inert solid waste items can be kept in this category. Fly ash generated from the thermal power plant accumulates in the ecosystem and move through ecosystem and ultimately enters biogeochemical cycling. Other examples include heavy metals, pesticides, and organo-mercuric compounds.

11.4.3 TOXIC WASTES

This includes waste items of chemical in nature such as Hg (mercury) and As (arsenic).

11.4.4 NON-TOXIC WASTE

This includes waste items which does not pose threat to human civilization or environment and mostly includes domestic waste items such as biodegradable paper rags, boards, etc.

11.4.5 BIOMEDICAL WASTES

Wastes generated in various healthcare units such as polyclinic, nursing homes, medical health care centers, microbiology, and biotechnology laboratories, hospitals, etc.

11.5 ENVIRONMENTAL IMPACT OF SOLID WASTE

Improper landfilling of waste items may pose significant threat for human health, environment, ground, and surface water pollution, spreading of diseases and odor pollution by various vectors such as insects, rats, etc. It may emit obnoxious gases through decomposition of waste items (Aljaradin and Persson, 2012). Waste are such items which creates vision pollution as well as all sorts of environmental pollution, causes water clogging leading to flash flood, causes serious public health hazard and causes alternate land uses practices. In country like India open burning is a common scenario (Thomas-Hope, 1998). Burning of waste releases toxic, obnoxious gases. Water may enter the waste dump and through percolation can generate toxic hazardous leachate contaminating the various sources of fresh water. Waste dumps becomes the active breeding site for mosquitoes < flies < cockroaches< rats and other diseases carrying vectors. Fedorak and Rogers (1991) mentioned three modes of microbial contamination from waste disposal site. Firstly through leachates in the groundwater, secondly through airborne particles and thirdly through biological life forms which come in contact with pathogenic microorganisms (Fedorak and Rogers, 1991).

11.6 NEED FOR MANAGEMENT OF SOLID WASTE

a. To regulate the various types of environmental pollution.
b. To prevent the spreading of contamination of infectious diseases.
c. To go for natural resource conservation such as forest, water, air, minerals, etc.

11.7 WHAT IS SOLID WASTE MANAGEMENT?

A SWM (solid waste management) system comprises of sequential steps of waste collection, storage, segregation, processing including recycling, treatment, etc., mobilization, and ultimately eco-friendly disposal (Annepu, 2012). Modern approaches include volume reduction of the waste items which requires land-filling as well as resource recovery from the waste item to the maximum possible extent.

11.8 WHAT IS WASTE DISPOSAL?

Comprehensive database regarding collection of waste disposal data is a challenging issue in the field of waste management. Most countries relies on waste disposal data produced income levels and regional difference. The methodology for calculation of disposal for each category was not consistent. In some places, waste disposal percentage for dumping, landfilling was only given or amount that belongs to other categories of disposal. In most cases compostable and recyclable materials were removed from disposable items (Hoornweg and Perinaz, 2012). The term disposal includes waste from all different processes such as production, processing, manufacturing as well as waste generated from various other sectors (Hoornweg and Perinaz, 2012).

11.9 METHOD OF WASTE DISPOSAL

11.9.1 OPEN BURNING

It is a process where waste items are combusted in air so that the gaseous emissions are directly released into the air. It is a widely practiced waste management strategy throughout India. Open burning is widely practiced in India due to some specific reasons: (a) rag pickers often burn waste for metal recovery from mixed waste items; (b) open burning of the waste items collected in bins to make them free of waste in order for future use; and (c) burning of waste items for getting warmness at cold winter night (Annepu, 2012).

11.9.2 INCINERATION

This is a process applicable for waste management only when land-filling operation is not giving good results. Moreover, waste items containing highly combustible material such as plastics, paper can be combusted through incineration.

However, incineration is always performed in the medical healthcare facilities such as hospitals and for other wastes of biological origin (Kansal, 2002). Air emissions from smokestacks of incinerator may have multifaceted effects such as emission of nitrogen and sulfur oxides may promote acid rain formation, release of heavy metals in gaseous form and carbon dioxide as global warming. However, modern day's incinerator is fitted with pollutant trapping devices but the entire process seems to be highly costly (Botkin and Keller, 2000).

11.9.3 COMPOSTING

Composting as a technological process has its origin since ancient times. It is simply decomposition of waste items mostly organic matter by microorganisms under moist wet both aerobic and anaerobic process. Compost made up from cow dung and agro-waste is being frequently used as compost by the farming community. Urban heterogeneous waste material tends to have higher nutrient content in comparison to compost used by the farming community. Plants usually use ammonium (NH_4) is the form of nitrogen which gets converted into nitrates (NO_3) by bacterial transformation-a process known as nitrification (Asnani, 2006).

11.9.4 PYROLYSIS

The process of destructive distillation is widely used to recover the chemical energy and constituents from organic wastes. The process takes place under anaerobic condition where organic waste materials are chemically disintegrated. The process operates at a temperature of about 430°C.

11.9.5 SANITARY LANDFILL

This is an engineering approach which is adopted through specially engineered structures in order to avoid the hazards of open dumping. In a pre-orderly manner site for building up off sanitary landfill requires proper site selection, proper preparation and management so that pollution due to leakage of sanitary landfill leachate can be prevented. Under this process waste items is disposed in a pit and covered with a layer of soil after each day's operation. In this way all the waste items are being isolated from the activities of insect and rodent population along with other soil-dwelling forms. Such configuration also helps to prevent leachate generation through movement of surface water within the waste items as well as emission of obnoxious toxic gases escaping from the waste items. It is the most scientific approach as all the other waste treatment process leaves some residue problem which can be effectively removed through sanitary landfilling process. As a consequence of this, landfilling would be the widely adopted option in India for upcoming times which requires technological up-gradation (Kansal, 2002).

11.9.6 VERMICOMPOSTING

Organic nature of MSW (Municipal solid waste) makes it a suitable candidate for vermicomposting process as most of the municipal solid waste is organic in nature. The process involves conversion of waste items into nutrient-enriched worm castings and has been widely manufactured and utilized in various big cities of India. During the process, waste items are used as food items which earthworm feed upon and convert them into vermicompost (Sharholy et al., 2008).

11.9.7 RECYCLING

It is the process of collection and reuse of waste items. Separate collection of recyclable materials by using separate beans and collecting equipments are known as carbide collection. Owner of the waste segregation unit may play active part in this waste segregation process. Recycling is such a process which involves conversion of discarded materials into newer product through proper processing, recycling process may lead to saving of money, cutting of trees and reduction of mining activities (Pappu et al., 2007).

11.10 THE POLICY OF 4R'S

1. **Refuse:** This is a strategy to control waste generation by avoiding purchasing new container from market or other items, and use that are already present in the house.
2. **Reuse:** Try to go for reutilization of waste product as much as possible. For example, soft drink cans and plastic bottles can be effectively used as pen stand or flower vases.
3. **Recycle:** Utilize items that can go through recycling process. For example, using jute bags or cloth bags can be used again and again we can ourselves take part in recycling process.
4. **Reduce:** Try to reduce the waste generation as much one can do. For example, carrying own bag in the market reduces the need of bags to be taken from others (Rajput et al., 2009).

11.11 SOLID WASTE MANAGEMENT SCENARIO IN INDIA

The population explosion with rapid industrialization is the biggest challenge towards sustainability of the world. Among the various challenges the biggest one is the management of municipal solid waste. Under municipal waste items it includes garbage material, rubbish; waste items from commercial, domestic, and industrial sectors. Garbage is the wet items such as vegetables, meat, and various other food items. Rubbish comprises of paper, plastic, textile, and glass items. Trash material includes appliances, furniture, and its parts, discarded mattresses. As a consequence effective monitoring and resource recovery needs to be done more effectively. One step towards such problem includes studying the role of increasing human population towards waste generation through system dynamics modeling in order to screen the best suitable alternative which could be applied to manage the waste (Pai et al., 2014). It is the need of the hour from the Indian perspective to achieve sustainable development.

In the sector of waste management, India is at the back front from world perspective. This is due to the fact that open dumping and uncontrolled burning are the most common practice of waste management in India. Such approaches cause significant level of pollution and health issues. India happens to be the largest emitter of CH_4 gas through its waste disposal activities. As methane is a potential greenhouse gas, therefore, it would have significant contribution towards global climate change (CC). It has been also observed that inappropriate treatment of waste items may affect the health of the people through contamination of toxic compounds emitted during burning along with leaching of toxic substances in the groundwater through landfill leachate (Axelsson and Theres Kvarnström, 2010).

11.12 SOLID WASTE MANAGEMENT SCENARIO IN SARGUJA, CHHATTISGARH

The municipality is the nodal agency to govern the management of solid waste facilities where proactive communities play a key role towards solid waste management planning. Integrated Solid Waste Management (ISWM) is an integrated system of prevention of waste generation, recycling of waste items, composting of biodegradable waste items along with their eco-friendly disposal. ISWM is an approach to protect human civilization and environment from pollution and other dangers (Lanjewar et al., 2014). It is a universal truth that waste in any form is not acceptable in this world. But in real situation we see wastage of many items such as water, energy, and power in our day-to-day life. Environment is the natural shield around us that protects us from all sorts of natural disorders. Anthropogenic activities day to day is creating pollution to our environment in terms of air, water, soil pollution. Therefore, we should take good care to protect our environment (Jharia, 2014).

11.13 MATERIALS AND METHODS

11.13.1 STUDY AREA

Study site-I is located to main road of Navapara. This area is very suitable of solid waste management. Management of the solid waste was done through green and red basket, in which wet and dry trash were kept. The ground area of this center is very low due to which the members face difficulty in making solid waste management. Study site-II Godhanpur SLRM center is a little away from the residential area, as many of the solid waste coming in it are from the Vasundhara colony and surrounding houses. Study site-III New bus stand the SLRM center is located beside the main road in which 23 members work. Here the solid waste is collected from the surrounding area. It is the only SLRM center in which organic waste is used to make biogas. Study site-IV Darripara is located at the route of new bus stand; its area is in the middle of residential area. The waste of the entire Darripara comes in this SLRM center, the area of this center is very small due to which the members face difficulty in making solid waste management. Study site-V D.C. Road is located on the roadside beside the Hanuman temple. The area of this center is very large, in which the area of the market falls short. Collection of the solid waste was done in green and red basket, in which wet and dry trash were kept respectively.

Study site-VI Khalpara is located in the middle of the settlement and the solid waste collected from these areas is brought to the SLRM center. Study site-VII Namnakala is located the ring road, where the waste of the entire Namnakala comes in this SLRM center. The ground area of this center is very small due to which the members face difficulty in making solid waste management. Study site-VIII Deviganj SLRM center is located in the middle of the market. In this center, a large amount of solid waste comes from the market, as well as a substantial amount of solid waste were collected from the hotels. The waste collection vehicles are available in the appropriate amount at this SLRM center. Study site-IX Thanganpara SLRM center in located near company market. In this SLRM center, highest amount of solid waste collection is happens because this center is surrounded by markets as well as hotels and house. In this center, solid waste in collected from ward number 30, 31, 27, 30, and commercial area as well as from company market. Study site-X Gangapur is located beside the Shiva temple. The area centers on residential area and market. In this SLRM center, solid waste is collected from the residential area only. Study site-XI Old bus stand is located in the middle of the old bus stand. The ground area of this center is very small because of this reason the management of solid waste is not properly managed. In this center, medical waste is most commonly found (Figure 11.1 to Figure 11.11).

11.14 RESULTS AND DISCUSSION

Among the eleven studied sites wastes of diverse nature were found during the present investigation. In Navapara center items such as plastic, organic matter, carton, paper, bottles, electronic waste, food waste, eggshells, hair, and medical items were recorded. In Godhanpur, center items such as thermocol, organic matter, paper, bottles, glass, food waste, eggshells, and hair were recorded. In new bus stand center items such as paper, plastic, bottle, carton, organic matter, food waste, eggshells, electronic waste were recorded. In Darripara, center items such as paper, plastic, bottle, carton, organic matter, disposal, food waste, food plastic, medical waste, glass were recorded. In D.C. Road center items such as bottle, carton, organic matter, disposal, food waste, eggshells, electronic waste, paper, plastic, hair, and clothes were recorded. In Khalpara, center items such as bag, plastic, mat, bottle, paper, carton, organic matter, shoes, food waste, sleeper, thermocol, and electronic waste were recorded. In Namnakala, center items such as clothes, bag, plastic, mat, bottle, shoes, sleeper, organic matter, cartoon, thermocol were recorded. In Deviganj center road items such as electronic waste, plastic, food waste, clothes, paper, carton, thermocol, glass, organic matter, and bottles. In Thanganpara center, items such as paper, carton, food waste, organic matter, plastic, bottles, medical items, mat, food plastic, and thermocol were recorded. In Gangapur, center items such as paper, plastic, organic matter, shoes, sleeper, disposal, food waste, carton, and food plastic were recorded. In old bus stand center items such as paper, plastic, food waste, disposal, carton, food plastic, bottles, medical items, electronic waste were recorded (Table 11.2).

FIGURE 11.1 **(See color insert.)** Study site-I Patelpara, Navapara. (Lat-23°C156'363," Long 83°C184'705")

FIGURE 11.2 **(See color insert.)** Study site-II Godhanpur. (Lat-23°C156'363," Long-83°C184'705")

FIGURE 11.3 **(See color insert.)** Study site-III New bus stand. (Lat-23°C115'237," Long-83°C189'829")

FIGURE 11.4 **(See color insert.)** Study site-IV Darripara. (Lat-23°C110'261," Long-83°C188'839")

FIGURE 11.5 **(See color insert.)** Study site-V D.C. road. (Lat-23°C128'679," Long-83°C201'949")

FIGURE 11.6 **(See color insert.)** Study site-VI Khalpara. (Lat-23°C143'365," Long-83°C188'564")

FIGURE 11.7 **(See color insert.)** Study site-VII Namnakala. (Lat-23°C122'914," Long-83°C171'402")

FIGURE 11.8 **(See color insert.)** Study site-VIII Deviganj. (Lat-23°C126'896," Long-83°C189'827")

FIGURE 11.9 **(See color insert.)** Study site-IX Thanganpara. (Lat-23°C126'126," Long-83°C199'959")

FIGURE 11.10 **(See color insert.)** Study site-X Gangapur. (Lat-23°C113'937," Long-83°C175'882")

FIGURE 11.11 **(See color insert.)** Study site-XI Old bus stand. (Lat-23°C113'937," Long-83°C175'882")"

TABLE 11.2 The Various Types of Solid Waste Generated in Various SLRM Centers at Ambikapur, Surguja, Chhattisgarh

Center Name	Name of Items
1. Navapara	Plastic, organic matter, carton, paper, bottles, electronic waste, food waste, eggshells, hair, medical items.
2. Godhanpur	Thermocol, organic matter, paper, bottles, glass, food waste, eggshells, hair.
3. New bus stands	Paper, plastic, bottle, carton, organic matter, food waste, eggshells, electronic waste.
4. Darripara	Paper, plastic, bottle, carton, organic matter, disposal, food waste, food plastic, medical waste, glass.
5. D.C. Road	Bottle, carton, organic matter, disposal, food waste, eggshells, electronic waste, paper, plastic, hair, clothes.
6. Khalpara	Beg, plastic, mat, Bottle, paper, carton, organic matter, shoes, food waste, sleeper, thermocol, electronic waste.
7. Namnakala	Clothes, Beg, plastic, mat, Bottle, shoes, sleeper, organic matter, carton, thermocol.
8. Deviganj road	Electronic waste, Plastic, food waste, clothes, paper, carton, thermocol, glass, organic matter, bottles.
9. Thanganpara	Paper, carton, food waste, organic matter, plastic, bottles, medical items, mat, food plastic, thermocol.
10. Gangapur	Paper, plastic, organic matter, shoes, sleeper, disposal, food waste, carton, food plastic.
11. Old bus stand	Paper, plastic, food waste, disposal, carton, food plastic, bottles, medical items, electronic waste.

Table 11.3 represents the various types of solid waste generated in various SLRM center at Ambikapur, Surguja, Chhattisgarh. Among the eleven studied sites wastes of diverse nature were found during the present investigation. In Navapara center, items such as plastic, organic matter, bottles, eggshells, food waste, medical items, glasses, paper, and electronic waste were recorded. In Godhanpur, center items such as Organic matter, bottles, glass, food waste, plastic, sleeper, garden waste. In new bus stands center items such as paper, plastic, bottle, carton, organic matter, food waste, eggshells, disposal, hair were recorded. In Darripara, center items such as paper, organic matter, food waste, food plastic, glass, and dust. In D.C. Road center items such as bottle, cartoon, disposal, food waste, eggshells, paper, hair, clothes were recorded. In Khalpara, center items such as bag, plastic, bottle, paper, cartoon, organic matter, shoes, food waste, sleeper were recorded. In Namnakala, items such as clothes, bag, plastic, mat, shoes, sleeper, organic matter, paper, and dust were recorded. In Deviganj road, items such as electronic waste, plastic, food waste, paper, cartoon, thermocol, glass, organic matter, bottles were recorded. In Thanganpara, items such as

paper, food waste, organic matter, plastic, mat, food plastic, bottles were recorded. In Gangapur, items such as paper, plastic, organic matter, shoes, sleeper, disposal, food waste, hair, carton, and food plastic were recorded. In old bus stands items such as organic waste, paper, food waste, food waste, disposal, food plastic, bottles, and glasses were recorded.

TABLE 11.3 List of Items Produced from Domestic Source

Center Name	Name of Items
1. Navapara	Plastic, organic matter, bottles, eggshells, food waste, medical items, glasses, paper, electronic waste.
2. Godhanpur	Organic matter, bottles, glass, food waste, plastic, sleeper, garden waste.
3. New bus stands	Paper, plastic, bottle, carton, organic matter, food waste, eggshells, disposal, hair.
4. Darripara	Paper, organic matter, food waste, food plastic, glass, dust.
5. D.C. Road	Bottle, carton, disposal, food waste, eggshells, paper, hair, clothes.
6. Khalpara	Beg, plastic, bottle, paper, carton, organic matter, shoes, food waste, sleeper
7. Namnakala	Clothes, beg, plastic, mat, shoes, sleeper, organic matter, paper, dust.
8. Deviganj road	Electronic waste, plastic, food waste, paper, carton, thermocol, glass, organic matter, bottles.
9. Thanganpara	Paper, food waste, organic matter, plastic, mat, food plastic, bottles.
10. Gangapur	Paper, plastic, organic matter, shoes, sleeper, disposal, food waste, carton, food plastic.
11. Old bus stand	Organic matter, paper, food waste, food waste, disposal, food plastic, bottles, glasses.

Table 11.4 represents the various types of solid waste generated in various markets at Ambikapur, Surguja, and Chhattisgarh. Among the eleven studied sites wastes of diverse nature were found during the present investigation. In Navapara, items such as plastic, bottles, food waste, thermocol, food plastic, and cartoon were recorded. In Godhanpur, no market items present. In new bus stands, items such as plastic, bottle, carton, and disposal were recorded. In Darripara, no market items presents. In D.C. Road items such as bottle, cartoon, disposal, food waste, and paper were recorded. In Khalpara, no market items were present. In Namnakala, items such as plastic, paper, and were recorded. In Deviganj road, items such as electronic waste, plastic, food waste, paper, carton thermocol, glass, bottles were recorded. In Thanganpara, items such as paper, food waste, organic matter, plastic, carton, medical items were recorded. In Gangapur, no market items are present. In old bus stand, items such as paper, food waste, disposal, food plastic bottles, glasses, thermocol, and medical items were recorded.

TABLE 11.4 List of Items Produced from Markets of Different Area

Center Name	Name of Items
1. Navapara	Plastic, bottles, food waste, thermocol, food plastic, carton.
2. Godhanpur	No items
3. New bus stands	Plastic, bottle, carton, disposal,
4. Darripara	No items
5. D.C. Road	Bottle, carton, disposal, food waste, paper.
6. Khalpara	No items
7. Namnakala	Plastic, paper, carton.
8. Deviganj road	Electronic waste, Plastic, food waste, paper, carton, thermocol, glass, bottles.
9. Thanganpara	Paper, food waste, organic matter, plastic, carton, medical items.
10. Gangapur	No items
11. Old bus stand	Paper, food waste, disposal, food plastic, bottles, glasses, thermocol, medical items.

Data presented in Figure 11.12 reflects the level of variation of different types of solid waste items at various SLRM center. Results reveal highest waste accumulation were seen in case of Thanganpara and lowest were recorded in Namnakala.

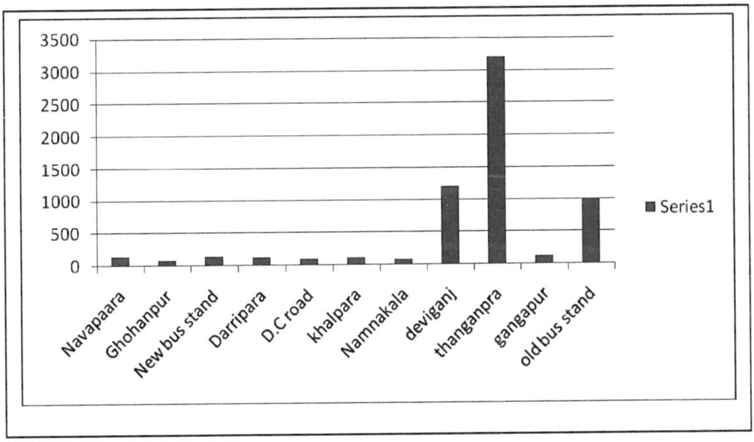

FIGURE 11.12 Variation of amount of solid waste collected from domestic source per day basis at various SLRM center.

Data presented in Figure 11.13 reflects the level of variation of different amount of solid waste collected from market on per day basis. Results reveal highest waste amount accumulation was seen in case of mark of place old bus stand and lowest were recorded in Navapara.

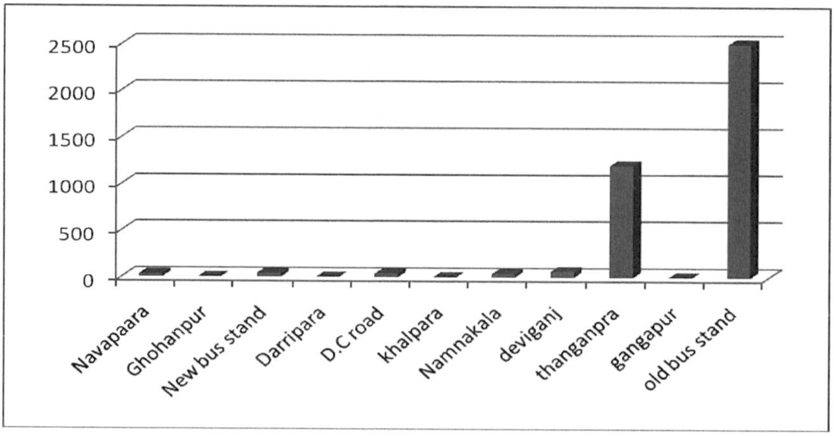

FIGURE 11.13 Variation of amount of solid waste collected from market on per day basis at various SLRM centers.

Data presented in Figure 11.14 reflects the level of variation of different amount of solid waste collected from dry waste on per day basis. Results reveal highest dry waste amount accumulation was seen in case of Thanganpara and lowest were recorded in Namnakala. The amount of dry waste range ranged between 34 kg to 1000 kg.

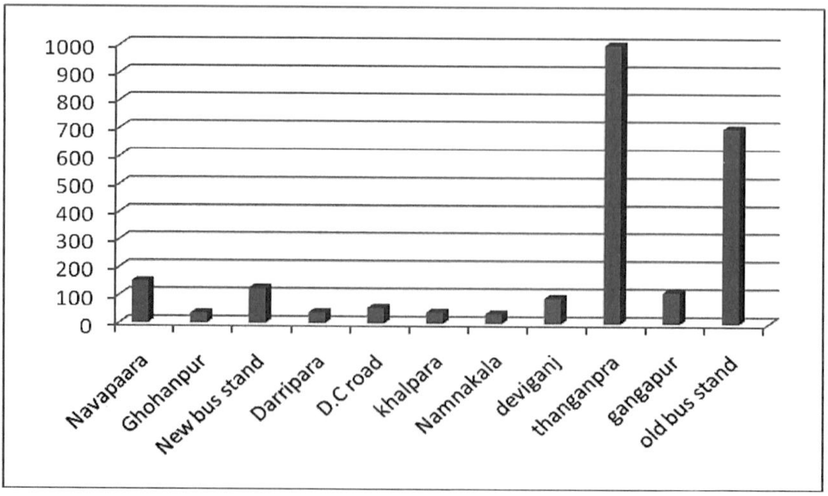

FIGURE 11.14 Variation of amount of generated of dry waste on per day basis of various SLRM center.

Data presented in Figure 11.15 reflects the level of variation of different amount of solid waste collected from wet waste on per day at various SLRM centers. Results reveal highest wet waste amount accumulation was seen in case of Thanganpara and lowest were recorded in Namnakala. The amount of wet waste range ranged between 45 kg to 2200 kg.

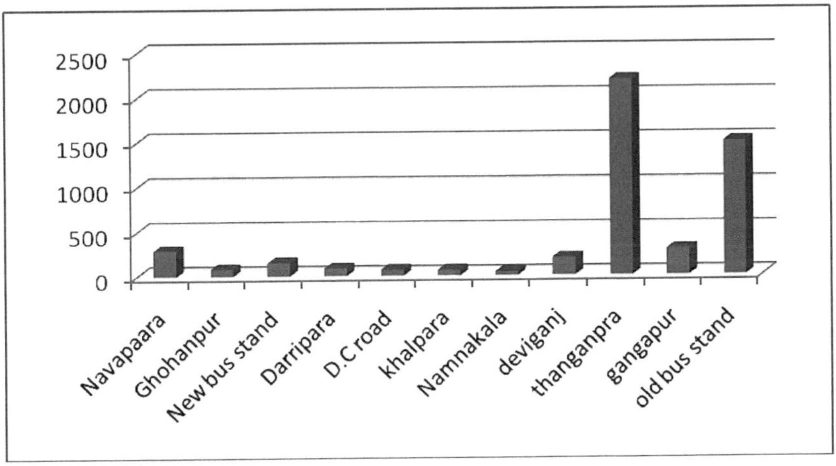

FIGURE 11.15 Variation of amount of collected wet waste on per day basis of various SLRM center.

Data presented in Figure 11.16 reflects the level of variation of different amount of dry solid waste collected on monthly basis at various SLRM center. Results reveal highest dry waste amount accumulation was seen in case of Thanganpara and lowest were recorded in Namnakala. The amount of dry waste range ranged between 1054 kg to 31000 kg.

Data presented in Figure 11.17 reflects the level of variation of different amount of wet solid waste collected per day/monthly at various SLRM centers. Results reveal highest wet waste amount accumulation was seen in case of Thanganpara and lowest were recorded in Namnakala. The amount of wet waste range ranged between 1395 kg to 68200 kg.

Data presented in Figure 11.18 reflects the level of variation of different amount of dry solid waste collected from dry waste on yearly basis at various SLRM centers. Results reveal highest dry waste amount accumulations were seen in the case of Thanganpara and lowest were recorded in Namnakala. The amount of dry waste range ranged between 12,648 kg to 372,000 kg.

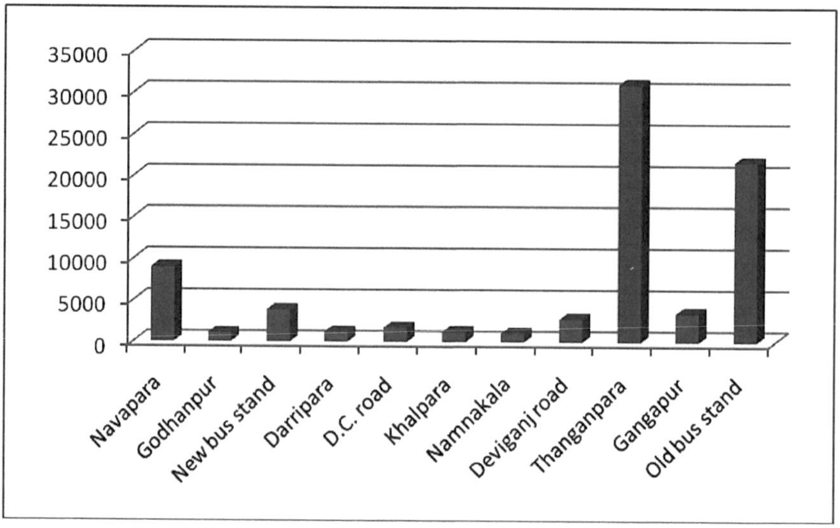

FIGURE 11.16 Variation of amount of collected dry waste on monthly basis at various study sites.

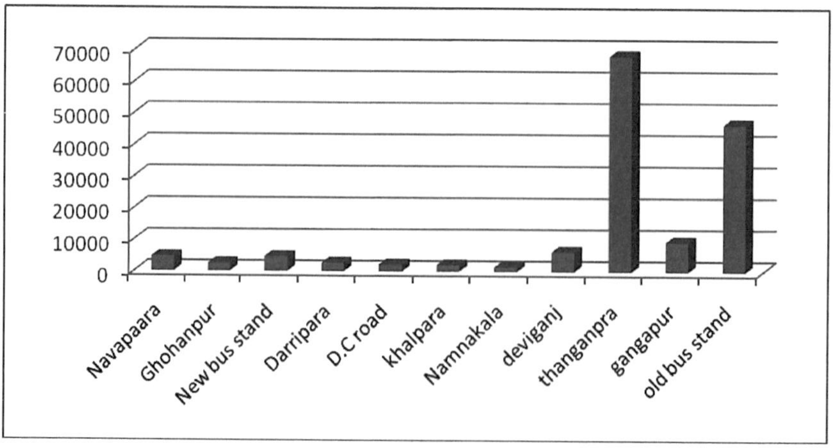

FIGURE 11.17 Variation of amount of collected wet waste on monthly basis at various study sites.

Data presented in Figure 11.19 reflects the level of variation of different amount of wet waste collected on yearly basis among various SLRM centers. Results reveal highest wet waste accumulation were recorded in the case of Thanganpara and lowest were recorded in Namnakala. The amount of wet waste ranged between 16,740 kg to 818,400 kg.

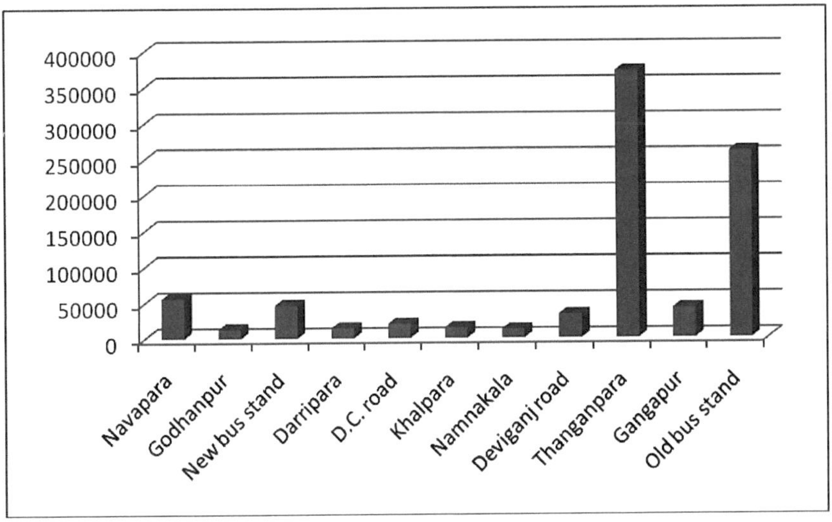

FIGURE 11.18 Variation of dry waste collected amount on yearly basis at various SLRM centers.

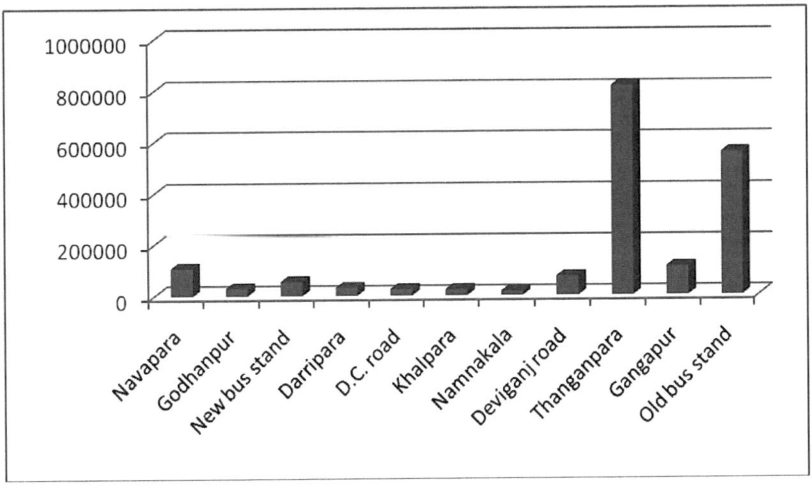

FIGURE 11.19 Variation of wet waste collected amount on yearly basis at various SLRM centers.

Data presented in Figure 11.20 reflects comparative level of dry and wet waste collected on yearly basis at various SLRM centers. Results reveal highest amount wet waste was higher in compare to dry waste. The amount of dry and wet waste percentage ranged between 39.56–60.43%.

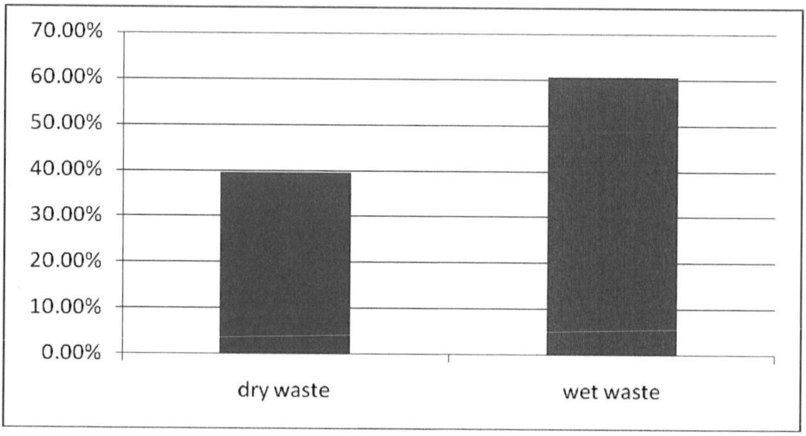

FIGURE 11.20 Comparative statement of annual amount of dry and wet waste on various SLRM centers.

Figure 11.21 represents the percentage contribution of waste items at various SLRM centers of Ambikapur town-ship. From the results, it appears that highest contribution were reflected by metal waste and electronic waste among the various waste items. Lowest contribution was recorded by the pipe items. Depending upon the contribution of different waste items they can be ordered as follows: Metals > Electronic > Hospital items > Glass> Domestic items > Disposal waste > Suttli > Leather items > other plastic items.

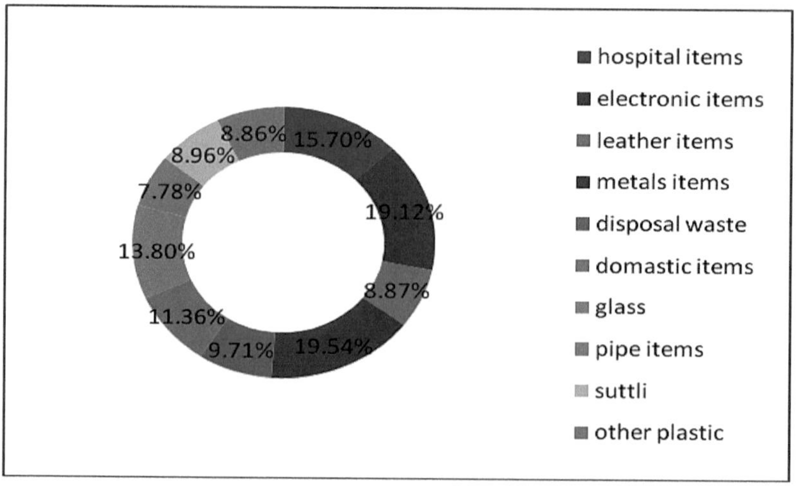

FIGURE 11.21 **(See color insert.)** Item wise contribution of waste material to the total waste load of Ambikapur township.

Data presented in Figure 11.22 reflects four small clusters of different SLRM centers depending upon the waste composition recorded in them. Cluster 1 comprises of Godhanpur and Namnakala, cluster 2 comprises of Darripara and Khalpara, cluster 3 comprises of Godhanpur, Khalpara, and D.C. road, cluster 4 comprises of new bus stand, Gangapur, Deviganj road and Navapara, Thanganpara, and Old bus stand is distantly related with other SLRM center.

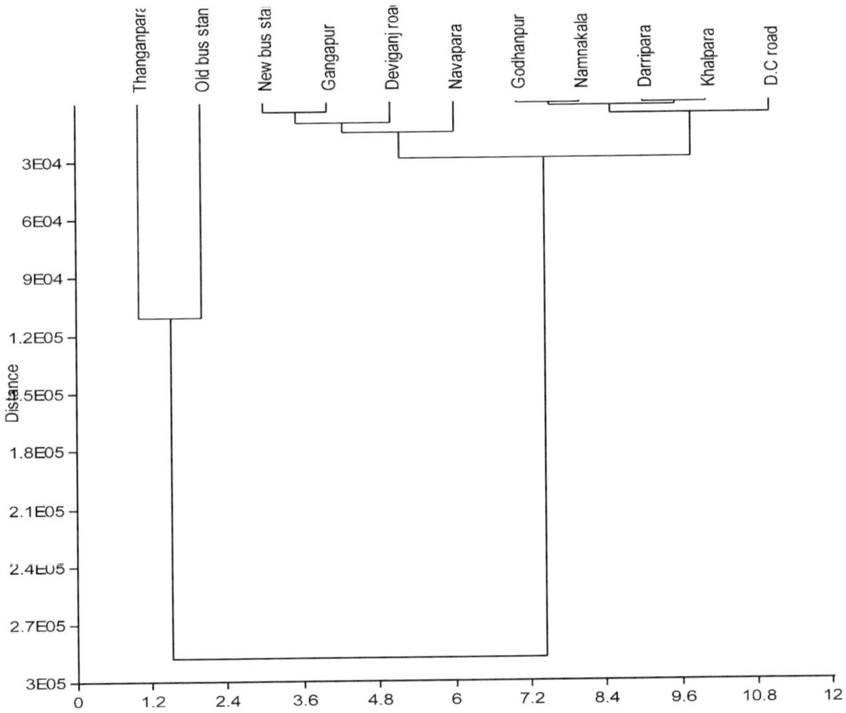

FIGURE 11.22 Dendrogram analysis dry waste of different SLRM center in Ambikapur township.

Data presented in Figure 11.23 reflects fourth small clusters of different SLRM centers studied during present investigation. Cluster 1 comprises of Gangapur and Navapara, cluster 2 comprises of D.C. road and Khalpara, cluster 3 comprises of Godhanpur and Darripara, cluster comprises of new bus stand, Gangapur, Deviganj road, and Navapara, Thanganpara, and old bus stand is distantly related with other SLRM center.

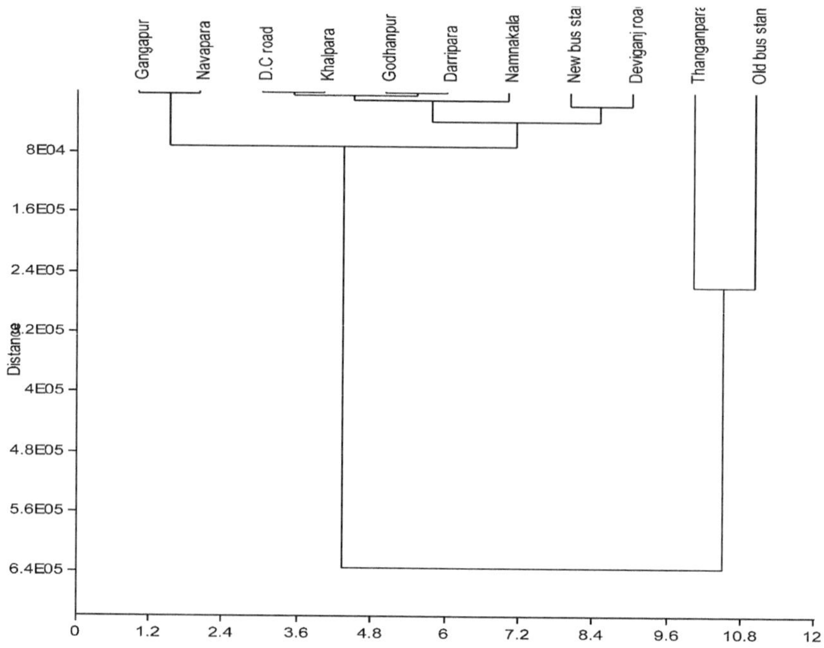

FIGURE 11.23 Dendrogram analysis dry waste of different SLRM center in Ambikapur township.

Data presented in Figure 11.24 reflects three small clusters for various SLRM centers. Cluster 1 comprises of glucose water bottle and syringe, cluster 2 comprises of glucose water bottle and syringe and glucose tube, cluster 3 comprises of glucose tube and plastic tonic bottle. Tablet cover is distantly related with other hospital items.

Data presented in Figure 11.25 reflects two small clusters. Cluster 1 comprises of mobile and varying, Cluster 2 comprises of mobile and wiring and light, CDs is distantly related with other electronic items.

Data presented in Figure 11.26 reflects two small clusters. Cluster 1 comprises of belt and shoes, cluster 2 comprises of bags, bag is distantly related with other Leather items.

Data presented in Figure 11.27 reflects three small clusters. Cluster 1 comprises of can and tin, cluster 2 of can, tin, and cable wiring. Bartan (utensils) is distantly related with other metal items.

Data presented in Figure 11.28 reflects three small clusters. Cluster 1 comprises of sketch pen and pen refill, cluster 2 comprises of sketch pen, pen refill, and pen plastic, cluster 3 comprises of white cap and ice cream plastic. Disposable plastic glasses are distantly related with other disposal items.

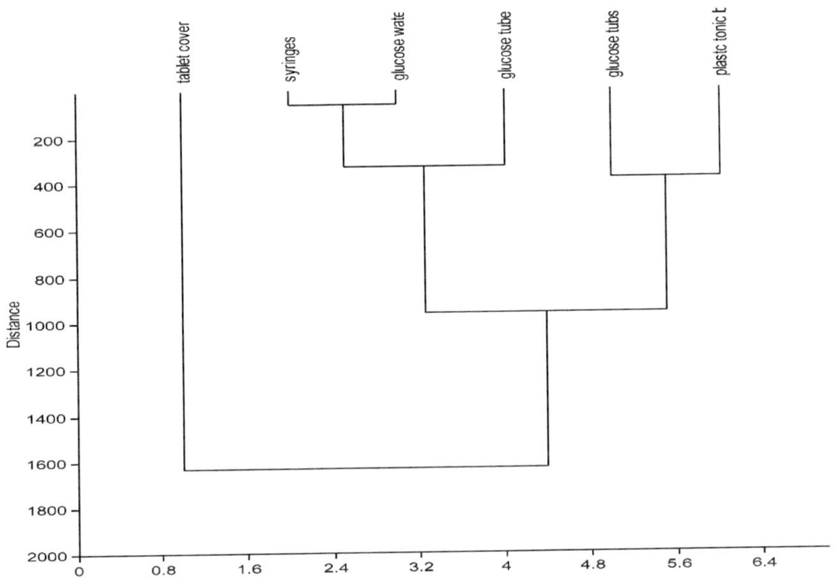

FIGURE 11.24 Dendrogram analysis of different hospital items of Ambikapur township, Chhattisgarh.

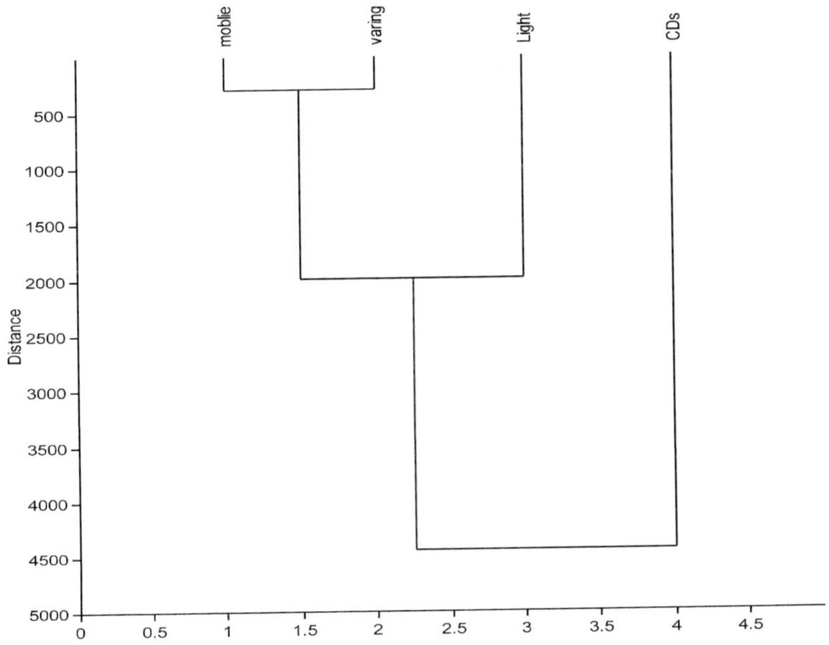

FIGURE 11.25 Dendrogram analysis of different electronic items of Ambikapur township.

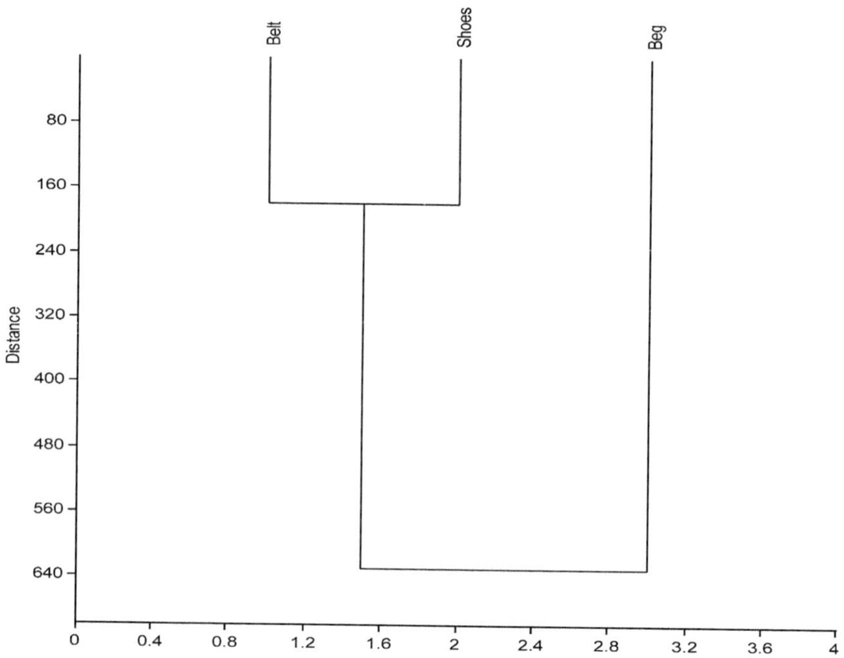

FIGURE 11.26 Dendrogram analysis of different leather items of Ambikapur township.

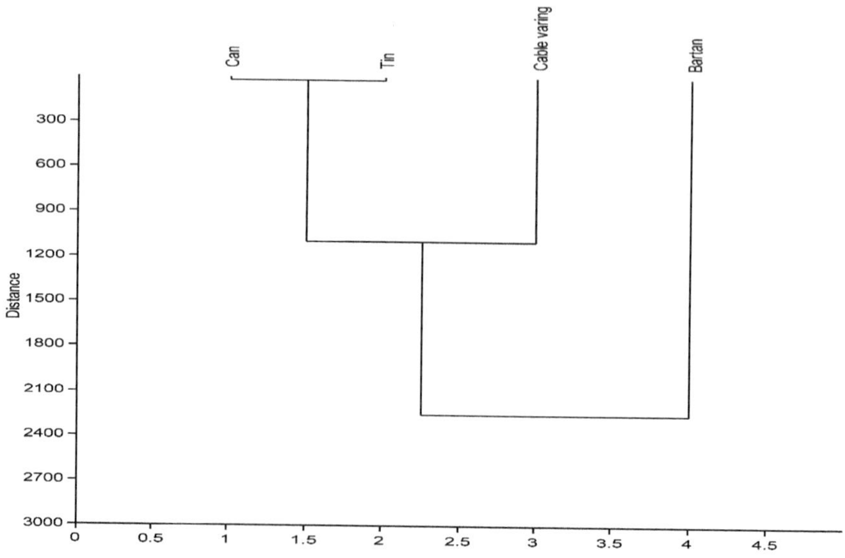

FIGURE 11.27 Dendrogram analysis of different metal items of Ambikapur township.

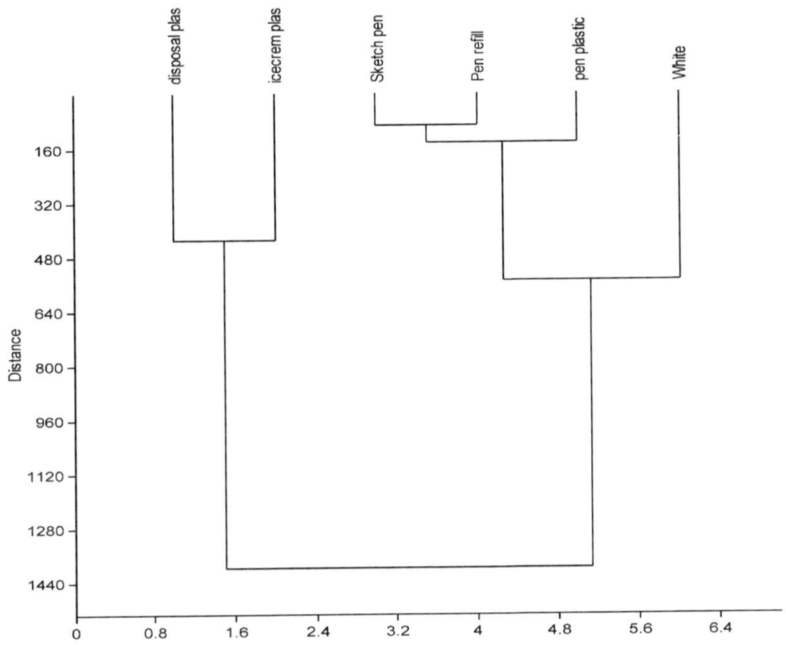

FIGURE 11.28 Dendrogram analysis of different disposal items of Ambikapur township.

- Data presented in Figure 11.29 reflects three small clusters. Cluster 1 comprises of Kanghi (comb) and toothbrush, cluster 2 comprises of floor mat and bucket. Toothbrush plastic is distantly related with other domestic items.
- Data presented in Figure 11.30 reflects two small clusters. Cluster 1 comprises of Bottle and Fridge, Glass items, Mirror is distantly related with other glass items.
- Data presented in Figure 11.31 reflects one single cluster. Cluster 1 comprises of water pipe, kishan pipe is distantly related with other pipe items.
- Data presented in Figure 11.32 reflects two small clusters. Cluster 1 comprises of Cable pack plastic and file document, cluster 2 Gudakhu dibba.

Table 11.5 represents the level of education of the people engaged in SLRM centers at Ambikapur. The results reveal that most of the workers have studied up to 10th standard, which reflects that they are not properly aware about environmental issues, not properly educated as well as untrained in terms of management of solid waste.

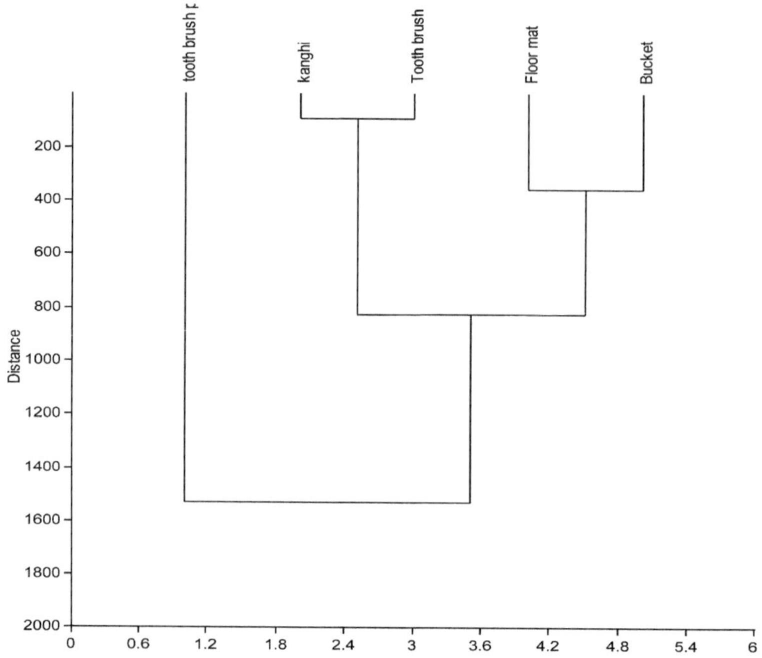

FIGURE 11.29 Dendrogram analysis of different domestic items of Ambikapur township.

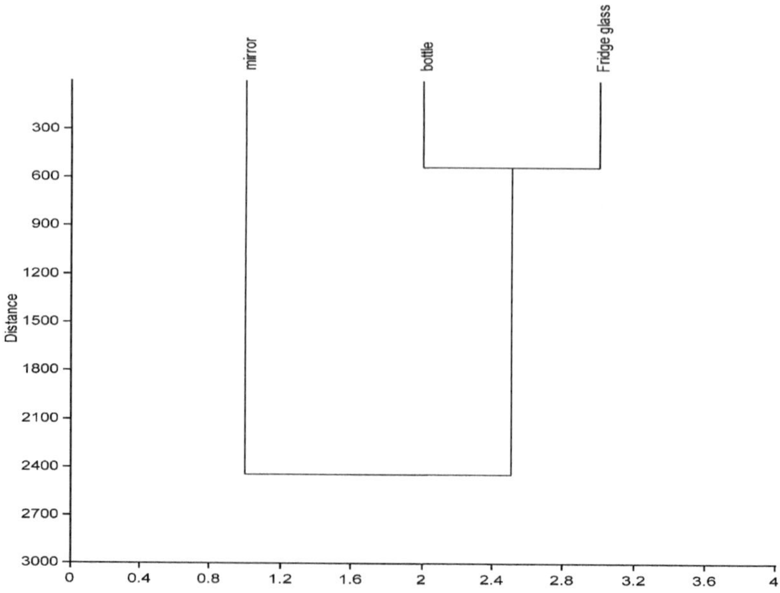

FIGURE 11.30 Dendrogram analysis of different domestic glass items of Ambikapur township

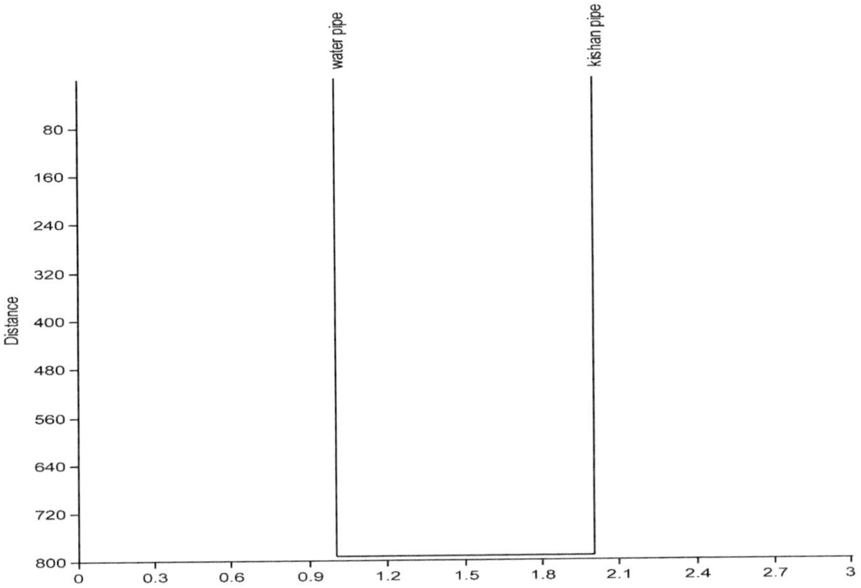

FIGURE 11.31 Dendrogram analysis of different pipe items of Ambikapur township.

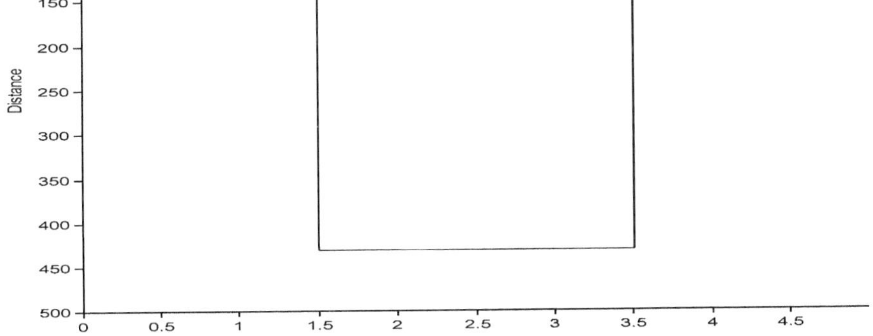

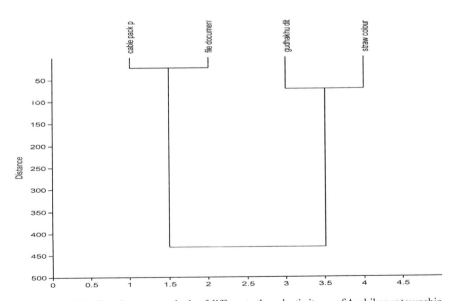

FIGURE 11.32 Dendrogram analysis of different other plastic items of Ambikapur township.

Table 11.6 represents the various problems associated with the different SLRM centers of Ambikapur township. The problems includes lack of proper infrastructure for proper functioning of solid waste management system, lack of adequate collection vehicle for waste collection, lack of adequate water and sanitation for proper management of organic and plastic waste, irregular disbursement of salary to the workers who are actively engaged in waste collection and its subsequent treatment in the SLRM centers.

TABLE 11.5 Educational Level of Workers Engaged in Various SLRM Center

Center Name	Educational
1. Navapara	5th, 8th, 10th
2. Godhanpur	5th, 8th.
3. New bus stand	10th, 12th, graduation
4. Darripara	5th, 8th, 10th
5. D.C. Road	5th, 10th, graduation
6. Khalpara	5th, 8th, 10th
7. Namnakala	5th, 10th, 12th, graduation
8. Deviganj road	5th, 8th, graduation
9. Thanganpara	5th, 8th, 10th, Post-graduation
10. Gangapur	5th, graduation
11. Old bus stand	8th, 10th, graduation

TABLE 11.6 Various Problems of SLRM Center

Center Name	Problems
1. Navapara	Raincoat, dumping, vehicles are not provided per day, rickshaw.
2. Godhanpur	Water, Rickshaw problems.
3. New bus stands	Lights are not provided, tap problems, salary problems
4. Darripara	Salary, light, waste collection problems.
5. D.C. Road	Low area, waste are not collected from all house.
6. Khalpara	Water problems, raincoat, and light.
7. Namnakala	Light problems, low rickshaw, water problems.
8. Deviganj road	Water, light, waste collection problems, salary.
9. Thanganpara	Water, rickshaw problems, low area.
10. Gangapur	Water, waste is not collect for all houses.
11. Old bus stand	Rickshaw problems, low area, salary not provide in time, water.

Ambikapur is the capital of Sarguja district in Chhattisgarh State in Central India. It has a population of 1.25 Lac. SLRM centers are aesthetically designed and functionally convenient work-sheds, used for secondary segregation of organic/inorganic refuse coming out of over 31,000 domestic and commercial units in the town. This model has been designed as part of an alternate approach to scientific disposal of the municipal solid waste. The model is technically correct, environmentally, and economically sustainable and socially very significant. It draws upon traditional wisdom and common sense. It rejects the profit-driven business models that require high-cost logistics, and instead relies upon community-based structures to manage municipal solid waste. It shifts the perspective on garbage from 'waste' to 'resource.' By doing so, it 'consumes' in the 17 SLRM Centers the entire refuse.

Livestock are actively involved in the management of organic 'resource' coming out of homes. The need for a trenching ground is thus totally eliminated. This, in turn, eliminates the environmental issues that trenching grounds pose. The SLRM Centers are manned by women self-help groups (SHGs) from urban poor homes. The model thus generates hundreds of green jobs, without putting financial burden on the State treasury. The SHGs are federated into a registered Society called Swachh Ambikapur Mission Sahakari Samiti Maryadit. Ambikapur Municipal Corporation has an agreement with this Society. Each worker earns around Rs.5,000/- a month. Dressed in uniform and geared up with cap, mask, shoes, and gloves, they now express a new and bold body language, a new identity.

Work in all the SLRM Centers can be centrally overseen, because they are all equipped with CCTV cameras. Every data pertaining to the operations starting with inflow of refuse two times a day, to sale of the various recycle-able products is digitized. The initiative, launched in June 2015, is in incubation in the technical sense. However, the sheer simplicity of the approach, combined with the fringe benefits of "District Collector," is an Indian term for the administrative head of the district. Originally meant to oversee revenue collection in the district, the role of this office has gradually shifted to marshaling of development activities in all sectors within the district.

SLRM solid and liquid resource management urban livelihood promotion and women's empowerment have impressed the State Government to order replication of the model in nearly 150 more towns. According to the size of the target town, the number of SLRM Centers can be increased or decreased. In the Ambikapur model, the SLRM Centers have been located on public land reclaimed from illegal occupation. This has hugely reduced the capital cost of the project. The model faces negative environmental impacts due to faulty handling of municipal solid waste can be observed in the Ambikapur city.

Open dumping of municipal solid waste is a common practice in the study area. Main streets, roads, railway tracks, open drains, and undeveloped plots in the study area have been seriously contaminated. In Ambikapur city, solid waste is being collected manually by the help of sweepers. They are normally using wheel borrows, hand carts and motorcycle rickshaws for the collection of solid waste from the streets. At most of the collection locations solid waste containers are not available and dwellers are dumping the solid waste on ground. The generated solid waste from these locations is being collected by the help of open body vehicles irregularly. The selected disposal site for the city is not suitable and producing negative environmental impacts on surrounding populations. Storage containers are not compatible with the existing system. It is also observed during the investigation that number of containers and collection vehicles are not sufficient to handle the generated waste. Considering the field analysis, the medium size containers with maximum 4.5 ft height may be more suitable under local conditions. The collection efficiency may be increased by the implementation of medium size containers, because at present large size containers are creating a lot of difficulties while handling of MSW in the town. As reflected from the study, the workers engaged in the project are not properly educated, and therefore often his handling of the waste taken place. Biogas production and compost production is often associated with very few SLRM centers.

11.15 CONCLUSION

Developing countries are seriously struggling to design useful and economical solid waste management systems. In Ambikapur city, municipal solid waste is being dumped openly along roadsides. In the study area, a diverse solid waste management system is in practice. Open dumps are responsible for so many negative environmental impacts in the study area. Due to lack of funding and proper management the existing solid waste management system is not working successfully in Ambikapur city. Due to the shortage of storage bins, collection efficiency is very low in the study area. Special wastes like hospital waste and other hazardous materials are being disposed along with municipal solid waste. Considering the overall negative impacts associated with open dumping and open burning, these practices must be strongly discharged.

Suggestions for proper disposal of solid waste- Considering the present solid waste management situation in the Ambikapur city the following suggestions are being presented for the improvement of system:

1. Deficiency of staff, crews, vehicles, and machinery must be recovered immediately.
2. Staff training programs must be initiated to motivate the workers.
3. Pen dumping and open burning of municipal solid waste are two major threats to the town environment, these kinds of practices must be discouraged through different awareness campaigns.
4. At present a reasonable amount of solid waste is being collected by the scavengers for recycling purposes, but they are not using any health and safety measures. These kinds of activities may be motivated while considering the proper health and safety measures.
5. Due to lack of available budget, the existing solid waste management system in not working effectively. The induction of small scale recycling plants by the municipal authority may generate a reasonable financial source in coming future.
6. Considering ethics and the town environment, the transportation of collected solid waste through open body vehicles must be discouraged.
7. At present, no proper sanitary landfill site is available for the final disposal of municipal solid waste. A proper sanitary landfill site while considering all environmental aspects may be designated immediately to avoid the environmental hazards due to illegal open dumps of collected municipal solid waste.
8. Encouragement of crews through proper incentives may be adopted to improve the sanitation practices.

KEYWORDS

- **electronic waste**
- **solid waste management**
- **municipal solid waste**
- **vermicomposting**
- **waste management**

REFERENCES

Aljaradin, M., & Persson, K., (2012). Environmental impact of municipal solid waste landfills in semi-arid climates-case study–Jordan. *The Open Waste Management Journal*, *5*, 28–39.

Annepu, R. K., (2012). Sustainable solid waste management in India. *M.Sc. Thesis*, Columbia University. pp. 56–58.

Asnani, P. U., (2006). *Solid Waste Management in India Infrastructure Report 2006* (pp. 160–189). New Delhi: 3i Network, Oxford.

Botkin, D. B., & Keller, E. A., (2000). *Environmental Science-Earth as a Living Planet* (6th edn.). John Wiley & Sons.

DCC, (1999). *Solid Waste Management by Dhaka City Corporation*. Dhaka City Corporation, Dhaka.

Dwivedi, P. R., Augur, M. R., & Agrawal, A., (2014). A study on the effect of solid waste dumping on geo environment at Bilaspur. *American International Journal of Research in Formal, Applied and Natural Sciences, 6*(1), 86–87.

Ejaz, N., & Janjua, N. S., (2012). Solid waste management issues in small towns of developing world: A case study of Taxila city. *International Journal of Environmental Science and Development, 3*(2), 167–168.

Fedorak, P. M., & Rogers, R. E., (1991). Assessment of the potential health risks associated with the dissemination of micro-organisms from a landfill site. *Waste Manage. Res., 9*(1), 537–563.

Hoornweg, D., & Perinaz, B. T., (2012). *A Global Review of Solid Waste Management*. The World Bank.

Jharia, B., (2014). Waste management: A study on Raipur waste management private limited, Raipur. *Recent Research in Science and Technology, 6*(1), 199–202.

Kansal, A., (2002). Solid Waste management strategies for India. *Indian Journal of Environmental Protection, 22*(4), 444–448.

Kumar, V., & Pandit, R. K., (2013) Problems of solid waste management in indian cities, India. *International Journal of Scientific and Research Publications, 3*(3), 1–5.

Lanjewar, S. S., Sharma, K., & Mahishwar, A., (2014). Solid waste management need and implementation in Raipur, Chhattisgarh. *Recent Research in Science and Technology, 6*(1), 61–62.

Pai, R. R., Rrdrigues, L. R., Mathu, A. O., & Hebber, S., (2014). Impact of urbanization on municipal solid waste management: A system dynamics approach. *International Journal of Renewable Energy and Environmental Engineering, 2*(1), 31–37.

Pappu, A., Saxena, M., & Asolekar, R., (2007). Solid wastes generation in India and their recycling potential in building material. *Building and Environment, 42*(6), 2312–2319.

Rajput, R., Prasad, G., & Chopra, A. K., (2009). Scenario of solid waste management in present Indian context. *Caspian J. Env. Sci., 7*(1), 46–52.

Saxena, A., & Oraon, R., (2016). Municipal solid waste management in Bareilly, India. *International Journal of Technical Research and Applications, 4*(3), 51.

Sharholy, M., Ahmad, K., Mahmood, G., & Trivedi, R. C., (2008). Municipal solid waste management in Indian cities-a review. *Waste Management, 28*(2), 459–467.

Singh, G. K. Gupta, K., & Chaudhary, S., (2014). Solid waste management: Its sources, collection, transportation and recycling. *International Journal of Environmental Science and Development, 5*(4), 347–349.

Thomas-Hope, E., (1998). *Solid Waste Management: Critical Issues for Developing Countries*. Canoe Press, Kingston.

Yadav, I. C., & Linthoingambi, Devi, N., (2009). Studies on municipal solid waste management in Mysore city- a case study. *Report and Opinion, 3*(1), 15–21.

CHAPTER 12

Environmental Education: An Informal Approach Through Seminar Talk Along with a Documentary Film

BUDDHADEV MUKHOPADHYAY[1] and JAYANTA KUMAR DATTA[2]

[1]*Research Scholar, Department of Environmental Science, The University Burdwan, West Bengal–713104, India, Mobile: +919635878593, E-mail: buddhadevmukhopadhyay@gmail.com*

[2]*Retired Professor, Department of Environmental Science, The University Burdwan, West Bengal–713104, India*

ABSTRACT

The present investigation deals with the effectiveness of media learning towards development of conservation attitude among students of undergraduate level – an indirect way of promoting environmental education (EE) at higher studies level. During the present investigation, the undergraduate students attended the traditional theoretical classroom teaching on mangrove ecosystem and their perceptions levels were evaluated. A seminar talk along with documentary film based on Borneo mangrove ecosystem was represented towards the same students by powerpoint presentation. Different factors of mangrove ecosystem were explained properly by interacting students and the perception levels were also evaluated again. By using questionnaires data was collected through close-ended questions and interviews and discussion with open-ended questions. One to one interaction was also followed. Effectivity of media learning was judged through statistical calculation (paired t-test). Results revealed that there is a significant difference ($p < 0.005$) in awareness level between pre- and post-period of promotion of media-based learning in students towards conservation. Therefore, interviewee response attitude changed towards positivity which was reflected through enhanced perception and awareness level about mangrove ecology along with conservation, which is a key for promoting education from environmental perspective at large scale.

12.1 ENVIRONMENT EDUCATION AN INTRODUCTORY REMARK

Environmental education (EE) has become the focal theme for different organizations including schools, government agencies since last 30 years. The planning of incorporating EE at school level is always present but no effective strategy is there to promote effective awareness generation through teaching-learning process in environmental issues. Ballard and Pandya (1990) mentioned that incorporation of environmental activities in syllabi has been the major theme of programmes of NAAEE (North American Association of Environmental Educators).

As per UNESCO's conference on environmental issues at intergovernmental level, the major aim was to generate awareness among people, to develop knowledge, to develop skills, and involve active participation towards EE (Athman and Monroe, 2000). Different authors have represented the approach related to environmental issues from education perspective in own way and as a result no unified approach regarding curriculum development took place from EE point of view. Gough (1997) gave a negative interpretation about EE through mentioning the abstractness in the concept leading to feebleness towards facing and solving problems related to environment. Some authors cited EE as an extra not as core or mandatory requirement (Orr, 1992). Such type of statements proves that imparting approaches that lead to sustainability and generating environmental awareness is biggest issue. Initially, the concept of education was thought as pathway of social reform towards society (Ornstein and Hankins, 1998). Social issues are very much intermingled with various aspects of racialism, sexual abuse, and environmental pollution, various diseases and its epidemic along with earth resource depletion. Under this circumstances teacher plays important role towards awareness of students, society, and human civilization who is posing severe threat. Teachers should stimulate students for developing problem-solving attitude for sustainable development of the society.

Madsen (1996) have emphasized role of education, knowledge, and awareness generation is very much crucial for environmental protection and conservation. He further emphasized about the basic clarity of common public about environmental issues. Public involved in such processes should bear extensive knowledge towards environmental pollution followed by problem-solving attitude.

12.2 ABOUT EDUCATION

It is a mixed term, which includes the traditional teaching-learning process in school colleges, classroom teaching and informal approaches which includes development of skill, knowledge, and attitude. Different people have different viewpoint regarding education. Therefore, education has various definitions, various issues without any universally accepted definition. It is a mechanism, which imparts knowledge, education, skill, and habits from one generation to next generation through knowledge gathering and research activities.

12.3 CONCEPT AND MEANING OF EDUCATION ACCORDING TO INDIAN THINKERS

Irrespective of various meanings of education along with their explanations, it is desirable to explore into the thinking of educationist, academicians in order to get a deeper understanding of the concept and the nature of education.

Human civilization, have attempted to define education in their own way of thinking and explanations. Therefore, we should more towards the exploration of great philosophers and thinkers, through different period of development of education by excellent academicians of India to which everybody is well known. It is the need to explore the past experience in ancient times regarding the mode of education in India.

12.3.1 EDUCATION IN ANCIENT INDIA

India has a rich heritage from the dawn of human civilization from an education perspective. This fact was supported by Gurukulas and Guru-Shishya parampara (teacher-disciple tradition), which was developed much a long time ago. Shiksha and Vidya are the two important words, which symbolizes the word of education. Shiksha stands for discipline, and Vidya stands for knowledge. At ancient times knowledge is considered for the benefit of society, not for own sake. It has got some religious relevance also. At that time, education means the gradual development of mind, recognition of own duty, spiritual bliss, and development physically, mentally. This, therefore, leads towards maintaining prosperity. From various aspects such as cultural, social, and political perspective the concept of EE has taken a dramatic change.

12.3.2 CHARACTERISTICS OF EDUCATION

Characterizing of education is big issue as it has different meaning and interpretations in different ways. Holistic terms such as education for character aims towards character and well-being of individual for societal development. Good individual implies man with proper civic sense, with discipline, good mannered. Learning concept related to EE involves logical skill development, developing reasoning approach along with education aspect in various dimensions. In modern perspective, numerous programmes were adopted by various institutions along with various business sectors. These approaches involve commercial activity; some are non-profit organization and other individual institutions. It involves incorporation of ethics, values through planned activities.

12.4 WHAT IS ENVIRONMENTAL EDUCATION (EE)?

It is a process of making world population aware about the environmental issues and to make the population work locally, globally, and at the individual level towards environmental sustainability. EE is a process that helps to understand the interrelationship between human beings and environment under natural, cultural, human-made environment. Major aim of EE includes generation of environmental responsible behavior (NEETF, 2000; Athman and Monroe, 2000).

EE is an organized approach towards environmental sustainability to establish a harmony between human behaviors along with ecological functions of ecosystems. The approach has multidisciplinary dimension encompassing diverse disciplines such as biological sciences, physical sciences, chemical sciences, earth atmospheric, and planetary sciences. It implies education from primary after secondary level and education informally as newsprint, media.

EE simply implies the knowledge exchange process to generate awareness about environmental issues through problem identification along with their solutions. Overall they are motivated towards environmental stewardship (Stapp et al., 1969). UNESCO (The United Nations Educational, Scientific, and Cultural Organization) mentioned about the obligation about the society towards awareness in the environmental issues along with development of environment consciousness towards protection of environment, poverty alleviation, to remove the disparities among the rich and poor countries and developed which would act as a future insurance towards sustainability. EE aims towards all round community participation towards environmental awareness, developing critical thinking attitude towards environment, develop decision making attitude towards environmental problems, responsibility development and skill towards sustainability, appreciation of environment by changing behavioral attitude towards positive direction (Bamberg and Moeser, 2007; Wals et al., 2014).

12.4.1 GOAL OF ENVIRONMENTAL EDUCATION (EE)

The major goals of EE include improvement of quality of environment, awareness generation of people regarding environmental issues, inculcate decision-making attitude for evaluating developmental programmes.

12.4.2 CHARACTERISTICS OF EE

To create environmental awareness EE plays significant influence towards public perception with skill development towards solving the problems related to environment leading to well-being and eco-friendly environment. The thinking related to EE was found to increase further by sustainability agenda keeping in view the various aspects of development, poverty irradiation, population, and gender biasness.

The joint venture of EE and sustainable development (EESD) were not supported and not properly executed by NGOs along with associated progressive organizations throughout India which is itself a big problem. However, this way of thinking is changing as the time is passing out. Under the banner sustainable education joint collaboration is gradually developing between the NGOs and governmental organization for environmental conservation and sustainability.

Organizations of international repute like UNESCO (United Nations Educational, Scientific, and Cultural organization), UNEP (United Nations Environment Program), MOEF (Ministry of Environment and Forest) and CEE (Center for EE) is promoting EE globally. However, future perspectives appear blink with numerous issues needs to be addressed.

EE helps to develop skill and personality development towards solving environmental problem. Such learning process can increase ecological perception of people at national and international level. EE can have multi-facets of advantages including literacy about environment; generate responsibility towards environment as an Indian citizen, moving towards sustainability lifestyle along with national and international level. EE provides stimulation towards personal improvement and involves project-based learning moving towards eco-friendly environment (Tal, 2004).

EE involves training of students towards development skill for basic classification. Additionally, the students can go for collaboration for solving problems. This leads to group co-operation among students. It also focuses on developing ability of independent and critical thinking, and critical-inquiry skills (Proulx, 2004). Therefore, EE has the following attributes: The attributes of EE involves holistic approach towards environment, interdisciplinary approach, lifelong sustainability, learning oriented, creation of thought, provide solutions for long duration, and develop accuracy and precision along with concerning approach relation to all round environmental development.

12.4.3 PRINCIPLES OF ENVIRONMENTAL EDUCATION (EE)

The main theme of EE is associated with the interaction of human being's with its surroundings that is his biophysical environment. Environmental pollution is the most significant level of problem globally faced by the human civilization. Over exploitation of nature is invariably done by human beings. Therefore, it is the urgent need of the hour to make people aware about the environmental degradation process and their likely consequences. The joint venture to aware public and their participation lead to environmental quality. UNESCO has mentioned EE towards promoting environmental protection. Multidisciplinary nature of EE improves environmental quality and well-being of humans.

12.4.4 OBJECTIVES OF ENVIRONMENTAL EDUCATION (EE)

EE includes multi-facets of objectives such as to aware the society and common people about environmental deterioration and pollution event, to promote awareness at trans-boundary scale, attitude development for environmental protection, skill development and capacity building among community people, incorporate public for community participation and environmental decision making.

12.4.5 PARTICIPATION

To provide social group and individuals with an opportunity to be actively involved at all leaves in environmental decision making.

12.4.6 AIMS OF ENVIRONMENTAL EDUCATION (EE)

UNESCO has mentioned about the different aims towards EE which includes interrelationship in relation to economic dimension, social, political, and ecological dimension from modernization perspective in which decision and action making regarding by various countries would have global implications. Such activities would promote gradual build up off responsibility and sensibility across countries at the international level for environmental betterment.

12.5 THE EVOLUTION OF ENVIRONMENTAL EDUCATION (EE)-GLOBAL PERSPECTIVES

EE was initiated by the IUCN (International Union for the Conservation of Nature) in 1970 at Neveda, USA. During the meeting in Nevada, USA–EE initiated through development of skill and attitude towards man and their surrounding environment. EE also aims towards development of proper functioning towards environmental protection (IUCN, 1998). EE was accepted by the global community as a protection measure for environment (UNESCO-UNEP, 1976).

12.5.1 BELGRADE CHARTER, 1975

In 1975, a conference was organized by UNESCO at Yugoslavia under which Charter of Belgrade was developed. This conference addressed the goals related to EE. It would promote general awareness globally regarding environmental issues along with skill, attitude, knowledge, technological expertise development and work collaboratively towards problem solving approach and prevention from the future ones (UNESCO, 1978).

12.5.2 TBILISI DECLARATION, 1977

After Belgrade, the world's first conference on EE at intergovernmental level was organized in Tbilisi, Georgia. Representatives of the Tbilisi Conference adopted the Tbilisi Declaration, which challenges towards EE for awareness and value generation among humankind for quality life and better environment.

The major outcome of Tbilisi conference includes formulation of objectives of EE which includes skills, attitude, knowledge, awareness, and overall community development. Environmental educationists thereafter adopted such objectives. Sensitization in relation to environmental matters among the public comes under the aegis of perception development in EE. Development of basic understanding in relation to environmental problems can take place through knowledge generation. Promote active participation towards environmental protection through attitude development and skill development.

12.5.3 THE 4TH INTERNATIONAL CONFERENCE ON ENVIRONMENTAL EDUCATION (EE), 2007

During 2007, the 4th International Conference in related to EE was organized by center for EE during 24th to 28th November. The Conference was represented by 1500 representatives from 97 countries globally. Center for EE on behalf of GoI hosted the conference. Ministry of Human Resource Development (MHRD) and MOEFCC (Ministry of Environment, Forests, and Climate Change (CC)) acted as the co-organizers of the department and the entire event was sponsored by United Nations Environment Programme and UNESCO. The mega event aimed towards finding out the outcome of EE in education for achieving sustainable development. The outcome of this event would also meet the joint collaboration between EE and ESD towards a sustainable future. Conferences of Tbilisi, Belgrade, and provided the framework for various EE programme.

12.6 THE EVOLUTION OF ENVIRONMENTAL EDUCATION (EE): AN INDIAN PERSPECTIVE

In 1991 the Honorable Supreme Court of India has made EE compulsory in the formal education system. NCERT has formulated a course curriculum for EE has been accepted by the court for textbook development at the basic level from 2004–2005 onwards. The judicial emphasis on promoting EE has framed at the grass route level in India. CEE (Center for Environment Education) is working with joint collaboration with education department and other academic bodies for teachers training for EE. Scope in service sector was provided for EE inclusion through projects and various courses. Some issues such as limitation of time, conventional approach creates problem for proper integration of EE at the formal area. Therefore, emphasis is to be given for all round development throughout the time span.

In this connection distance learning is the pathway. 42nd amendment of Indian constitution in 1974 has incorporated two new articles. According to the amendment within our Constitution Article 48A mentions the state's obligation towards environmental protection. Further Article 51A(g) mentions that protection and conservation of the environment should be fundamental duty of every citizen of India. Article 21 of Indian Constitution mentioned about the life and personal liberty as fundamental right for citizens. The necessity of improving living standard, health has been narrated as a duty of State Government under Article 47 of Indian Constitution. National policy on education emphasized on sensitization for all towards environment. It mentioned that all age groups starting from different age group awareness generation through paper about the facts and issues of environment. National Curriculum Framework (2005) recommended the infusion model for imparting EE at Upper Primary and Secondary level of formal system of education. In this matter policy formulation has started from 1968 and therefore highly awareness generation was the outcome. Policy formulation regarding this issue is being formulated since 1968. The output of such exercise would lead to more awareness generation among local community. However, still there are lacunae in such action plan.

From sustainability perspective, India is striving across various sustainability issues like poverty alleviation, population growth, illiteracy rate and others along with degradation of the environment. Sustainable development demands harmony among socio-cultural and economic dimension along with ecological system.

From Indian perspective integration various sustainability aspects like conservation and protection of environment is always integrated with socio-economic and ecological dimensions. India possesses a rich heritage by means of its diverse values to work for environmental protection. Under such concept, human beings forms component of environment which is inseparable. Therefore, enjoyment comes from co-existence of life. Each species across world should not deprive the other species for its existence.

12.6.1 FORMAL ENVIRONMENTAL EDUCATION (EE)

EE is the backbone of education at higher level in Asia and the Pacific. Formal system provides a broad platform for the student community to become aware about various issues related to EE and develop their own interpretation through analysis and intellect. Various factors have influenced the progress of EE in the concerned region. In this connection educational policy at national level and policy matter related to population and environment are the driving factors. Actually, the policies are the manifesto of the existing socio-cultural knowledge across various countries. As a consequence of the aforesaid matter development of policy related to EE paves the way of broader dimension (UNESCO-PROAP, 1996). As for instance major academic reforms were observed in Australia since 1970s. In this case, a migration towards nature and eco-based educational programme to fulfill the various dimensions of sustainability were occurred. The progression of education has migrated from school level to adult education towards community development (Fien, 1996).

Besides going for a new curricular aspect most of the countries across the world have attempted incorporate EE in their existing syllabus. However, some countries have invested themselves to upgrade the curricula by incorporating EE along with development of separate curriculum. As most of the EE learning perspective is practical oriented therefore it helps the school education to promote value education for the well-being of the society (UNESCO-PROAP, 1996).

It was observed that syllabus and curriculum spread across nationally through coordination in various schools across the countries at primary and secondary level. For national framework perspective there is huge lacunae or gap which prevents the development of EE programme.

12.6.1.1 PRIMARY AND SECONDARY LEVELS

A lot of difference at a different level of education was observed across developing regions considering various vulnerable issues of various countries. In the Maldives, EE, and programmes related to awareness gives major importance to marine environment originating from National Environment Action Plan of 1989 (IUCN, 1998). In Nepal optimum use of available resources and heritage were given special priority depending upon problem solving approach for specific problems and integration of resource, its use, and population explosion and associated environmental problems under EE plan (IUCN, 1998).

From 1990 onwards in the Republic of Korea, issues related to environment are incorporated at school level to promote environmental awareness in children with an attitude build-up for solving environmental pollution problems.

In China, integrative work related to EE takes place through public participation and academic bodies. Such a concept was the child of Conference of National EE in 1992 by the National Bureau. At pre-school and primary level promotion of EE is mediated through non-formal systems. Integration of environmental courses into the basic courses at higher level is the need of the hour. Local examples were cited by the teacher for better environmental perception (UNESCO-PROAP, 1996).

For most South Pacific island the sub region efforts have been emphasized for integration at lower and higher level of school education. The curriculum at the secondary stage has given top priority on curriculum related to environment.

12.6.1.2 TERTIARY LEVEL

Since 1990s EE has gained increasing demand among the environmental managers. There are some specific trends that has been developed such as incorporation of environment as a subject issue at UG and PG level of curriculum with formulation of new modules or units for in-depth study of environmental issues, foundation, and other courses on environment training in-service people and understanding of eco-friendly practices, publication of relevant print material as textbook and audiovisual aids, developing research capabilities in environmental field, focusing research towards educational policies and practices, dialogue, and information exchange programmes both in governmental and non-governmental sectors, promoting adult and community education which would increase the literacy on environmental issues.

Emphasize on environmental issue is the need of present realities, environmental crisis situation. Different countries have catered the issue of EE at a different way. For example, country like Viet Nam since 1995 has made man and environment course compulsory in various academic disciplines with specialized courses at the university level (UNESCO-PROAP, 1996).

In Thailand, degree courses related to EE were implemented at various academic programmes with various schemes. Such courses rose perception in environmental issues which is also integrated with other academic disciplines (UNESCO-PROAP, 1996).

China and Pakistan has already incorporated environmental courses at various college and university level for proper training of officials, technicians, and academicians. In Mongolia, master plan has been formulated for promoting EE through various pathways (IUCN, 1998).

In developing region the EE has been effectively implemented from grass route level to higher level but the major problem lies between improper development of conceptual skill among the learners as well as improper implementation and lack of management (UNESCO-PROAP, 1996).

12.6.2 NON-FORMAL EDUCATION

Non-formal systems of environmental educational include actions related to activities at curricular and extracurricular levels in the form of occupational training and create public perception through mass media, and active involvement NGOs.

12.6.2.1 LEARNING BY DOING

In various countries, practical approach has been made to correlate students with the local problems and generate a problem-solving approach. For example, an education programme named *Muktangan Siksha* implemented in Bangladesh promotes field-based studies. In Myanmar, EE is implemented informally through *patwinkyin* programme at pre-school and lower primary level without traditional textbook teaching. In Sri Lanka, WWF has initiated EE programme through greening of learning approach. The programme would involve 750 schools. Under such scheme students are motivated for plantation, gardening activities and other eco-friendly approaches within their school premises. Success in such schemes has widened up of application of such schemes in various countries globally.

Various research foundations and institutions are gradually developing environmental activities in non-formal way. From the Indian perspective, DA (Development Alternatives) has promoted the scheme of Community-Led Environmental Action Network (CLEAN), which promotes "four r" among school student. Globally in various countries government and non-government organizations has initiated development of eco-clubs in school to motivate the students to act in an informal way for environmental protection.

12.6.2.2 OUTDOOR ACTIVITIES

Government sectors and various non-governmental agencies have promoted various outdoor activities among the youths as well as among adults for betterment of life. In this connection Environmental Camps for Conservation Awareness (ECCA) an NGO in Nepal is promoting outdoor activities among children, adults, and even disabled individuals. Camps are organized in areas of environmental importance and raise awareness towards conservation. Such environmental camp has been widely adopted by other countries also. In this connection, environment department of Malaysia is an example. Under such schemes children are given duty of adopting and protecting beaches. In this way they can become aware about the ocean ecosystem.

The China Association for Science and Technology (CAST) actively promotes informal education in science among young generation. Scientists, technologists of CAST help towards developing human intellect as resources for EE.

Also in China, the State Environmental Protection Administration (SEPA) is playing an active role in-formal mode of learning leading to promote learning process globally to provide benefit under Global Learning and Observations to Benefit the Environment (GLOBE) programme and plays leading role in non-formal activities to educate the world globally. It is actually a worldwide programme integrating students, researchers, scientists in the global platform in science and education. In Australia, Water watch–a community based programme were initiated in 1990s at country level which monitoring hydrosphere and environmental quality which would bring all round development globally (Palmer, 1995). In this, both the students and their parents participated in monitoring of water quality.

12.6.2.3 INNOVATIVE APPROACHES

As the time passes away new developments has been initiated in EE covering formulation of newer strategies and development. In this direction formation of eco-clubs under various banner is a fruitful step toward successful implementation in Singapore, India, and other countries.

Celebration of National Environment Days and Weeks worldwide provide another footstep for various environmental activities workshop. Education materials on eco-friendly farming in schools are facilitated to young learners to aware those about the ongoing changes in environment, agriculture, and how to cope up with such problems. Japan is a leading front in this perspective of awareness generation, skill development through various programmes, campaigning towards conservation of energy and resource. At junior level eco-club programmes has given good results. Such eco-club activities have been supported by government agencies throughout the country in the form of festival within a school year. About 4000 clubs and 70000 people actively participating in EE programmes in Japan.

In India, challenge of creating awareness has moved forward through Act Now' process by CEE (Center for Environment Education). Under the scheme initiatives were taken for active participation in EE. CEE further addresses the issues of resources through video story entitled *Dhraki*. The programme was further supported by booklet publication which is widely circulated in classrooms teaching, nature experiences. Further, CEE promoted EE Bank to promote community learning. The programme encompasses films, videos, poster publications and computer-based databases for more than 700 environmental concepts, more than 2000 activities with 600 case studies dealing with environment. Similar types of innovativeness were recorded from Thailand in the form of Science and Environment Management curriculum at Hard Amra Aksornluckvittaya School, situated at Samutprakan Province. A very innovative programme where local problems were given consideration and students actively engage themselves towards solving the problem. As a consequence of such activities, the academic community involves themselves in plantation programme within the Asokaram Temple.

EE with its multiple dimensions addresses social issues, focuses on community-based programmes and their solutions (Rasmsey and Hungerford, 1989). This, therefore, creates the necessity towards development of individuals working towards environmental protection and conservation in the area of pollution and its mitigation both in present and future time frame. Active participation and thinking is required for human being to maintain eco-friendly lifestyle for environment conservation (Veeravatnanond, 2010). Visual media could play an important role in this aspect.

Various reports related to informal mode of imparting knowledge in EE happen to far better than the traditional classroom teaching. Visual media plays the active role from this aspect. Electronic and print media has proved their efficiency in imparting knowledge of EE. However, through books, magazine happens to provide updated reference whereas audio-visual aids in terms of radio, transistors, and television mediates better and quick knowledge transfer (Prathap and Ponnusamy, 2006). For everybody total education along with good environment was the major aim for Millennium Development Goal needs to be achieved through innovativeness in the learning process. For mankind major challenges such achieving sustainability through food security, creating employment opportunities, improving productivity and yield can be mediated through related video programmes (Isiaka, 2007).

Knowledge regarding natural landscape was more effective studied and understood through nature-based videos. This provides a very suitable condition to children to know more things, to move across the world and move across their world of imagination. Information regarding areas to which the individual doesn't belong was easily obtained through video learning. Video learning also increases the individual capability for learning more. This actually provides real life situation in agriculture and environment for students. Economic burden of field visits can be reduced by VCD and VHS. Creating awareness in relation degradation of environment can be easily mediated through video learning. Advantage of video learning includes gaining of knowledge in a particular issue again and again (Spencer, 1991; Ahmad, 1990; Adedoyin and Torimiro, 1999).

It has been proved that video has the potentiality to inculcate knowledge among students. Based on theoretical principles in video learning it was observed that it is highly effective. Video learning mediates group discussions leading to personality development (Isiaka, 2000; Dopemu, 1990; Ahmed, 1990; Talabi, 1989). Very popularly known fact is that visual media learning process is very much effective in relation to gaining knowledge along with public attitude and behavior development. Visual media may occur both in the form of print or electronic media. Visual media is acting significantly towards imparting knowledge in students and common public. Different mass media tools were found to be effective towards awareness generation among public. Therefore, knowledge, and information transfer takes place in short span of time. Attractive features of visual media make the learning process student-friendly (Akpabio, 2004).

A study at Turkey on role of various electronic media technology by counselors through distance learning was reported by Kumar et al., (2003). Results reveal effectiveness of videocassette towards knowledge inculcation among students beside telephone and computer. Moore and Kearsly (1996) gave a comparison of learning between face-to-screen study and traditional teaching methods. Previous researches have necessitated the use of video as instructional medium. Low cost, suitability of application, better knowledge imparting capability among students has made video learning friendly. It leads to social skill development among students (Isiaka, 2000; Dopemu, 1990).

Mangrove ecosystem is unique in nature which is fragile in nature. General awareness regarding mangrove ecosystem is often blink in student community at school, colleges, and Universities. The major ecological function of this unique ecosystem and their characteristics of, the reasons behind mangrove ecosystem degeneration, methods of mangrove ecosystem conservation are yet to be explored for common people by the scientist. Therefore, during present study mangrove ecosystem were considered as the discussion topic for proper knowledge dissemination regarding mangrove ecosystem.

Burdwan town were chosen as study site which is an important educational center of West Bengal having many schools, colleges, University, etc. Throughout West Bengal gradual awareness regarding environment is gradually developing up by incorporating environmental issues in school, colleges, and Universities curriculum. But generating awareness in environment is blink through classroom teaching, text book knowledge, etc. due to lesser effectively of teachers in imparting knowledge among students as well as reluctant attitude of students towards this subject. Therefore, creating importance and interest among the people is a major challenge to us under the Indian context.

Several research reports on this aspect were recorded from different European countries but very few reporting regarding the influence of visual media, print media towards disseminating knowledge of environment and generating awareness. Under this context the study were carried out to evaluate the influence of alternate mode of knowledge dissemination along with awareness generation in students of higher class through the visual media under Indian context where the matter of environment were very easily overlooked by other issues and a big lacunae in strict implementation of policies laws, acts for conservation and protection of environment.

12.7 ENVIRONMENTAL EDUCATION (EE) SCENARIO IN INDIA

In India, a wider dimension of EE scenario exists. Our country has a rich heritage of diversity in terms of socio-cultural aspects, geographical locations, biodiversity along with climatic variability. This, therefore, indicates that the dimension of EE is area specific. At primary level, proper care should be given at the level of children and women, which comprises half of the population. Awareness in relation to health, hygiene, family planning, population control, food security issues should be imparted. Regarding these NGOs should lead from the front. Reports reveal that one hundred and fifty NGOs are working in EE field. Children need to be made aware of about wildlife species. EE should develop environmental conscious behavior (NEEAC, 1996).

Bhandari (2001) reports about the agendas that may be formulated for the Asia-Pacific region for up-gradation of promotion of EE which is yet still to be implemented properly without adequate framework, traditional classroom teaching as well as exam-based learning without adequate resources. Total lacunae exist in learning by doing. Action plan for effective EE implementation includes capacity building, partnership development, academic reforms through curricular buildup, developing collaboration and support.

A comparative study was done by Bartosh (2003) in order to assess the influence of EE programmes over traditional formal subjects. Comparisons were made between two schools EE has been successfully implemented within one school and without EE on the other. Results reveal positive results for school that have successfully incorporated EE in course curriculum in comparison to that school who haven't done that. The study further revealed that schools adopting EE is getting better performance in terms of knowledge impartment as well as greater support from parents< students and overall from administration. More value is added towards promoting EE.

From the context of India environment is of wider dimension. India is diverse in nature by means of its climate, geologic features, geographical orientation, edaphic features, and biodiversity, ethnic, and socio-cultural aspects. Therefore, EE is case specific. Women and school children are the primary focal point which is half of the human population. Awareness regarding personal health hygiene, population control needs to be considered properly. NGOs should lead ahead in this aspect and play the key role. EE has four component parts in its execution. Firstly the lower secondary stage which aims towards awareness generation for real life situation leading towards sustainability and developing conservation attitude. The contents are supplemented through practical and field oriented study. Secondarily through higher secondary school stage aiming towards skill development through solving problems. Mostly they are science oriented accompanied by theory and applied work (Sharma, 2004).

De La Vega (2004) reported that EE is an inherent component for the curriculum of Southwest Florida public schools for last 30 years by which major emphasis were given towards awareness generation, attitudinal build-up and knowledge development among students in order to develop future warriors against environmental degradation. The major target people for this assessment were the high school instructors, subject experts, and students' parents. Significant variations were observed among the various target groups for the study variables such as attitude, awareness, and knowledge in environment. Positive response was obtained for subject experts and the other people of the target group lowest awareness and skill development were observed. Socioeconomic condition, culture, and other activity involvement positively influenced the difference across the groups. It is a good exercise to evaluate district level prevailing EE system. As per Kuo and Taylor (2004) EE tends to reduce symptoms of attention-deficit/hyperactivity disorder (ADHD). Some researchers reported EE programme on urban set up. The main theme of research was EE programme influence evaluation in generating awareness in children about their surrounding environment. Results reveal positive attitude regarding awareness about surrounding environment in students coming from higher economic neighborhood. Result was little bit negative for students coming with low socioeconomic condition. Ozden (2008) considered that today's young generation as future warriors for sustainable world and EE can be a solution for mankind in front of environmental pollution.

Another investigation carried out by Berman et al., (2008) revealed positive result for improved cognitive functioning in EE with respect to human health. Jickling and Wals (2008) study reported that under the globalization process EE concept has been reoriented towards education to achieve sustainability.

Petrillo and Demchik (2008) mentioned about the Rio Summit in 1992 produced *Agenda 21*. Global agreement addresses sustainability through community participation which addresses social, economic, and environmental issues affecting the community. Sustainable development was indeed a debatable issue but at the global scale improved condition for developing skill, communication, and critical thinking capability in students is a worldwide need. Educational reform demands environmental issues in skill development. By developing courses having natural resource interaction and environment surrounding man it would be possible to generate future concern student community. In this context community resources would play important role in student empowerment leading to skill development.

Programmes of EE were aims towards reform in the prevailing systems of education and sustainability which depends upon variable factors such as geographical location, culture, and community perspective. Generalized factors can be applicable for wider dimensions.

A survey conducted at primary and pre-primary schools in Extremadura (Spain) on environmental audits (eco-audits) of school reveal about the progress of EE through its curriculum integration. Such initiative promotes participation process by students and teaching community (Conde and Sanchez, 2009).

Young and Lafollette (2009) evaluated the mode and way of applying EE in classroom teaching during 2005. Response obtained during the study revealed above 90% response reflected 22–100 minutes of exposure towards EE or at least single exposure for the year. However, teaching community response were mixed type with 49% people reflected their own interest in environment, and 47% revealed shortage of time for EE class. In Malaysia, Green strategies has been formulated to be incorporated in national policy. Non-formal pathway may work as complimentary system for different level of education with updated knowledge and skill base. The informal mode is indeed advantageous due to its flexible nature, different nature of target group through various activities. In this aspect various investigation were carried out at Sabah, Malaysia. Results reveal the campaigning related to environmental awareness were participated and actively supported by NGOs (non-government organizations). Hassan et al., (2009) mentioned that work-shops, seminar, symposium, and other functions improve public perception about environment and their protection measures.

Under the directives of Honorable apex court of India, EE has been incorporated in regular course under formal system of education in school. Prior to such legal step various programmes such as Environmental Orientation to School Education (EOSE) related to EE were implemented throughout the country which involves the school children towards solution of local problems. From the Government part initiatives includes formation of three nodal agencies for effective implementation of the scheme (Sonowal, 2009). EE programs aims towards development of cognitive abilities among individuals and become sensitize about what they are learning. For better reasoning and decision making students should not just be a subject expert only but also build a connection with the environment by involving with local environmental problems. Cheng and Monroe (2010) reported that connection of children with environment promotes them to actively participate in nature based studies. This leads to eco-friendly behavior and attitude development in children. The work of Donald Burgess and Jolie Mayer-Smith mentioned biophilia concept which emphasizes the biological interrelationship between human being and environment and enhanced quality of life may be influenced by condition of interrelationship (Burgess and Mayer Smith, 2011).

EE can be thought of as an open system of awareness and sensitivity development among the local community towards sustainability. The overall change across the globe due to science and technological progress has emphasized the necessity of reorientation of EE in school system to educate our youth about environment, its problems and overall how to solve them (Agnes, 2011). Various studies on this particular aspect were done across the globe which reveals Malaysia and Nigeria are effectively implementing them throughout their country. Within the school system conducive environment, community, and social development programmes related to social development have been organized to effectively implement EE programmes.

Budvytyte (2011) mentioned the vital role of EE about developing attitude and knowledge towards protecting environment which also serves the environmental protection issue. The problem of EE includes the incorporation of EE in formal education system.

Halder (2012) mentioned that EE happen the key pathway for solving environmental problems and move towards sustainability from global perspective. During their study they considered frequency of environmental classes, field-based study, and nature-based study, type of methodology of teaching adopted and evaluation system. Results reveal blink fate of EE and therefore reorientation of the entire education system needs to be done from EE perspective and sustainability. Jena (2012) mentioned the role of EE towards involving mass people in movement towards conserving environment and sustainability. The major goal of EE is to sensitize people about environment and its associated problems along with developing skill, attitude towards solving problems. It also includes development of inquisitiveness to work from environmental perspective with a collaborative approach for solving problems related to environment. In this regard school frames the platform to sensitize the mass about environment.

According to some researchers, EE concept and sustainability should be under the root of education philosophy (Sund and Lysgaard, 2013). Ghosh (2014) mentioned about the perception, awareness of secondary students of school in EE. Among 200 students the study were conducted comprising of one hundred males and females. The descriptive survey method was used in the investigation. Significant result was obtained for both genders of city and village set up but it was insignificant for school students at the secondary level. A positive interaction was obtained regarding developing perception about environment among students.

Kopnina (2014) projected future visions regarding implementation of EE and its effectiveness. Limitation of growth, modernization, and ecology along with sustainability was addressed with an anthropocentric dimension. Such issues were intricately related with sustainability and EE. Berberoglu (2015) mentioned the field education importance in relation to EE. These studies aim to assessing the effectivity of EE towards imparting awareness regarding biodiversity, environment, and sensitization towards environment. Results reveal that nature-based EE is much more effective than formal way of EE. However, generating awareness about biodiversity is lacking.

Rudy (2015) mentioned about multidimensional of EE for primary, secondary, higher secondary, undergraduate, and postgraduate level. Such functions are based upon structured and tiered analysis with an integrated along with monolithic curriculum. The major problem for effective implementation includes lack of understanding, awareness, and low public participation.

Siddiqui and Khan (2015) addressed about designing working model for EE under formal condition as well subsequent analysis of the dataset of formal sector to evaluate perception power in relation to EE. The methodology for assessment includes structured questionnaire survey among teaching community belonging to school and colleges. The major finding of this study includes formulation of framework and strategy for EE promotion. From the overall scenario it is evident that scenario of EE is not promising but its future perspective seems blossoming up. Development of school curriculum on conservation issues requires planning.

Kasimov et al., (2015) mentioned about the incorporation of sustainable development in the curriculum of EE in Russian Universities. Pavlova (2013) has provided co-existence nature of EE and education for sustainable development (ESD) for the educationists and policymakers. This co-existence has created problems of overlapping between EE and ESD concept. No clear cut demarcation EE and ESD. Absence of clarification often hinders achieving sustainable development goals.

12.8 METHODOLOGY

12.8.1 STUDY SITE DESCRIPTION

The study site for present investigation comprises of MUC women's college, a state Government aided College at undergraduate level affiliated to University of Burdwan, West Bengal, India. It comprises students of three different faculties, science, arts, and commerce in which environmental studies are compulsory subject in the course curriculum. The group of students who were involved as a sampling unit during the present study was the final year students of different undergraduate programmes with age class between 21–23 yrs.

12.8.2 METHODOLOGY ADOPTED

1. **Section A:** Evaluation of perception about mangrove ecosystem and its importance at undergraduate level before implementation of seminar talk along with documentary film (SDF) were assessed. Hundred students attended the theoretical classes on mangrove

ecosystem before evaluation. Different tools and instruments were considered like short questionnaire with closed ended questions, open discussion along with open-ended question, one to one interaction, etc.

2. **Section B:** Seminar talk along with documentary film based on Borneo mangrove ecosystem was represented towards the students. Before playing the documentary film a short seminar was given to the students by powerpoint presentation. A 100 students participated in the powerpoint presentation regarding mangrove ecosystem. Different factors of mangrove ecosystem were explained properly. Emphasis was given specially on viviparous germination, tilt root, key stone species of mangrove ecosystem, edaphic factor, mangrove conservation, socioeconomic influence of mangrove ecosystem, etc. After that, the documentary film was played to the students. This approach was too much attractive towards the students. Students were very much interested about the representation. It seems that when formal education in terms of classroom teaching as well gaining of knowledge from reference textbooks ends, media become the most available and sometimes only source for the public to gain information about scientific discoveries (Nisbet et al., 2002).
3. **Section C:** It consists of post-evaluation of perception of visual media learning. Data for this study was collected using questionnaires with close-ended and interviews and discussion with open-ended questions. One to one interaction was also followed. A 100 students participated in the interview session after seminar talk and documentary film. The interview questions, discussion with open-ended questions, one to one interaction (qualitative data), etc. and further the data analysis were undertaken were through interpretative explanation where responses were categorized by theme. Results obtained from questionnaires were subjected to statistical analysis and responses were counted and the frequencies converted into percentages.

12.8.3 TOOLS USED

Evaluation of perception of students' attitude measurement was done by various means by which various data (quantitative and qualitative) were obtained. To achieve this, the help of different tools and instruments were taken like short questionnaire of survey sheet, open discussion along with open-ended question, one to one interaction, etc.

12.8.4 QUESTIONNAIRE FOR MEDIA LEARNING (SDF) SEMINAR TALK ALONG WITH DOCUMENTATION

The survey instrument (questionnaire) consists of eighteen statements or parameters that solicited information about different aspect of mangrove ecosystem study concerned with viviparous germination, tilt root, key stone species of mangrove ecosystem, edaphic factor, mangrove conservation, socioeconomic influence of mangrove ecosystem, food chain and many others having significance in terms of conservation and awareness attitude development. The responses to each statement were tallied and converted into percentages under the categories of strongly agree (SA), agree (A), neutral (N), disagree (D), and strongly disagree (SDA). The questionnaire was applied before and after application of the proposed approach of the model (Tables 12.1 and 12.2).

12.8.5 VALIDITY TESTING OF THE STRUCTURED QUESTIONNAIRE

The survey instruments for each of the proposed model were self-structured keeping the view to measure the environmental knowledge, view, awareness, perception, etc. within the students which is the prime aim of the EE. The questions of the questionnaire were framed in such a manner so that the students' knowledge, view, awareness, and perception, etc. are predicted. The interviewee were asked for various types of responses (agree, SA, neutral, disagree, and SD) with a given statement. For the validity testing three expert members apart from supervisor checked those instruments. They have made the necessary corrections and provided some suggestions. Statement number 9: 'Key stone species are getting affected by different anthropogenic causes' and Statement number 16: Cultural aspects are also getting affected due to loss of diversity of living organisms in mangrove ecosystem are incorporated to add on the survey sheet based on media learning (Tables 12.1 and 12.2).

Expert members also suggested treating the statements of the questionnaire or survey sheet as open-ended questions also so that students can express their own view apart from the prescribed format. They suggest to follow open discussion and one to one interaction for that purposes and

TABLE 12.1 Responses in Percentage Before Implementation of SDF

SL No.	Conceptual Clarity **A case study of Mangrove ecosystem study through a documentary film.**	A	SA	N	DA	SDA
1	Concept about species diversity has been developed.	30	40	12	10	8
2	Concept about habitat/ecosystem diversity has been developed	35	30	8	10	17
3	Concept regarding species interaction in the natural habitat has been developed	29	28	19	11	13
4	Importance of wildlife has been understood	30	20	5	20	25
5	Effectiveness of wildlife conservation has been clarified	35	28	10	10	17
6	Importance of ecosystem of mangrove have been understood	8	50	2	10	30
7	Functions of different important species attribute in mangrove ecosystem	40	30	8	15	7
8	Concept of the Key stone Species has been developed	30	35	15	10	10
9	Key stone species are getting affected by different anthropogenic causes	30	36	6	10	18
10	Other species dependent on species of keystone would also get affected	32	30	10	15	13
11	Can predict the reasons of its degeneration	45	16	16	15	8
12	Reasons for mangrove habitat loss have been understood	30	40	14	13	3
13	Impact of mangrove forest degradation from socioeconomic and environmental aspect of human beings has been understood	20	48	14	10	8
14	Sociological impact tends to be degenerated due to loss of mangrove ecosystem	25	40	6	10	19
15	Economic condition of the surroundings is getting affected for degradation of mangrove ecosystem	35	26	18	11	10
16	Cultural aspects are also getting affected due to biodiversity loss in mangrove ecosystem	30	40	4	12	14
17	Spiritual aspect is also declined through biological diversity loss.	30	30	20	15	5
18	Overall environmental aspect is badly affected due to loss of biodiversity	30	40	2	6	22

N = 100; A – Agree; SA – Strongly agree; N – Neutral; D – Disagree; SD – Strongly disagree.

TABLE 12.2 Responses in Percentage After Implementation of SDF

SL. No	Statement	A	SA	N	DA	SDA
1	Concept about species diversity has been developed.	44	50	6	0	0
2	Concept about habitat/ecosystem diversity has been developed	50	44	4	2	0
3	Concept regarding species interaction in the natural habitat has been developed	42	44	14	0	0
4	Importance of wildlife has been understood	40	54	4	0	2
5	Effectiveness of wildlife conservation has been clarified	44	48	4	2	2
6	Importance of mangrove ecosystem have been understood	12	80	4	4	0
7	Functions of different important species attribute in mangrove ecosystem	48	46	4	2	0
8	Concept of the Key stone Species has been developed	40	48	10	0	2
9	Key stone species are getting affected by different anthropogenic causes	42	46	6	0	6
10	Other species dependent on keystone species are also getting affected	46	48	4	0	2
11	Can predict the reasons of its degeneration	60	28	10	0	2
12	Reasons for loss of mangrove habitat have been understood	38	56	6	0	0
13	Impact of loss of mangrove forest in socioeconomic and environmental aspect of human beings has been understood	32	62	6	0	0
14	Sociological impact tends to be degenerated due to loss of mangrove ecosystem	32	54	10	0	4
15	Economic condition of the surroundings is getting affected due to mangrove ecosystem degeneration	48	34	14	0	4
16	Cultural aspects are also getting affected due to loss of biodiversity in the mangrove ecosystem	40	54	6	0	0
17	Spiritual aspect is also declined with the loss of biodiversity	36	42	12	10	0
18	Overall environmental aspect is badly affected due to loss of biodiversity	38	56	4	2	0

N = 100; A – Agree; SA – Strongly agree; N – Neutral; D – Disagree; SD – Strongly disagree.

the students' feedback of responses may be analyzed to get the qualitative information.

12.9 RESULTS AND DISCUSSION

12.9.1 EFFECTIVENESS OF VISUAL MEDIA LEARNING

Information may be provided in non-textual form in a much more attractive way through video (Hampapur and Jain, 1998). Video provides multi-sensory learning environment, which enhances the mode of accumulation of information (Syed, 2001). Heterogeneous nature of results was obtained under various treatments of visual and verbal information in video. Nugent (1982) reported a comparative assessment of several forms of video presentations and revealed that combined application of visual and auditory information were much more fruitful to propagate the information among people in comparison to other single source. Some others reported nullified impact of video on learning process (Mbarik et al., 2001). Research reports suggest that video promotes learning interest among students and therefore becomes a motivation for them to move forward (Wetzel et al., 1994).

The documentary film was very much acceptable for the students. Acceptability of the film by the students indicates a promising result. 40% students were in opinion that acceptance level of the documentary film was excellent whereas 28% as very good, 28% as good, 4% as average and 0% as bad (Figure 12.1). 50% students were in opinion that level of interest of the documentary film was excellent whereas 36% as very good, 10% as good, 4% as average, and 0% as bad (Figure 12.2). 58% students were in opinion that level of knowledge imparting activity of the documentary film was excellent whereas 30% as very good, 6% as good, 6% as average, and 0% as bad (Figure 12.3).

Therefore, documentary film plays an excellent role towards the motivation of students in the environmental activity.

12.9.2 QUANTITATIVE ANALYSIS OF MEDIA LEARNING

The effectiveness of media learning was assessed through survey sheets (Tables 12.1 and 12.2). Development of concept about species diversity has been developed within the students by SDF approach or media learning. Before implementation of media learning the percentage of Agree (A), SA, Neutral

(N), Disagree (DA), and SDA level against the development of concept about species diversity were 30%, 40%, 12%, 10%, and 8% respectively but after implementation of media learning the said responses became 44%, 50%, 6%, 0% and 0% respectively.

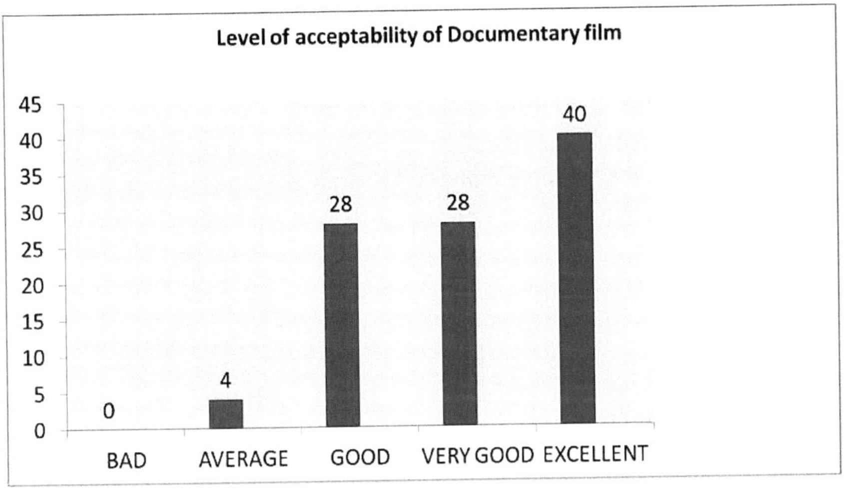

FIGURE 12.1 Responses towards acceptability of SDF in percentage.

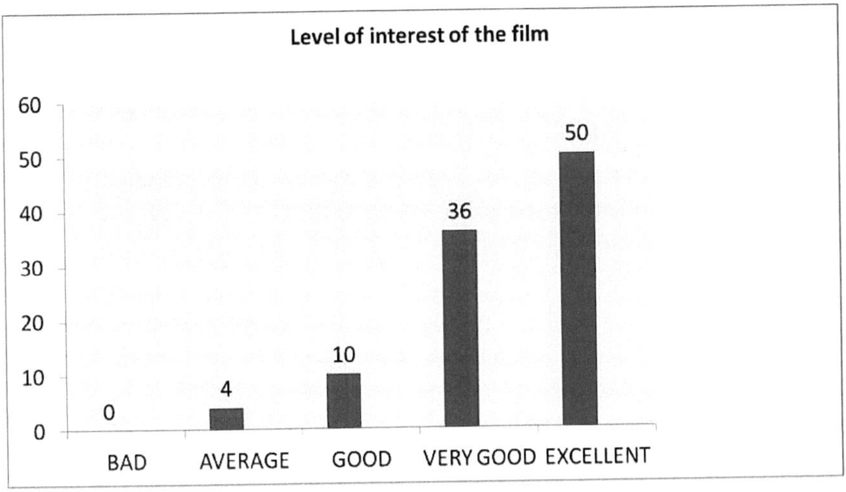

FIGURE 12.2 Responses towards interest of SDF in percentage.

Student's perception about species diversity has been developed after power point representation of documentary film based on Borneo mangrove ecosystem. Students can understand well about habitat diversity following the documentary film. Disagree and SDA level towards the development of concept of habitat diversity have been reduced from 10% to 2% and 17% to 0% after integration of media learning. Concept of species interaction in the nature has been well understood as agree, and SA level have been increased from 29% to 42% and 28% to 44%, respectively, whereas ignorance level (N), DA and SDA level have been reduced from 19% to 14%, 11% to 0% and 13% to 0% respectively.

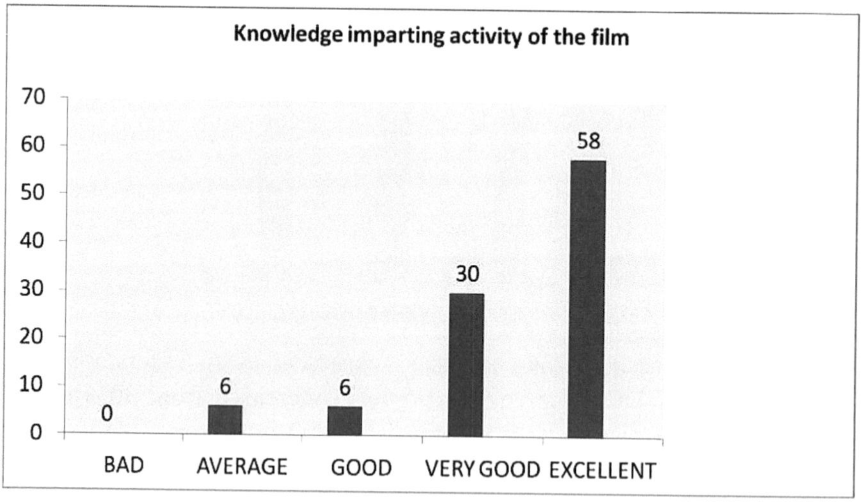

FIGURE 12.3 Responses towards knowledge imparting activity of SDF in percentage.

This is a good response towards the awareness development of the students. Habitat conservation is the prime initiative of the conservation of the biological diversity and this also helps to check the species extinction. In the present study response regarding biodiversity reduction, i.e., the local, regional or global extinction of species and their support towards conservation will depend on their knowledge about biodiversity, their conceptions about species number present and an awareness regarding the threat of extinctions (Lindemann-Matthies and Bose, 2008).

Average perception about importance of wild life is being elevated through media learning. Towards this parameter Agree and SA level improved from 30% to 40%, and 20% to 54%, respectively. Similarly, disagree and strongly DA reduced from 20% to 0% and 25% to 2% only. Towards the function of different species attributed in ecosystem of mangrove student's perception is better after application of media learning. Students must understand key stone species concept as key stone species regulate the ecosystem entirely. The documentary film based on mangrove ecosystem was played to the students. After implementation of media learning a very good feedback was obtained about understanding of concept of key stone species as agree level increased from 30% up to 40% and SA level has also increased from 35% to 48%. Ignorance level (N) has also reduced from 15% to 10%.

Key stone species are getting affected by manmade causes. Students can understand this fact following seminar talk along with documentary film as agree and SA level increased from 30% to 42% and 36% to 46% respectively. Disagree and strongly DA also declined from 10% to 0% and 18% to 6%. Other species dependency on key stone species was properly understood after implementation of media learning. Documentary film also predict well about reasons of degeneration and habitat loss mangrove ecosystem. Agree level and SA level have been increased for this respect which reflects effectiveness of media learning. Actually, any documentary film based on wild life helps to increase knowledge about concerned matter than general ways. A marine mammal documentary increased the knowledge about marine mammals in the treatment (documentary viewers) compared to an untreated control group (Fortner, 2006).

Reasons behind mangrove habitat loss have been well understood by the students by the SDF as agree and SA level increased from 30% to 38% and 40% to 56%, respectively. Neutral level, DA and strongly DA reduced from 14% to 6%, 13% to 0% and 3% to 0% respectively. Documentary film can also predict well about socioeconomic impact of mangrove ecosystem. Agree and SA level increased from 20% to 32% and 48% to 62%. Neutral level, DA and strongly DA also reduced from 14% to 6%, 10% to 0% and 8%to 0%, respectively. It is a good indication for understanding ecosystem influence on human society.

Visual media has a capability of making complex matter into simpler one. This capability helps to mitigate the conflict man and nature. Environmental problems are scientifically complex; the media have to find an easy way to cover such pieces in a compelling that engages common people (Nisbet et al., 2002). Economic and sociological condition of native population is also getting affected due to loss of mangrove ecosystem. This fact is understood by the students as agree and SA level have increased by more than 10% (35% to 48%) and 26% to 34% respectively whereas Neutral disagree and strongly DA have been reduced from 18% to 14%, 11% to 0% and 10% to 4%, respectively.

Cultural and spiritual aspect has negative impact in terms of the loss of biodiversity. Applying the SDF it is well understood by the students.

Agree and SA level have been increased from 30% to 36% and 30% to 42%, respectively. Neutral, disagree, and strongly DA have been reduced from 20% to 12%, 15% to 10% and 5% to 0%.

Total score for each student for survey sheet on visual media learning was calculated before and after application of proposed model. Paired t-test were done with the observed data (score of the students) set and significance was tested at 0.005 level of significance. Results reveal that students score has been increased after implementation of proposed models. An upward curve comparison was done with the previous one (Figure 12.4). The effectiveness of visual media learning was judged through statistical calculation (Paired t-test). Results revealed that there is a significant difference ($p < 0.005$) in achievement level between pre and post-application of visual media learning within the students (Table 12.3).

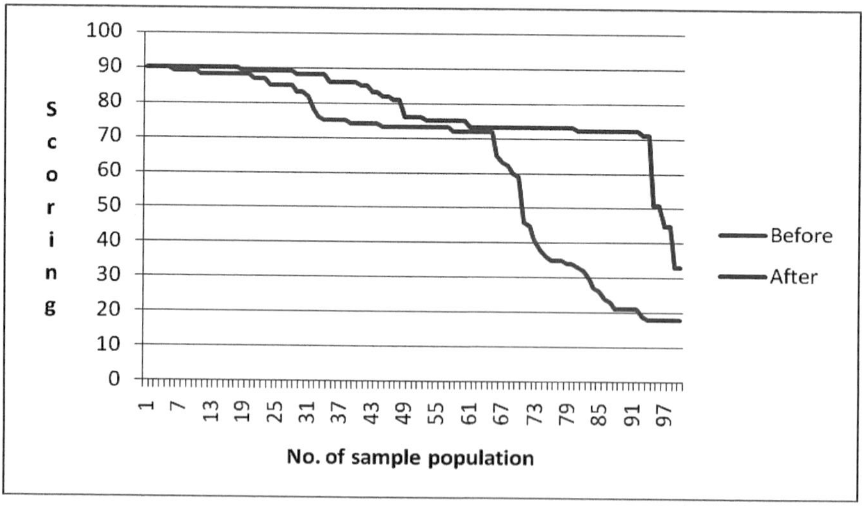

FIGURE 12.4 **(See color insert.)** Scoring of sample population of media learning.

TABLE 12.3 Paired T-Test of Media Learning (SDF)

	Paired Differences					t	df	Sig. (2-tailed)
	Mean	Std. Deviation	Std. Error Mean	95% Confidence Interval of the Difference				
				Lower	Upper			
Before SDF–After SDF	15.040	17.131	1.713	18.439	11.641	8.779	99	.005

Therefore, the student's attitude has been changed positively which is the goal of EE. Integration of documentary film into teaching and learning in EE is essential for meaningful interaction between learners and educators. Visual media learning can be used to advance cognitive skills such as comprehension, reasoning, problem-solving, and creative thinking related to EE. It is one of the objectives of the EE that aims at promoting the students' critical thinking, problem solving, and decision making that made the students understand the relation and the connection of the content studied (Ratanapeantamma, 2005). Therefore, over all perception about ecological aspect may be developed by visual media learning.

12.9.3 QUALITATIVE ANALYSIS OF MEDIA LEARNING

Students are very much interested about visual media learning. All the students were in opinion about effectivity of visual media learning would be more than others. A representation of different aspects of mangrove ecosystem through the visual media, i.e., the documentary film was attractive to most of the students. Students at least understood different factors of mangrove ecosystem and the ways by which they are getting degraded. Students suggested that to mitigate the conflicts between man and mangrove ecosystem a buffer zone must be created and restored if present. Students also suggested providing alternative livelihood to the people who are dependent on mangrove ecosystem directly or indirectly. Poly bags use in the adjacent area must be stopped. Human interferences should be minimized so far as possible. Students also suggest including political and administrative initiative to conserve the mangrove ecosystem.

12.10 CONCLUSION

Earlier research findings revealed heterogeneous results of visual media learning on knowledge and awareness generation in student community. Our study reveals impact of visual media learning through SDF has positive influence over the student's attitude towards gaining EE. Integration of visual media learning would help in traditional knowledge exchange in student community to expand the area of knowledge beyond classroom. Students become capable of conceptualizing and visualizing the facts more clearly through observation of facts directly learning from textbooks and classroom teaching. This would give better results towards higher learner satisfaction. The study has also tried to explore heterogeneous outcome of such types of studies done previously offered an explanation for inconsistent findings reported in previous studies. It suggests that the visual media learning process can be a valuable means for improving effectivity of learning under traditional classroom teaching process. It would promote development of comprehension, reasoning, problem-solving, and creative thinking. Overall perception about ecological context can be developed through visual media learning.

KEYWORDS

- **awareness level**
- **communication**
- **environmental education**
- **mangrove ecosystem**
- **global learning and observations to benefit the environment**

REFERENCES

Adedoyin, S. F., & Torimiro, D. O., (1999). *A Manual of Children-in-Agriculture Programme in Nigeria, Ago-Iwoye, Communication, Extension and Publication Component*. CIAP National Headquarter.

Agnes, A. M., (2010). Implementation of environmental education: A case study Malaysian and Nigerian secondary school. *International Conference on Biology, Environment and Chemistry IPCBEE, 1*. IACSIT Press, Singapore.

Ahmed, N., (1990). The use of video technology in education. *Educational Media International, 27*(2), 119–123.

Akpabio, E., (2004). Nigerian home video films as a catalyst for national development. *Journal of Sustainable Development, 1*(1), 5–10.

Athman, J., & Monroe, M., (2000) *Elements of Effective Environmental Education Programs*. Retrieved from: *Recreational Boating Fishing Foundation*. http://www.rbff.org/educational/reports.cfm (Accessed on 10 October 2019).

Ballard, M., & Pandya, M., (1990). *Essential Learnings in Environmental Education*. Troy, OH: North American Association for Environmental Education.

Bamberg, S., & Moeser, G., (2007). Twenty years after Hince, Hungerford, and Tomera: A new meta-analysis of psycho-social determinants of pro-environmental behavior. *Journal of Environmental Psychology, 27*(1), 14–25.

Bartosh, O., (2003). Environment education: improving student achievement. *Master Thesis*. The Evergreen State College, Washington.

Berberoglu, E. O., (2015). The effect of ecopodagogy-based environmental education on environmental attitude of in-service teachers. *International Electronic Journal of Environmental Education, 5*(2), 86–100.

Berman, M. G., Jonides, J., & Kaplan, (2008). The cognitive benefits of interacting with nature. *Psychol. Sci., 19*(12), 1207–1212.

Bhandari, B. B., & Osamu, A., (2001). *Environmental Education in the Asia-Pacific Region: Status, Issues and Practices*. Kanagawa, Japan: Institute of Global Environmental Strategies, Environmental Education Project.

Budvytyte, A., (2011). Environmental education at secondary school system in Lithuania. *Master Thesis*. Lund University, Lund.

Burgess, D. J., & Mayer-Smith, J., (2011). Listening to children: Perceptions of nature. *Journal of Natural History Education and Experience, 5*, 27–43.

Cheng, J., & Monroe, M., (2010). Connection to nature: Children's affective attitude toward nature. *Environment and Behavior, 44*, 31–49.

Conde, M. D. C., & Sánchez, J. S., (2010). The school curriculum and environmental education: A school environmental audit experience. *International Journal of Environmental and Science Education, 5*(4), 477–494.

De La Vega, R. M. C. L., (2004). Awareness, knowledge and attitude about environmental education. *PhD Thesis*, University of Central Florida.

Dopemu, Y. A., (1990). The effect of color on students cognitive achievement in video instructional. *Educational Media International, 27*(12), 64–69.

Fien, J., & Tilbury, D., (1996). *Learning for a Sustainable Environment: An Agenda for Teacher Education in Asia and The Pacific*, Printed by UNESCO Principal Regional Office for Asia and the Pacific, Bangkok, Thailand.

Ghosh, K., (2014). Environmental awareness among secondary school students of Golaghat district in the state of Assam and their attitude towards environmental education. *Journal of Humanities and Social Science, 19*(3), 30–34.

Gough, A., (1997). *Education and the Environment: Policy, Trends and the Problems of Marginalization*. Australian Education Review No. 39. Melbourne, Australia: The Australian Council for Educational Research Ltd.

Halder, S., (2012). An appraisal of environmental education in higher school education system: A case study of North Bengal, India. *International Journal of Environmental Science, 2*(4), 2223–2233.

Hampapur, A., & Jain, R., (1998). Video data management systems: Metadata and architecture. In: Klas, W., & Sheth, A., (eds.), *Multimedia Data Management*. McGraw-Hill, (Chapter 9). New York.

Hasan, A. A., Noordin, T. A., & Sulaiman, S., (2010). The status on the level of environmental awareness in the concept of sustainable development amongst secondary school student. *Procedia Social and Behavioral Science, 2*, 1276–1280.

Isiaka, B. T., (2000). *Assessment of Video Tape Recordings as an Instructional Media in Teaching Agricultural Science in Lagos State*. Public Secondary Schools, Unpublished M. Agric. Thesis, University of Agriculture, Abeokuta, Nigeria.

Isiaka, B., (2007). Effectiveness of video as an instructional medium in teaching rural children agricultural and environmental sciences. *International Journal of Education and Development Using ICT, 3*(3), 105–115.

IUCN, (1998). *Report on the End of the Project Workshop on the National Conservation Strategy Implementation Project*. Kathmandu: NPC/IUCN NCS Implementation Project.

Jena, A. K., (2012). Awareness, openness and eco-friendly (AOE) model teaches pre-service teachers on how to be eco-friendly. *International Electronic Journal of Environmental Education, 2*(2), 103–117.
Jickling, B., & Wals, A. E. J., (2008). Globalization and environmental education: Looking beyond sustainable development. *Journal of Curriculum Studies, 40*(1), 1–21.
Kasimov, N. S., Malkhazova, S. M., & Romanova, E. P., (2005). Environmental education for sustainable development in Russia. *Journal of Geography in Higher Education, 29*(1), 49–59.
Kopnina, H., (2014). Future Scenarios and environmental education. *Journal of Environmental Education, 45*(4), 217–231.
Kumar, A., Sharma, R. C., & Vyas, R. V., (2003). Impact of electronic media in distance education: A study of academic counselor's perception. *Turkish Online Journal of Distance Education-TOJDE, 4*(4), 1–9.
Kuo, F., & Taylor, A., (2004). A potential natural treatment for attention-deficit/hyperactivity disorder: Evidence from a national study. *Is J. Public Health, 94*(9), 1580–1586.
Lindemann-Matthies, P., & Bose, E., (2008). How many species are there? Public understanding and awareness of biodiversity in Switzerland. *Human Ecology, 38*, 731–742.
Madsen, P., (1996). What can universities and professional schools do to save the environment? In: Callicott, J. B., & Da Rocha, F. J., (eds.), *Earth Summit Ethics: Toward a Reconstructive Postmodern Philosophy of Environmental Education* (pp. 71–91). NY: Albany State University of New York Press.
Mbarika, V. W., Sankar, C. S., Raju, P. K., & Raymond, J., (2001). Importance of learning-driven constructs on perceived skill development when using multimedia instructional materials. *The Journal of Educational Technology Systems, 29*(1), 67–87.
Moore, M. G., & Kearsley, G., (1996). *Distance Education: A Systems View*. Boston: Wadsworth Publishing Company.
NEEAC, (1996). *Report Assessing Environmental Education in the United States and the Implementation of the National Environmental Education Act of 1990*. NEEAC, Washington, DC.
NEETF, (2000). *Environment-Based Education: Creating High Performance Schools and Students*. Washington DC: National Environmental Education Training Foundation.
Nisbet, M. C., Scheufele, D. A., Shanahan, J., Moy, P., Brossard, D., & Lewenstein, B. V., (2002). Knowledge, reservation or promise? A media effects model for public perceptions of science and technology. *Communication Research, 29*, 584.
Nugent, G. C., (1982). Pictures, audio and print: Symbolic representation and effect on learning. *Educational Communications and Technology: A Journal of Theory, Research and Development, 30*(3), 163–174.
Ornstein, A., & Hunkins, Fr., (2004). *Curriculum Foundations, Principles, and Issues* (4th edn., p. 117). Boston, MA: Pearson Education, Inc.
Orr, D., (1992). *Ecological Literacy: Education and the Transition to a Postmodern World*. Albany, NY: State University of New York Press.
Ozden, M., (2008). Environmental awareness and attitudes of student teachers: An empirical research. *International Research in Geographical and Environmental Education, 17*(1), 40–55.
Palmer, J. A., (1995). Environmental thinking in the early years: Understanding and misunderstanding of concepts related to waste management. *Environmental Education Research, 1*(1), 35–40.
Pavlova, M., (2013). Towards using transformative education as a benchmark for clarifying differences and similarities between environmental education and education for sustainable development. *International Research in Geographical and Environmental Education, 19*(5), 656–672.

Petrillo, W., & Demchik, M., (2008) *Environmental Education's Role of Sustainable Development: Three Case Studies from India, South Africa and the United States*. Retrieved from: www.uwsp.Edu/forest/StuJournals/Documents/IRM/Reilly.pdf (Accessed on 10 October 2019).

Prathap, D. P., & Ponnusamy, K. A., (2006). Effectiveness of four mass media channels on the knowledge gain of rural women. *Journal of International Agricultural and Extension Education, 13*(1), 73–81.

Proulx, G., (2004) Integrating scientific method and critical thinking in classroom debates on environmental issues. *American Biology Teacher, 66*(1), 26–59.

Ramsey, J. M., & Hungerford, H. R., (1989). The effects of issue investigation and action training on environmental behavior in seventh grade students. *Journal of Environmental Education, 23*(4), 35–50.

Ratanapeantamma, W., (2005). *Environmental Management Administration.* Bangkok: Wattanapanit.

Rudy, P. C., (2015). The perspective of curriculum in Indonesia on environmental education. *International Journal of Research Studies in Education, 4*(1), 77–83.

Siddiqui, Z. T., & Khan, A., (2015). Environment education: An Indian perspective. *Research Journal of Chemical Sciences, 5*(1), 1–6.

Sonowal, C. J., (2009). Environmental education in schools: The Indian scenario. *J. Hum. Ecol., 28*(1), 15–36.

Spencer, K., (1991). Modes, media and methods. The search for educational effectiveness. *British Journal of Educational Technology, 22*(1), 12–22.

Stapp, W., Bennett, D., & Bryan, W., (1969). The concept of environmental education. *Journal of Environmental Education, 1*(1), 30–31.

Sund, P., & Lysgaard, J. G., (2013). Reclaim "education" in environmental and sustainability education research. *Sustainability, 5*(4), 1598–1616.

Syed, M. R., (2001). Diminishing the distance in distance education. *IEEE Multimedia, 8*(3), 18–21.

Tal, R. T., (2004). Community-based environmental education a case study of teacher-parent collaboration. *Environmental Education Research, 10*(4), 523–543.

Talabı, J. K., (1989). The television viewing behavior of families in Kwara State, Nigeria. *British Journal of Educational Technology, 20*(2), 135–139.

UNESCO, (1978). *Intergovernmental Conference on Environmental Education.* Tbilisi (USSR), 14–26 October 1977. Final Report. Paris: UNESCO.

UNESCO-PROAP, (1996). *Celebrating Diversity, Cultivating Development, Creating Our Future Together: UNESCO in Asia and the Pacific,* UNESCO, Bangkok.

UNESCO-UNEP, (1976). The Belgrade charter. Connect: UNESCO-UNEP. *Environmental Newsletter, 1*(1), 1–2.

Veeravatnanond, V., (2010). Teacher education for sustainable development AEE-T. *Journal of Environmental Education, 1*(1), 1–4.

Wals, A. E. J., Brody, M., Dillon, J., & Stevenson, R. B., (2014). Convergence between science and environmental education. *Science, 344*(6184), 583.

Wetzel, C. D., Radtke, R. H., & Stern, H. W., (1994). *Instructional Effectiveness of Video Media.* Hillsdale, NJ, Lawrence Erlbaum Associates.

Young, R. M., & Lafollette, S., (2009). Assessing the status of environmental education in Illinois elementary schools. *Environ. Health Insights, 9*(3), 95–103.

Index

D